Safety of Historical Stone Arch Bridges

Dirk Proske · Pieter van Gelder

Safety of Historical Stone Arch Bridges

 Springer

Dr.-Ing. habil. Dirk Proske, MSc.
University of Natural Resources
and Applied Life Sciences, Vienna
Institute of Mountain Risk Engineering
Peter-Jordan-Street 82
1190 Vienna
Austria
dirk.proske@boku.ac.at

Dr.-Ir. Pieter van Gelder
Delft University of Technology
Section of Hydraulic Engineering
Stevinweg 1
2628 CN Delft
The Netherlands
p.h.a.j.m.vangelder@tudelft.nl

ISBN 978-3-642-42552-3   ISBN 978-3-540-77618-5 (eBook)
DOI 10.1007/978-3-540-77618-5
Springer Heidelberg Dordrecht London New York

Cover design: eStudio Calamar S.L.

Printed on acid-free paper

Springer is part of Springer Science+Business Media (www.springer.com)

Preface

Stone arch bridges are special technical products in many aspects. Two of the most important aspects are their very long time of usage and their landscape changing capability. First, for more than two millenniums, stone arch bridges have been part of the human infrastructure system, and some of them are still in use. Most of the stone arch bridges now in use are older than the first century. The only type of structures reaching the same duration of usage are tombs and other religious structures. However, in contrast to those, arch bridges are much more exposed to changes in usage conditions. There exist Roman bridges that were crossed not only by Roman legions but also by tanks in World War II. When most stone arch bridges were constructed, motorized individual car traffic was yet unknown. This load now has to be borne by these historical bridges. We should probably much more esteem the farsightedness and endeavour of our ancestors, which we often count on nowadays without perception.

Or perhaps we do notice as some common attitudes indicate, don't we? In many children's books, landscapes often include stone arch bridges. And if people are asked whether arch bridges are disturbing or accepted, in most cases people consider arch bridges as part of our man-made landscape and not necessarily as human artefact. Painters such as Paul Cézanne have included arch bridges in their landscape paintings as early as the 19th century, which refutes the theory that arch bridges are now just accepted because they have been part of the landscape for centuries.

Stone arch bridges are considered beautiful because they apply some simple rules of aesthetics. First of all, they use building material from the vicinity and therefore are embedded in the landscape. Furthermore, the genius idea to arrange stones geometrically in such a way that the mechanical properties of stones are used in a nearly perfect way gives the impression of harmony, whereas beam bridges made of reinforced or prestressed concrete are often felt as strange.

Besides beauty, the bridges show in a very clear way one of the biggest conflicts of our human civilisation. In the untiring trial to rationally describe all elements of our world, we have seen the limits of this concept in the last decades. Even though arch bridges have been built and used for more than two millenniums, we still face problems in numerically describing their behaviour. Only in the last decades have appropriate tools been provided. Such tools are presented in this book. However, the book embeds these procedures in an even wider concept. Not only are computation strategies and strengthening techniques for arch bridges given, but adaptations of today's loads to preserve the bridges are also presented.

However, strengthening of arch bridges is often not required: The major cause of the destruction of arch bridges is the insufficient width of the roadway, which means not the safety but the usability has limited the lifetime of the bridge. Perhaps we could live with this limitation and give respect to the arch bridges. They still provide us with the lowest maintenance costs of all bridge types.

Expression of Thanks

This book is a translation and extension of a former German book primarily written during a visit of the first author at the TU Delft in 2005. Therefore, first of all we thank the TU Delft for the financial support. Additional support was given to translate the book.

Furthermore, many persons have contributed by proof-reading and giving suggestions. Therefore, the authors thank Prof. Konrad Bergmeister, Prof. Han Vrijling, Prof. Jürgen Stritzke, and Prof. Udo Peil. Additionally, Mrs. Angela Heller did the proof-reading of the German version. We thank her for the intensive work. Last, but not least, we thank Springer and their assistants for the opportunity to publish the book in English, and for their strong support.

Contents

1 Introduction

"The first bridges men built were in wood, which were suited to their requirements at the time. But then they began to think about the immortality of their names. And because their richness gave them heart and made better things available to them, they began to build bridges in stone, which lasted longer, cost more, and brought glory to those that built them."

A. Palladio 1570 (taken from Corradi 1998)

1.1 General Introduction

Only natural phenomena exist in the world of materials. The concept of technology, which is seen as a real characteristic of human culture, already shows the following fact by its definition: under technology, one understands the method and capacity to use the natural phenomena in a practical way.

This statement also applies to the technical product of bridges. Nature is easily able to create bridges without human involvement: bridges, as physical features, already existed for millions of years, created from geological formations by wind and water or as fallen trees that cross a creek. One only has to travel to the Arches National Park in the United States to see examples.

However, at the moment in which nothing except the forces of nature create bridges without the forces of intellect, the situation changes fundamentally. An artificial phenomenon then originates. The artificial distinguishes itself from the natural by the agreement of intention. Thus, a line, for example, becomes an artificial phenomenon, when the lines are shaped into a symbol. This symbol, however, requires an agreement in advance. Based on this hypothesis, a bridge is considered to be an artificial phenomenon since it is designed to span something and to ease progress.

Next to that, it seems as if the concept of art is related to the notion of "artificial". But the notion of art is actually more strongly linked to the notion of technology. Originally, the word "art" describes a high degree of

D. Proske, P. van Gelder, *Safety of Historical Stone Arch Bridges*, DOI 10.1007/978-3-540-77618-5_1,
© Springer-Verlag Berlin Heidelberg 2009

skill with which a human accomplishes a task. Later on, the term is expanded to the product itself.

An example of a product that deserves the notion of artwork is the world atlas "Atlas Maior" (later edition 2005) that appeared between 1662 and 1665. The atlas showed the geography of the known world in a quality that had never been achieved until that time. The publisher from Amsterdam Joan Blaue writes in the preface of the enclosed maps: "[With them ...] we tread inaccessible mountains and traverse oceans and rivers without risk" (FAZ 2005).

The human insistence to explore, which drives one to expand his own living space, eventually cannot be satisfied by maps or by pure conceptual, context only. But these maps can be a movement towards it. The spatial movement of humans and objects does have a substantial meaning, not only in science, but also in the present day-to-day world.

Since the beginning of humanity, humans could only cover large spatial distances in a very long time frame. The individual could not separate himself too far away from his place of origin. Yet the settling of humans for 10,000 years after the last Ice Age on the necessity of agricultural grounds once again contradicts the human wish for mobility. However, settlement led to an unexpected effect: it paid off by making "inaccessible mountains" crossable and "oceans and rivers" traversable without risk, since this has to be done on a regular basis.

The broadening of the skill of the human movement apparatus with regard to the development of speed, with regard to the development of dynamic mingling, and with regard to the reach has probably lead to the development of improved movement and transport systems, respectively, with the start of settlement.

The invention of the wheel, as well as The taming of horses, is often regarded as one of the greatest inventions of humankind, because there exists no example in nature of a wheel that rotates around its own axis. The wheel is the basis for the construction of vehicles that allow the transport of goods and people in a very economic way. The basis for the contribution of wheels in transport is, however, a proper condition of the surface of the roads. One of the requirements is a certain wheel straightness and horizontalness of the roll face. Such wheel straightness also relieves the human and animal movement apparatus. This is meaningful, because pulling animals like horses served as the power for vehicles for almost 5,000–6,000 years after the invention of wheel.

The free and efficient movement of vehicles, people, and animals is, for example, not given for the movement through rivers. Also, very steep roads quickly result in the exhaustion of both humans and animals and are unsuitable for wheeled vehicles. Therefore, the wish to ease the movement of people and goods arose probably very early. The best way to ease movement is shortening and horizontally or vertically detouring to avoid unsuitable stretches of the road. Bridges follow this idea. They are structures that serve to cross obstructions. They not only can be used for various means of transportation (road and railway bridges) for people (pedestrian bridges) and animals, but also for passing over water.

When one compares bridges with the above-mentioned maps, they are both invitation and tool for movement in a double sense: not only do you show the design of the roads, just like on the map, but you also offer a physical entrance. The first examples of primitive bridges are the stone bridges of Dartmoor (Brown 1994), and the stone beam bridges in Gizeh (2,500 B.C.) in China (500 B.C.) (Heinrich 1983).

The physical achievements brought about by the creation of bridges are substantial. Even in present times, with the elaborate technical resources available, many people *sense an inner feeling* when they marvel at the beauty and size of the extraordinary historic bridge structures. This is especially true for the mighty arched bridges of the Romans—for example the Pont du Gard in France or the Bridge of Alcántara in Spain. Many of these over 80-generations old structures, like the Ponte Milvio, were crossed by Roman legions as well as by German and American armoured vehicles in World War II (Heinrich 1983, Straub 1992). These military operations basically always have the objective of demolition or destruction.

In contrast to this, a bridge is a construction – a synthesis. In this passage, it is noted that humankind is also a synthesis. The translation of the concept "synthesis" in Latin is composition, also additio (von Hänsel-Hohenhausen 2005). The written work at hand is the composition of an analysis. An analysis again is a derivation, since the Latin word for it is *reductio* (von Hänsel-Hohenhausen 2005). The objective of this analysis, however, is the progress of the actual structure – the progress of the synthesis.

Progress can reach far over the everyday present. It will and must encompass future generations. Our ancestors, for example, encouraged us as they erected bridges that are useful even today.

The preservation and acknowledgment of the skills needed to erect historic bridges is, in the view of the writers, a duty we have to the up coming generations. One can best carry out this duty when one defines a use for the historic bridge structures and extends their utilization time. Just as work is an inextricable criterion for the merit of a human being, the utilization of a bridge is an inextricable criterion for the justification of its existence.

These structures, however, are only utilized when the advantages from its utilization are greater than the possible disadvantages. An elementary disadvantage of a historic structure can be the poor capacity for present day loads. Complying with modern safety codes is not negotiable for any product, even for historic structures. The mentioned safety requirements of technically created structures are transposed through safety concepts. In recent years, a new safety concept for structures has been introduced both in European context and also on a national level in Germany.

The successful integration of historic arch bridges of natural stone into these safety concepts is the foundation for a reutilization of these types of bridges. This book takes up that task.

Figures 1-1 to 1-15 not only provide an impression of the great diversity of masonry arch bridges, but also arch bridges of other materials. However, many further fine examples are known. The reader can consult the homepages by Bill Harvey (2006) or Janberg (2008). Another interesting example is the Minzhu Bridge in China (NN 2005), where three half arches meet in the arch crown. Some better modern examples are the completion of the stone arch bridge Pont Trencat in Spain by a steel arch with closed spandrel walls, a bridge in the inner port of Duisburg with a flexible lane that is lifted up in case of ship traffic and thus can perhaps count as an arch bridge (Bühler 2004), the Gateshead Millennium Steel arch bridge in Newcastle by Chris Wilkinson with a curved lane that rotates along its alongside axis in case of ship traffic, the Puente La Barqueta (Langer's Beam), the steel arch bridge with suspension above Ebro in Logrono, the Leonardo Bridge in Norway, or the Juscelino Kubitschek Bridge in Brazil (Goldberg 2006). Besides the success of steel and concrete arch bridges, in recent years, some stone arch bridges have again been erected in Great Britain and Portugal.

As the above-mentioned examples demonstrate, the types of arch bridges change and live on by continuous variation of the original idea. The presently applied mathematical optimization for defining optimal bridge variants, leading to uniform standard solutions in the end, will and has already partially failed. One such optimization requires a multitude of entry sizes, which are not yet known at the time of the optimization calculation. Next to this uncertainty, various optimal solutions are often possible. A magnificent example of this is the variety of living organisms on Earth. One occasionally happens to find similar organic solutions, for example for the limbs, but even so one frequently finds differences for the same boundary conditions. The constant variation of the arch bridges is one key element of their success.

Fig. 1-1. Sweden Bridge in Dresden, built in 1845 (side view)

Fig. 1-2. Sweden Bridge in Dresden, built in 1845 (view on the carriageway)

Fig. 1-3. Ponte Vecchio and Ponte St. Trinità in Florence, Italy

Fig. 1-4. Ponte Vecchio in Florence, Italy

Fig. 1-5. Bridge in Toscana, Italy

Fig. 1-6. Bridge in Toscana, Italy

Fig. 1-7. Viadukt de Saint-Chamas in France, built in 1848

Fig. 1-8. Bride in the Saxon Switzerland, Germany

Fig. 1-9. Göltzschtal Bridge, Germany

Fig. 1-10. Stone arch bridge, Delft, The Netherlands

Fig. 1-11. Arch bridge, Valencia, Spain

Fig. 1-12. Railway stone arch bridge, Melk, Austria

Fig. 1-13. Concrete arch bridge, Valencia, Spain

Fig. 1-14. Steel arch bridge, Valencia, Spain

Fig. 1-15. Steel arch bridge, Vienna, Austria

1.2 Advantages and Disadvantages of Arch Bridges

"Like the back of a tiger, the bridge curves from Jade."

G. Mahler 1860–1911: The Song of Earth, of the Youth

All biological solutions and technical systems have advantages and disadvantages. An advantage of arch bridges is that their beauty is certainly not to be underestimated. The beauty of arch bridges is often traced back, not only to the use of natural materials (stone look), but also to the high aesthetical measure. Birkhoff (1933) has introduced such a concept. He defines this aesthetical measure as follows,

$$A = \frac{O}{C} \tag{1-1}$$

in which the aesthetic measure A is between zero and one, the O stands for the number of relations of order, and the C stands for the complexity. An application of the measure can be found in Staudek (1999). An essential assumption of this measure is the relationship between beauty and effectiveness. According to Piecha (1999), humankind possesses assessment mechanisms by nature and nurture that assess the effectiveness of a structure. Objects are aesthetically surveyed after such an observation and assessment mechanism. Birkhoff assumes that a high measure of aesthetic satisfaction exists with a balanced relation between the observation effort to identify orders within an object and the complexity of the object—for example, the new elements. In case of arch bridges, the number of ordering relations, as well as the complexity, is very limited, so arch bridges are considered aesthetic in accordance with this consideration.

This fact fits very well with some observations. For example, in Switzerland, arch bridges actually act as tourist attractions. For the largest part of the population, they are regarded as an element instead of an interference with nature. This could admittedly also be because many historic arch bridges have high seniority and with that a customary right. The above-mentioned consideration is also strengthened, however, by the fact that arch bridges are constructed on many historic sites and with that are then already regarded as aesthetic. The French painter Paul Cézanne included arch bridges in his paintings and considered the bridges as part of nature (Becqué 1983)

But at this point it should be pointed out that the ultimate numerical description of beauty has not been achieved up to now. However, developments of the Birkhoff measures have taken place, for example, by Bense (Ebeling and Schweitzer 2002, Klein 2008).

Further advantages of arch bridges, in addition to their indisputable beauty, are summarized by Weber (1999):

- Limited deformations under traffic loads (some tens of millimetres in case of railroad bridges)
- Usability and fatigue are irrelevant (the total strains are often in the cyclic pressure load region)
- Application of uninterrupted rails based on limited deformations (no rail fissures required)
- A high failure safety and robustness (insensitive to unplanned impacts)
- A high damage tolerance (recently, the notion of fitness is also used for that: a system has high fitness when it stays functional despite a large number of occurring faults)
- Early indication of malfunctioning
- A long lifetime and period of utilization
- Arch bridges, like all deck bridges, guarantee an undisturbed view for the travellers
- The construction materials can be disposed of and re-used as environmentally compatible material, respectively
- Excellent insertion into the landscape

However, there are also disadvantages, according to Weber (1999):

- Considerable reduction of the loading capacity by large support displacements (this assumption is however valid for all bridges)
- The clearance diagram under the bridge is not constant
- Complex renaturation

Obviously, such a summary of pros and cons is always subjective. It appears, however, as if the advantages prevail over the disadvantages. That would support the preservation of these structures, and with that, the realization of safety assessments with the aim of conservation.

1.3 Structure of the Book

Petryna (2004) has displayed a very good description of the basic elements of damage-oriented safety analyses of structures in his paper (Fig. 1-16). The description shows the following five basic elements: load models, material models, damage models, carrying models, and the formulation of verification equations. These elements can be found in this book, as well.

Figure 1-17, according to Mori and Nonaka (2001) and Mori and Kato (2003), is another way of describing the time-dependant probability of failure of a structure as a safety measure substitute without, however, considering the load and carrying models.

The relatively abstract description by Petryna (2004) can be transformed into a four-staged assessment scheme according to Diamantidis (Fig. 1-18) and ICOMOS (2001) (Fig. 1-19). Then, the single points of the process as well as the personal allocation are mentioned in this scheme. There is also a less detailed approach for that purpose by Czechowski (2001) (Fig. 1-20).

Figure 1-21 classifies the safety assessment of bridge maintenance procedures. However, the maintenance procedures for historic monuments under protection include some further restraints (ICOMOS 2001, Vockrodt 2005, Vockrodt et al. 2003, Yeomans 2006, Žnidarič and Moses 1997, Rücker et al. 2006, Jensen et al. 2008, and SIA 269 2007).

That the reliability theory-based analysis can ultimately only be a part of an over-organized and discipline-crossing inspection and maintenance strategy is shown in Fig. 1-22. It clarifies the various historic developments of risk-based inspection concepts for different industrial areas. Whereas in some areas, (for example risk-based inspection and maintenance) is standard practice in the monitoring of offshore platforms, this approach has not yet been implemented into other areas such as the construction industry. This is for various reasons—i.e., different owners.

Whereas the proprietor of bridge structures is generally the state, other structures such as oil platforms normally belong to commercial firms. For the state, meeting safety responsibilities is of utmost importance, whereas commercial firms are also constrained by market-based forces. Therefore, safety responsibilities are only a subtask. This is subsequently shown in cost-efficiency considerations for safety precautions.

Dealing with protective measures, however, is and stays a political decision. The practically active engineer cannot argue with such political decisions in his daily job. Structuring as in Fig. 1-18 would certainly be ideal for this situation. Despite this, the systematic by Petryna (2004) is chosen in this chapter, since it is only slightly bounded by organizational limitations. Before the single emphases are discussed, however, this chapter first gives some general information regarding arch bridges.

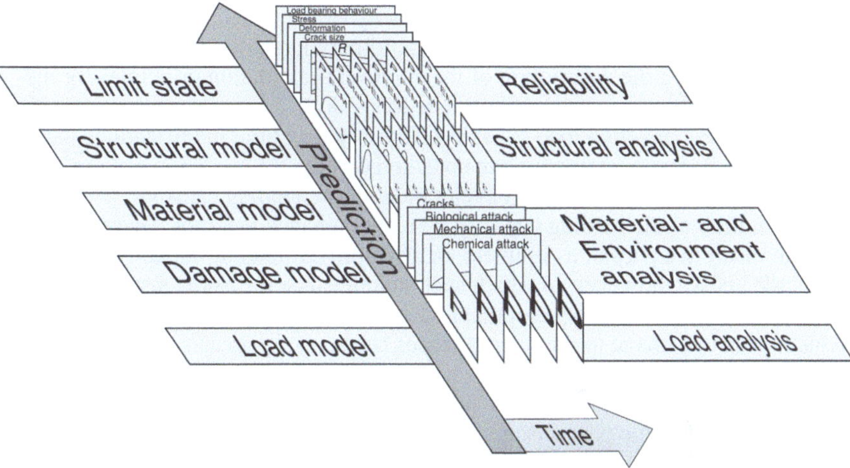

Fig. 1-16. Basic elements of damage-oriented reliability analysis of structures according to Petryna (2004)

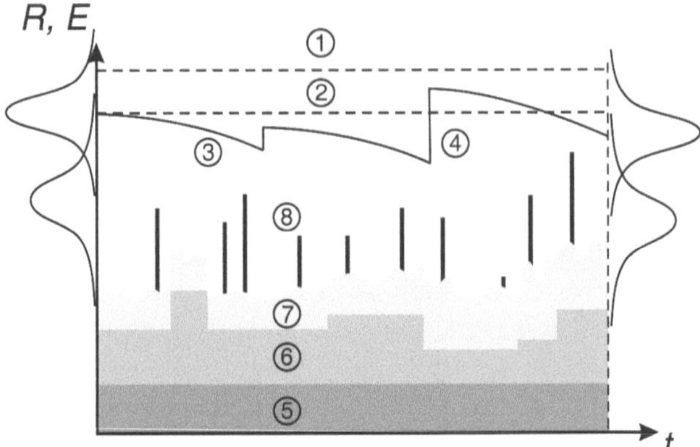

① Corresponds to the targeted loading capacity of the structure according to the design. Just as for historic structures, extraordinary large distinctions can also exist here, insofar as the description of the empirical calculation principles for historic arch bridges is necessary to estimate this value.
② Structurally converted load capacity. Not only structural modifications, but also deficiencies during construction are considered here.
③ Time and damage-dependent development of the loading capacity.
④ Restoration of the loading capacity by a maintenance measure. One such measure can result in a partial, a complete, or an improved loading capacity. Some cases are known, however, in which such maintenance measures resulted in an acceleration of the damage development—i.e., by the use of the wrong mortar—which leads to damage to the natural stones because of a greater stiffness.
⑤ Effects of dead load.
⑥ Effects of upgrading loading. A simplified assumption is made here that in time, for example, the parapets are converted, backfills are exchanged, and spare vaults are filled up.
⑦ Continuously changing effects, such as wind, These effects are normally not decisive for the massive arch bridges. However, traffic loading for continuously used bridges can also be reckoned among this type of impact. These can then definitely become dominant.
⑧ Impulse type effects, such as bumps and hits. Normally, exceptional effects are dealt with here.

Fig. 1-17. Representation of the time-dependent probability of failure for safety assessment based on Mori and Nonaka (2001) and Mori and Kato (2003)

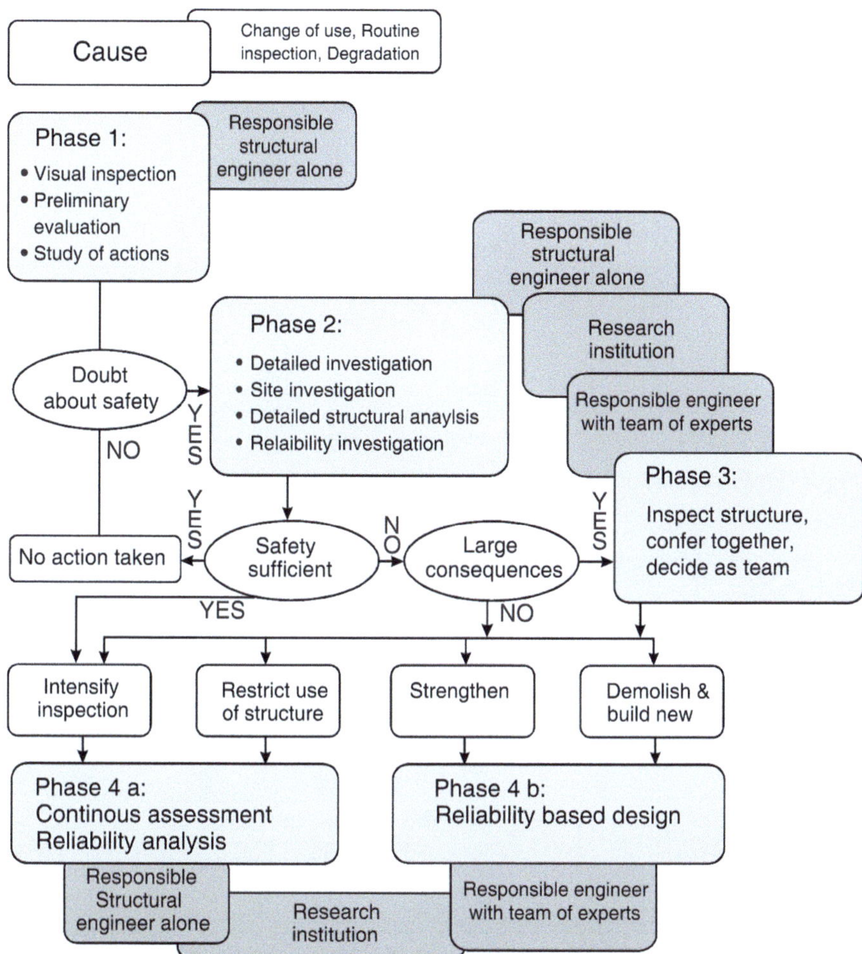

Fig. 1-18. Evaluation chain of historical structures according to Schueremans et al. (2003), Diamantidis (2001) and Schneider (1996)

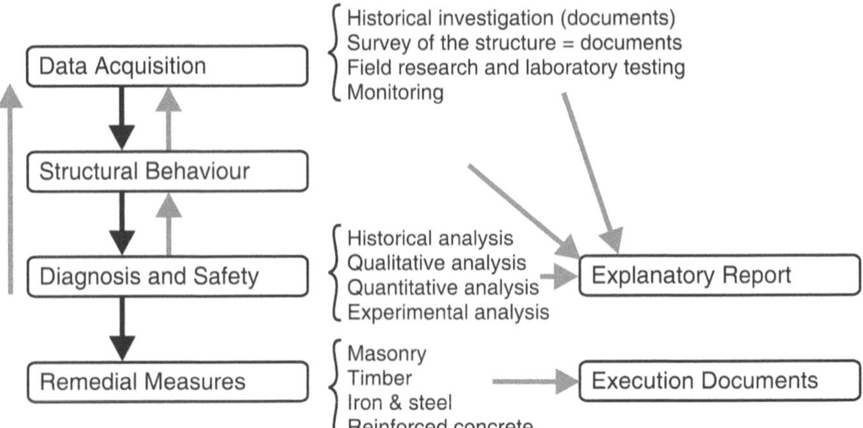

Fig. 1-19. Flowchart of structural interventions (ICOMOS 2001, Lourenço 2002)

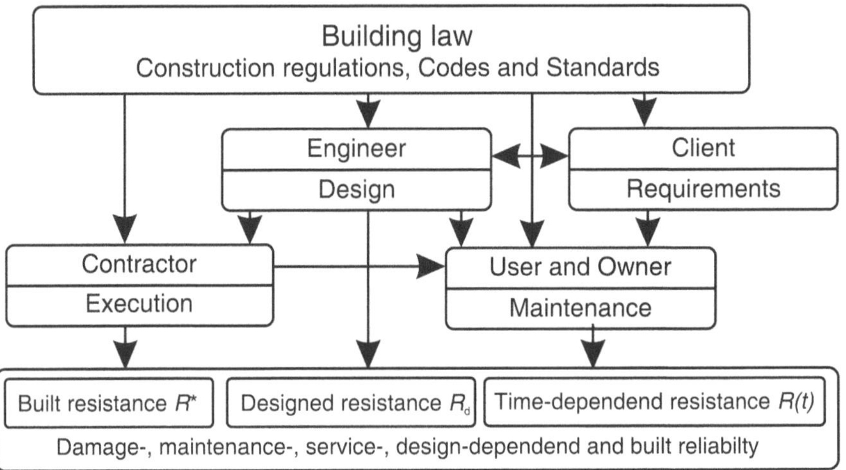

Fig. 1-20. Time dependence of the structures' resistance based on Czechowski (2001)

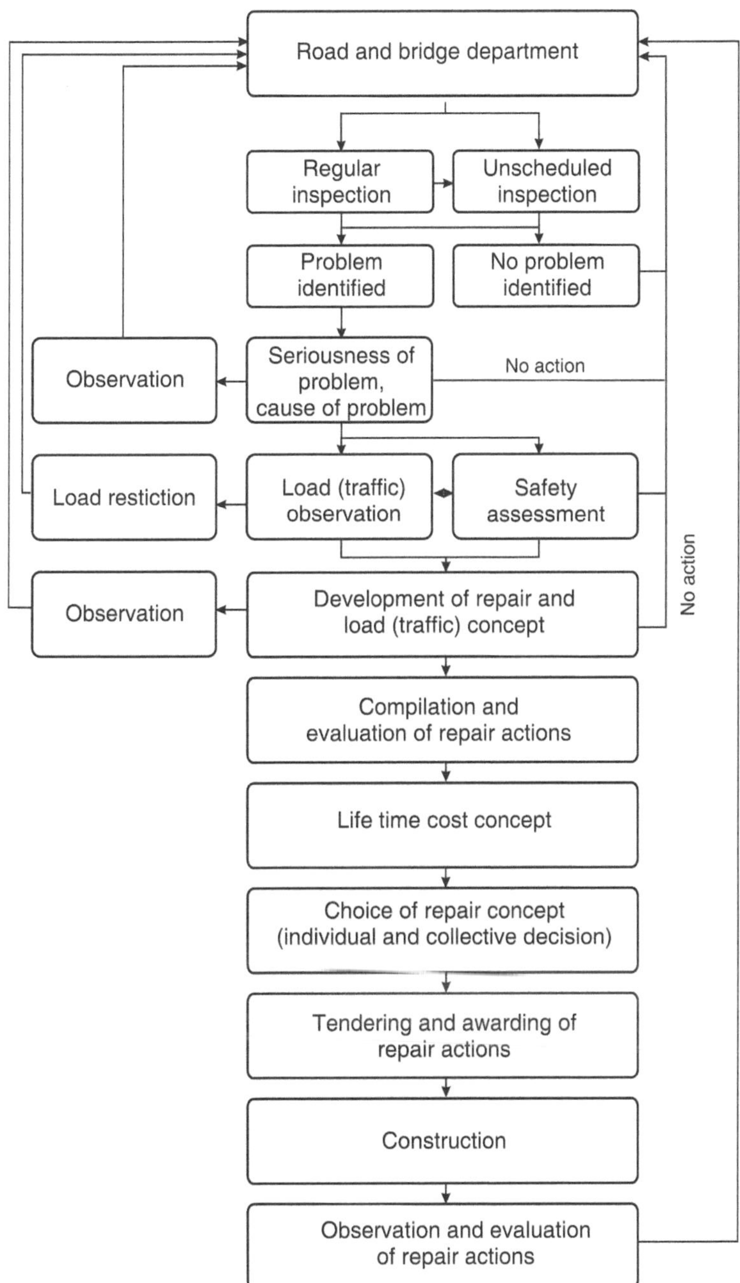

Fig. 1-21. Flow chart of bridge inspection, evaluation, and strengthening according to REHABCON (2000)

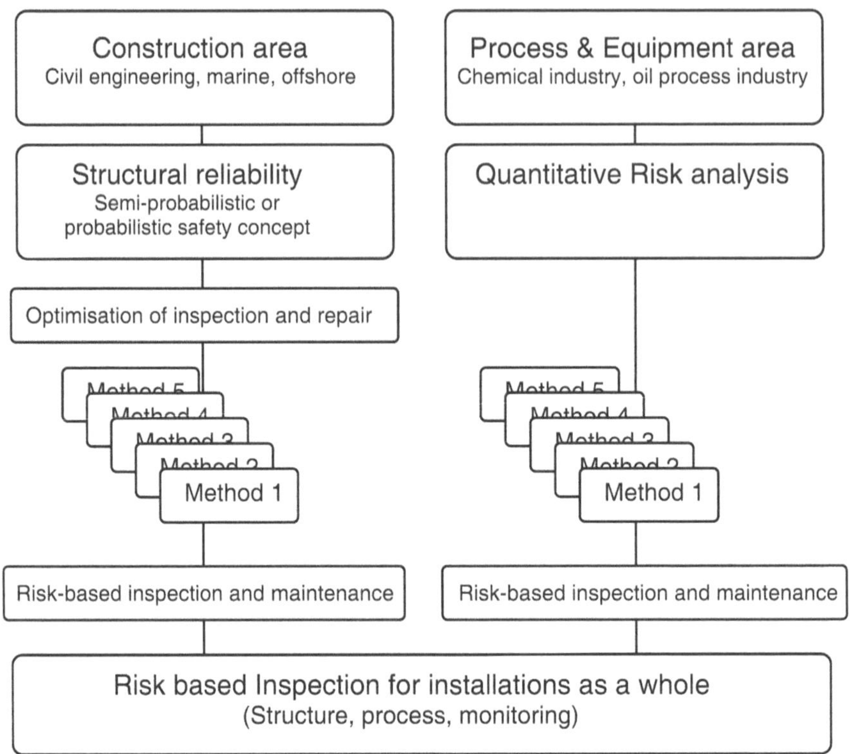

Fig. 1-22. Development of risk-based inspections and observation concepts in different industries according to Goyet (2001)

1.4 Terms

The German term for arch "bogen" can be traced back to the old high German "bogo." It can also be found in the Dutch "boog" and the English "bow." In the German dictionary by the Grimm brothers from the 1800s, it is called, following Kurrer (2002): "Arch now is the curved, bended, twisted." The root of the German term for arch (bogen) lies in the verb bending. "According to Kurrer (2002), an arch is a concave curved support framework from double flexural rigid building materials, in a structural sense."

Pauser (2003) writes on arch bridges: "Arches derive their high structural efficiency from the utilization of the compression cross section." Pauser (2002) defines an arch as "a curved compression member with a transverse load."

Weber (1999) defines: "An arch appears when a rod shaped (line shaped) support framework, whose system line is located underneath its tangents, is also concave shaped. The support framework experiences two translatory movement restraints in each support in its main curvature plane. Its construction materials are pull, push and pressure capable."

In contrast to that word exists the term "vault." The origin of the term "vault" (German: *Gewölbe*) probably lies in the Roman term "camera." This term is applied to curved ceilings and eventually not only to the ceilings themselves but also to the space below the ceiling: "... camera became the broad term for the entire room that is covered by the ceiling." This is how it is called in the Grimm dictionary (following Kurrer 2002). Later on, the term vault was ascribed back to the support structure again. Already in 1735, the following term definition can be found: "a ceiling shaped from an arch of stones." In 1857, the term was extended to other materials as well (Kurrer 2002).

The definition that domes carry out their support function only by compression-capable construction materials with negligible tensile resistance was occasionally disputed. But as the following examples prove, the definitions of the term "vault" display a great divergence.

Haser and Kaschner (1994) define: "With vault bridges, whelmed support structures are considered with an in front view curved support axis, which geometrically demonstrate a surface shape based on their limited cross-sectional height compared to their large cross-sectional width and with that distinguish themselves from the rod shaped considered arch bridges."

Lueger (from Weber 1999) defines: "A vault is a stone ceiling assembled from wedge-shaped stones that as a result impends freely and transfers its operating loads and own weight to walls and columns."

Mörsch (from Weber 1999) defines: "The vault bridges can be regarded as arched girders from a static point of view, because they exert a horizontal push as a result of vertical loads."

Dimitrov (from Weber 1999) defines: "Vaults are arched girders, whose statue is based on the pressure line."

Kurrer (from Weber 1999) defines: "A support structure is a vault when the support function, required as safeguard for traversing a space, is only realized by compression proof construction materials with a negligible tensile resistance."

Weber (1999) defines: "A vault bridge appears as a support structure to cross over roads and obstacles. This support structure is characterized by a curved system surface with either only parabolic or parabolic and elliptic points out of the sides plus a clearance of at least 2.0 m. Its material is compression capable with a negligible small tensile resistance."

In the common parlance, the distinction between arches and vaults is hardly noticed. There are various definitions for culverts as special types of vaults and arch bridges, respectively. According to Mörsch (1999), culverts can be distinguished by small spans (<8 m), by a high earth fill above the key, and by a high rise–span ratio ($f/l > 1/3$). Other publications state a span of 2 m (Orbán 2004) or 3 m (Bién and Kamiński 2004). Mörsch (1999) furthermore discerns river and valley bridges from vaults. River bridges thus distinguish themselves by flattened arches and greater spans, and valley bridges by high columns and semi circular-shaped arches.

The terms for the single components of a vault bridge are displayed in Figs. 1-23 and 1-24. English terms and definitions of arch bridge elements can be found in Griefe (2006). A radial array of stones is regarded as a real vault, whereas a false vault exists of cantilevers.

The integration of vaults and arch bridges, in the classification of bridges respectively, follows in the next section. The various kinds of vault bridges are displayed in Fig. 1-25.

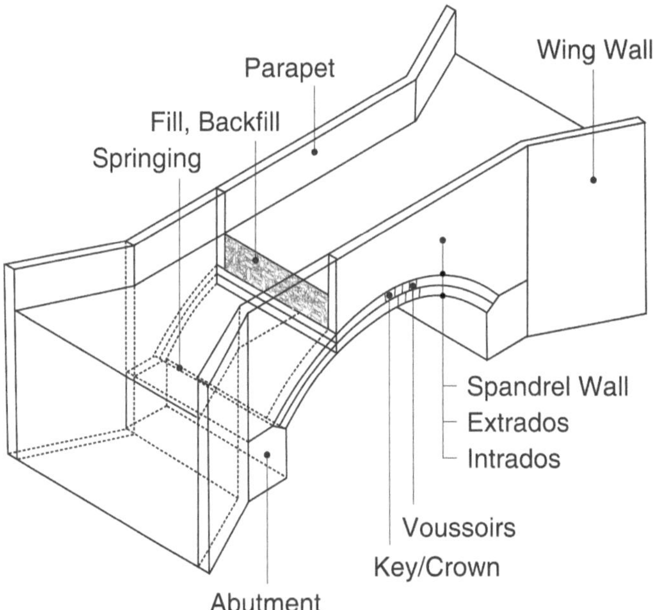

Fig. 1-23. Elements of stone arch (vault) bridges according to Huges and Blackler (1997)

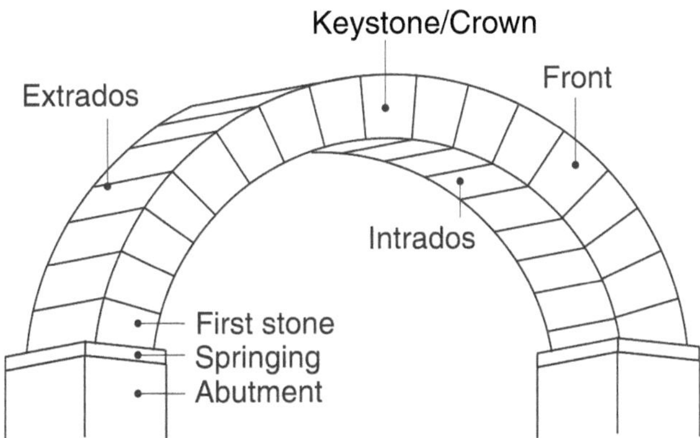

Fig. 1-24. Terms of stone arch (vault) bridges according to Koch (1998)

Further terms for special types of bridges are only touched upon hereafter, because they are reserved for education on window arches. A stilted arch has an elongation between curvature and springing. The springings lie at different heights in case of rising or single-hip arches. Further notations for arch bridges are round arches, flat bow, rising, segmental arches, elliptical arches, basket arches, shoulder, collar plunge, panel, pointed, lancet arches, clover leaf or trefoil arches, fan, jagged, keel, saddle-backed coping, flame arches, curtain arches, tudor arches, horseshoe arches, and plunge arches. The arch shapes for vaults bridges will still be treated later on.

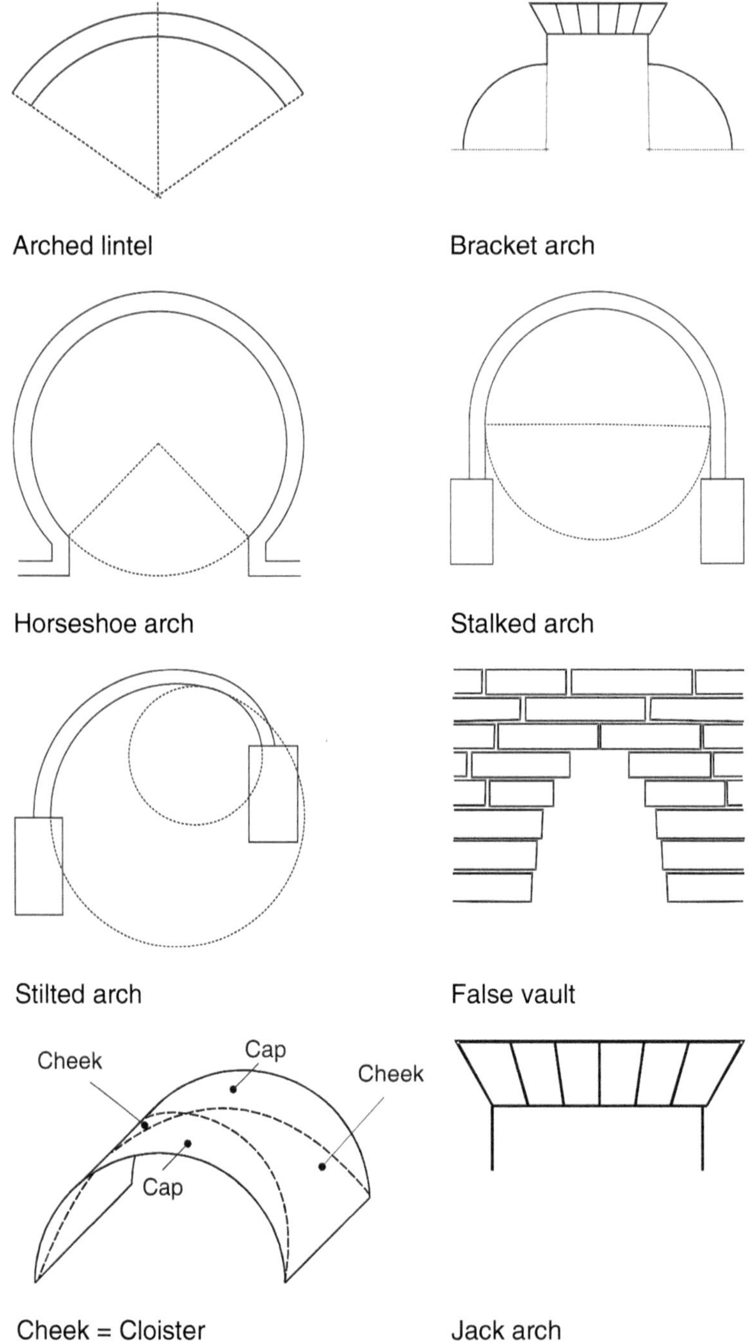

Fig. 1-25. Types of arches according to Koch (1998)

An exception is illustrated by the Roman description of the bridge for overpassing: "Aquaduct." This term stems from the Latin term *aquae ductus* = water conduit. This refers to antique Roman constructions that, as arch bridge, often transport an open or a closed water channel to a settlement.

The fillings of a vault bridge lying between the vault and the roadway are denoted as backfilling. Backfilling can be developed differently, both constructively and in the choice of materials. The backfilling can thus consist of

- unbound imported fill (loose material)
- reinforced soil by means of injection
- concrete or walled support
- hollows for weight saving, alongside arranged fittings (front and between alongside the walls) (Haser and Kaschner 1994)

With many bridges, one has attempted to achieve weight saving by hollows and openings in the backfilling. Thus, spare vaults are incorporated in numerous bridges. These spare vaults can run both in longitudinal and transverse directions. While an example of spare vaults in longitudinal direction is the Orleans Bridge by Perronet (Fig. 1-26), an example of spare vaults in transverse direction is the Verde Bridge in Italy (Fig. 1-27). The spare vaults can be arranged differently, often parallel, sometimes on top of each other, or even irregularly. In many cases, the spare vaults are covered by spandrel walls. As a rule, fully closed spandrel walls are, however, not incorporated for spans greater than 70 m to avoid unwanted stiffening by the spandrel walls. Melbourne and Tao (1995, 1998, 2004) are mentioned as further literature for bridges with open spandrel walls.

Towards the end of the 19th century, however, spare vaults were again increasingly disguised by spandrel or front walls. In Italy, one presumes that all bridges with a span over 20 m contain spare vaults in an arbitrary form. Towards the end of World War II, spare vaults were only very seldom incorporated. As a rule, the bridges were entirely backfilled, because the labour costs for the construction of spare vaults were greater than the material costs (Brencich and Colla 2002).

One of the first bridges with longitudinal spare vaults was the Westminster Bridge in London. In 1748, it was decided, after settling occurred at Pillar 4, to construct adjacent arches with smaller vaults and with the application of spare vaults with a lower deadweight to decrease the loads on the foundations (Brencich and Colla 2002).

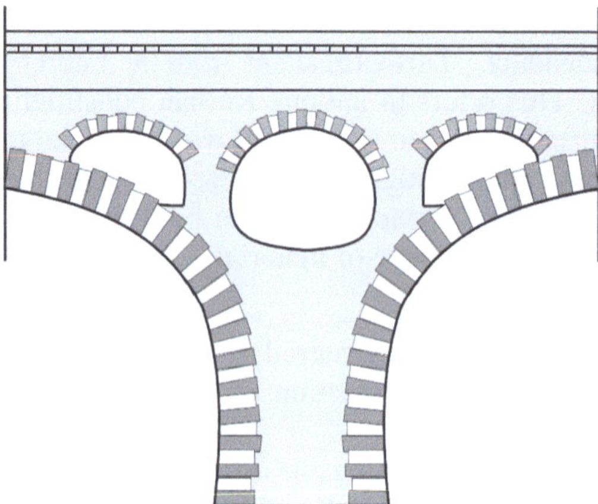

Fig. 1-26. Longitudinal spare vaults after Brencich and Colla (2002)

Fig. 1-27. Transverse spare vaults in the Verde Viaduct in Italy, constructed 1883–1889, span 18.5 m, after Brencich and Colla (2002)

Although spare vaults with very flat arches, with elliptical arches, and even with pointed arches can occasionally be found, spare vaults are usually constructed as half-circular arches to limit the horizontal loads on the walls of the spare vault, especially with the front walls as the outer wall. In the case of very flat arches (segmental arches) in the spare vault, metal chains are used halfway to transfer the horizontal loads. The layers of the spare vault are chosen in such a way that the loads on the walls of the spare vault generate a compression arch line that follows the main arch (Schaechterle 1937, 1942).

The application of multi ring arches in the main arch at the connection to the spare vaults occasionally leads to separation of the arches. This construction solution was frequently applied at the end of the 19th and the start of the 20th century. An original solution to the problem of separation of arches can be found at the Badstrassen Bridge in Berlin. This bridge was constructed between 1861 and 1864. It involves a skewed multi ring bridge across six spans. In two spans, the skewness is achieved through the shifted string of straight sectors of arches. These arch sectors are delimited in contact sector to the nearest row of arches by the front walls. Additionally, a front wall is located in the middle of such a sector of arches, as well. The front walls end at the springing sector. The connection between the front walls of the single arrays is achieved through spare vaults across the springing. These spare vaults are also filled with scrap rock and mortar. A minimal weight for simultaneous regular support of the main arch is accomplished by the combination of stiffening walls and filled spaces. Next to the application of spare vaults in the backfilling, they are also regularly used in the pillars (Brencich and Colla 2002).

The possibility of multi ring arches must also be checked when no multi ring arches are visible from the outside of the arches. Especially in the second half of the 19th century through the thirties of the 20th century, the multi ring arches were often disguised by natural stones in the edge region. Thereby, the impression of a thicker arch is created. Actually, significant distinctions can appear between the apparent thickness of the arches and the actual thickness. The Cornigliano Bridge (Italy) from the year 1932 is mentioned here as an example. Additionally, the thickness of the arches in the inner sector can often indicate significant variances, since stones are arranged with varying accuracy and rise up to various depths in the backfilling (Brencich and Colla 2002).

Also, the springing construction outside and inside often points out distinctions, as shown by the example of the Verde Bridge in Italy in Fig. 1-28. Further examples of the distinction between visible and structural compositions are shown in Figs. 1-29 and 1-30.

After this rough introduction to some of the arch bridge elements, this bridge type shall be integrated into a general bridge system.

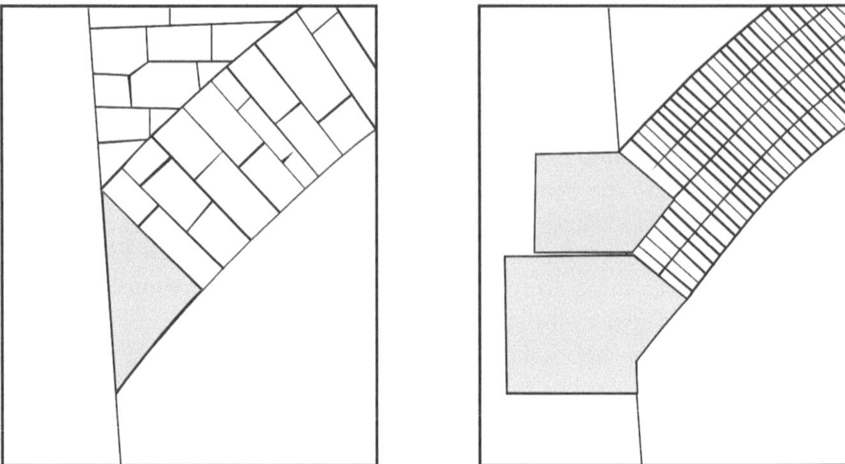

Fig. 1-28. Construction of abutments in the Verde Viaduct. The *left image* shows the visible springing and the *right image* shows the actual inner construction as multiple shelled stonework arches with overhanging springing stones (Brencich and Colla 2002).

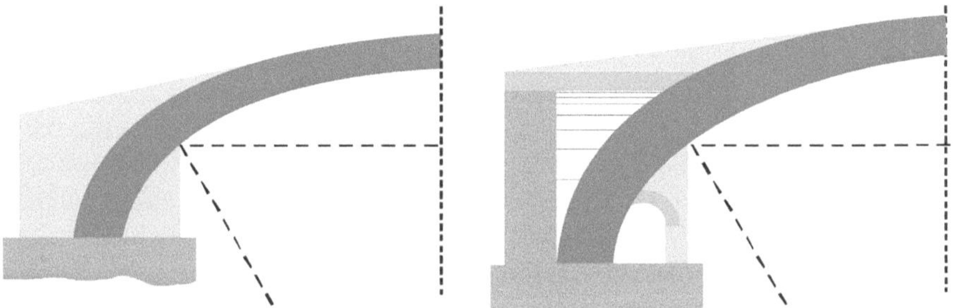

Fig. 1-29. Construction from the springings at semi elliptical arches (from Brencich and Colla 2002). Visible springings must not coincide with the static springing

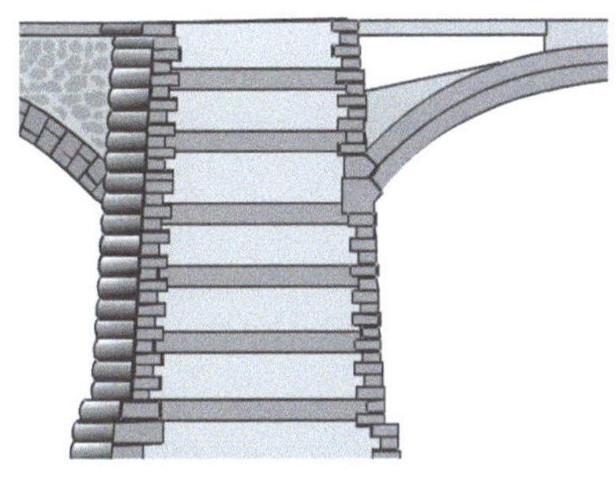
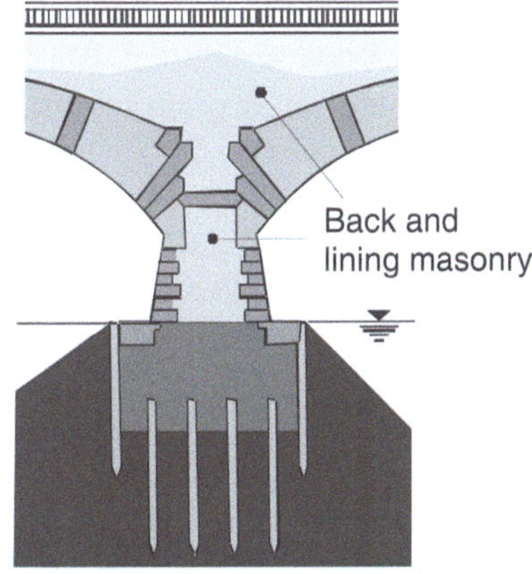

Fig. 1-30. Examples of a pillar construction after Brencich and Colla (2002)

1.5 Classification of Static Bridge Types

The ways to design bridges in terms of statical systems are limited. Schlaich (2003) has introduced a typology of bridges as shown in Fig. 1-31. Historically, the statical systems were determined by the mechanical properties of the building material, usually taken from the vicinity. However, with the development of new building materials, especially steel and composite materials such as reinforced concrete, the possibilities for bridge design increased and traditional limits were exceeded. The longest arch bridge worldwide is the Lupu Bridge in Shanghai, China, with a span of 550 m (Chen 2008). Dubai plans to build the Bur Dubai-Deira arch bridge with a span of 667 m (Chen 2008). In China, arch bridges with an incredible span of 1,000 m are planned (Čandrlić et al. 2004, Martínez 2004). Such spans can only be achieved with advanced construction materials. Whereas in historical times location and shape of the bridges were strongly limited, nowadays architects and engineers experience more freedom in the choice of bridge type by choosing from the tool box of materials.

Historical building materials were axial tensile force-capable ropes, axial compression-capable masonry, and bending and axial force-capable wood. However, wood geometries were limited by biological limitations, for example the size of trees. Using these materials, suspension bridges with ropes, arch bridges with masonry, and beam bridges with wood were constructed for probably more than two millenniums.

An arch bridge and a suspension bridge in these concepts are the assembly of several single structural elements in such a way that the experienced force types comply optimally with the capable force types. If such an arch structure is designed, the location of the roadway can either be up on the arch or suspended from the arch. The latter is not found for stone arch bridges but for steel arch bridges. Furthermore, the so-called Langer's Beam is distinguished by not transfering horizontal loads to the foundation but keeping the horizontal forces inside the roadway by tensile structural elements.

However, in the context here, a more appropriate classification of arch bridges is shown in Fig. 1-32, originating from Bién and Kamiński (2004). The classification considers the building material, the arch geometry, the arch thickness, the number of spans, and the type of the front or spandrel wall.

Bridges can cross the obstacle either in rectangular form or in another angle. Bridges crossing differently than in rectangular form are called skewed bridges. Skewed masonry arch bridges are classified according to the joints pattern (Figs. 1-33 and 1-34).

A detailed discussion of skewed, multi ring arch and multi span bridges is not part of this book. For skewed bridges, please refer to Chandler and Chandler (1995), Choo and Gong (1995), Melbourne (1998), and Hodgson (1996); for multi ring arch bridges, please refer to Gilbert and Melbourne (1995), Gilbert (1998), and Drei and Fontana (2001). For multi span bridges, detailed information can be found in Molins and Roca (1998) and Fanning et al. (2003).

Figure 1-31 shows that the arch shape is an important property for the classification of stone arch bridges. Therefore, this property will be discussed in detail in the next section.

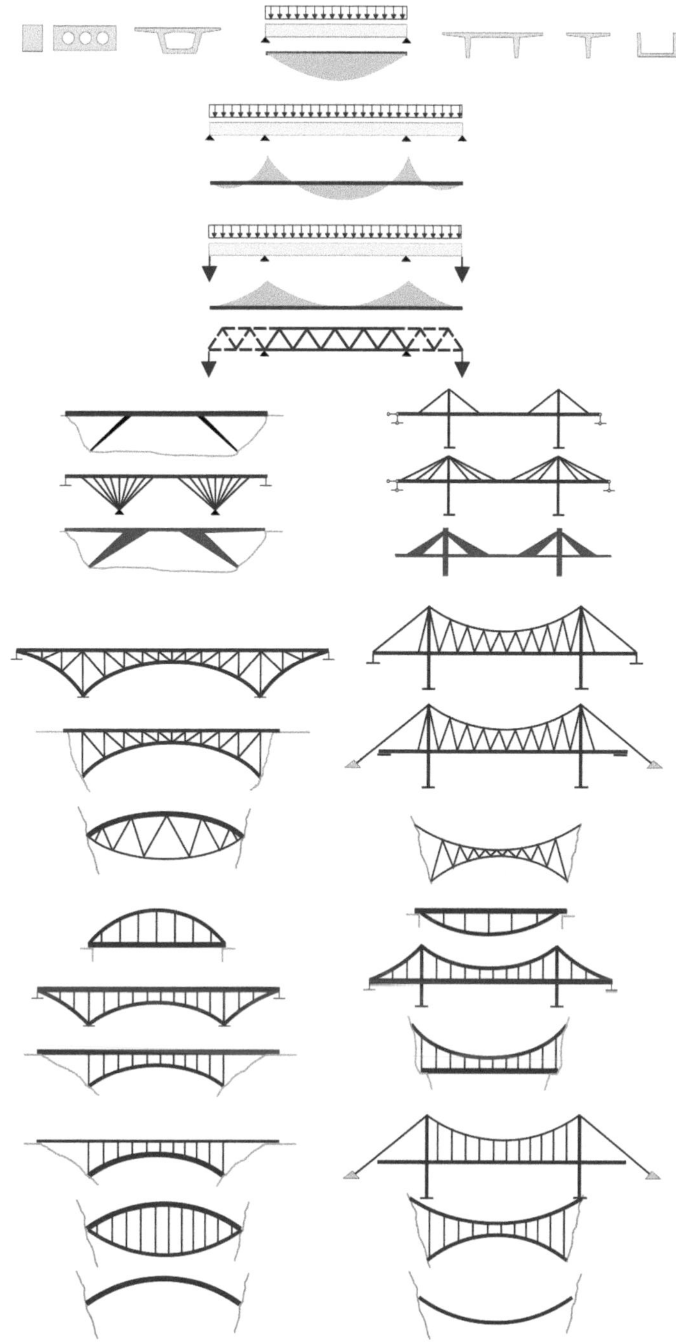

Fig. 1-31. Typology of bridges according to Schlaich (2003)

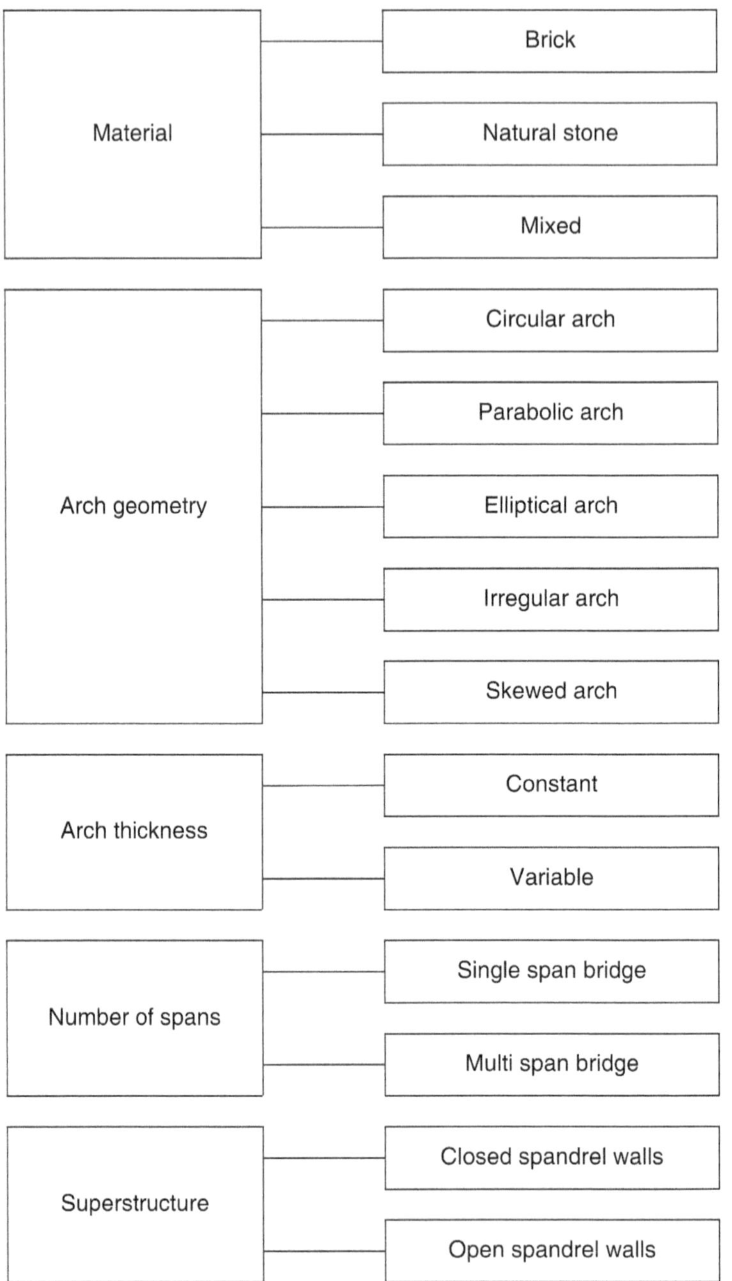

Fig. 1-32. Typology of stone arch bridges according to Bién and Kamiński (2004)

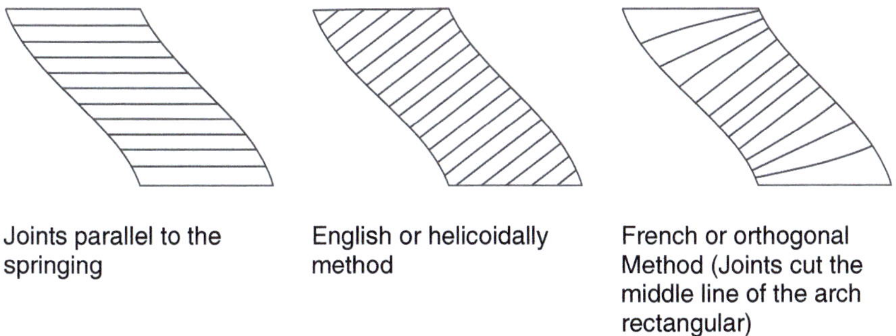

Fig. 1-33. Typology of joint patterns in skewed masonry arch bridges according to Melbourne (1998)

Fig. 1-34. Example of protruding stones to achieve skewness

1.6 Types of Arch Geometry

In the section on terms, it was indicated that arches are curved and possess a curvature. The selection of the curvature and the geometry of the arch is based on a maximum match between the line of thrust and the arch geometry. The line of thrust depends heavily on the type of loading. Therefore, for example, a circular arch is the optimal choice for a constant radial load (Fig. 1-35 *left*), a parabolic arch results from a constant vertical uniform load (Fig. 1-35 *middle*), and a catenary arch is the optimum for constant dead load of the arch (Fig. 1-35 *right*) (Petersen 1990).

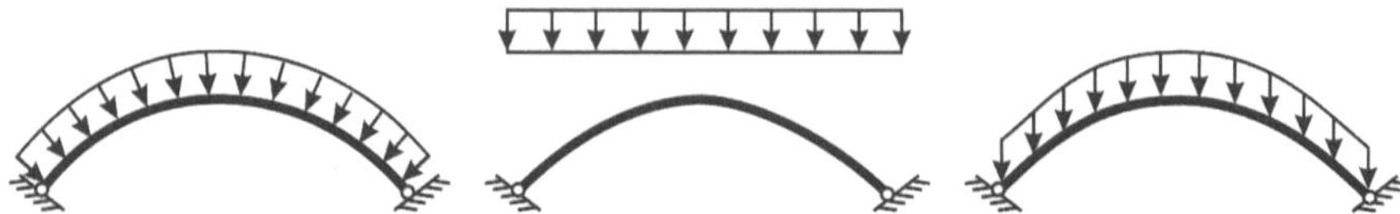

Fig. 1-35. Optimal arch shape according to different loading patterns (Petersen 1990)

The term "line of thrust" describes the geometry under which the load is only transferred by axial forces, here it is being transferred by compression forces. It can also be compared to ropes, which transfer only axial tensile forces. The affinity between ropes and arches also becomes visible by the choice of the catenary arch geometry.

However, under realistic conditions, load changes and some further conditions such as construction boundaries have to be considered when designing the arch shape. Table 1-1 gives an overview about possible arch shapes. The table mentions the circular arch, the parabolic arch, the elliptical arch, and the basket arch. The basket arch is a particular case of the circular arch since it is assembled from several circles with different radii. A further arch shape is the cycloid. A cycloid is created as the trace of a point inside a circle, when a circle is rolled on a certain line, usually a straight line. It is therefore related to the circular arch. Formulas for the computation of certain geometries can be found in Petersen (1990). Weber (1999) furthermore refers to Kammüller and Swida, mentioning a parable fourth order for an earth backfilled arch bridge. And, last but not least, Weber (1999) also mentions a sinus curve and a half wave as arch shape.

The identification of the exact arch curvature is often difficult, as shown in Fig. 1-36. In Fig. 1-36, a historical reinforced concrete vault is shown. The Santa Trinità bridge in Florence is chosen to illustrate the difficulties in a second example. Ferroni assumed in 1808 that the arch geometry followed a basket arch with six circular segments. Brizzi suggested two parabolic arches in 1951, whereas Torricelli guessed a logarithmic curve (Corradi 1998).

Table 1-1. Types of arch geometries

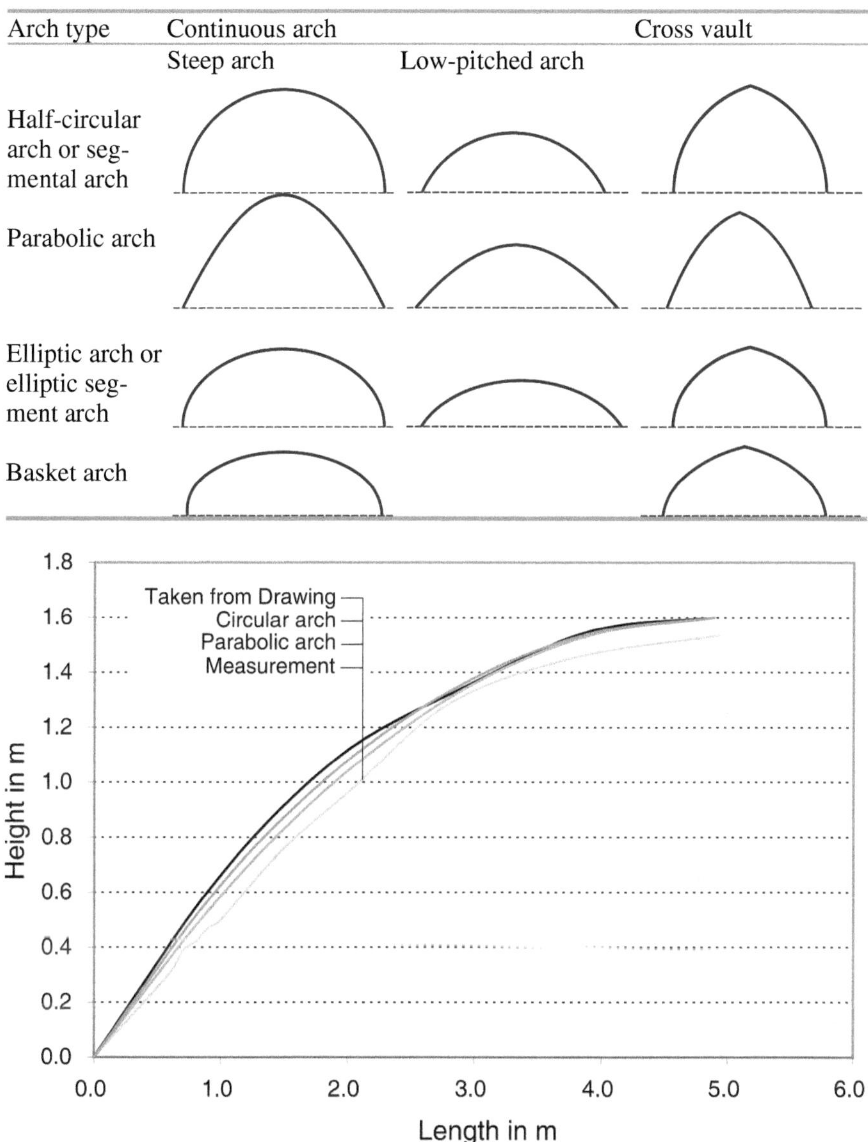

Fig. 1-36. Arch shape of a historical concrete vault

The application of different geometries was strongly related to different historic epochs. For example, the Romans mainly used the half-circular arch. However, the required rise-to-span ratio yielded a major bridge height and consequently to long driveways with a significant slope. This caused certain problems, especially in cities. After the 16th century, more and more basket arches, elliptical and centenary arches became popular, since this shape of arches avoided the long driveways and slopes.

1.7 History of Stone Arch Bridges

The application of arches and vaults for bridging space is probably several thousand years old. Barrel vaults with a span of more than 1 m were already built about 5,000 years ago in Mesopotamic burial chambers (Kurrer 2002). Von Wölfel (1999) mentions the first known vault in the royal grave of Ur about 4,000 B.C. Also, the Sumerians and the Old Egyptians knew the vault (Heinrich 1983, Martínez 2004, von Wölfel 1999).

There are many different theories on how this type of structure was invented. However, final proof of these theories is virtually impossible. One theory claims that the overturning of false vaults yielded to the first arches. Other theories consider the refinement of support stone elements or the subdivision of stone beams into single elements as shown in Fig. 1-37 (Kurrer 2002, Heinrich 1983).

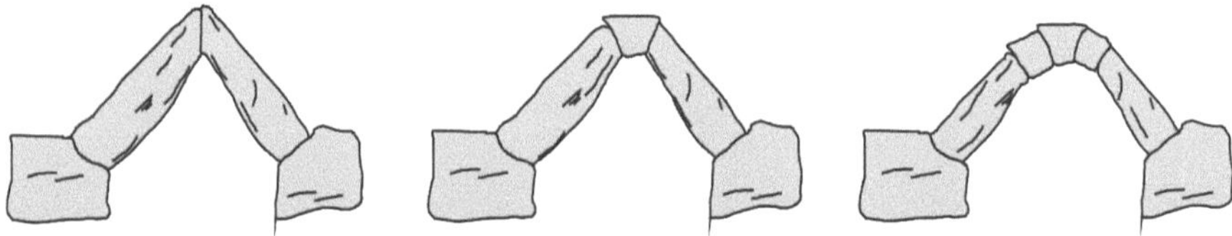

Fig. 1-37. Subdivision of stone beams into single elements

Van der Vlist et al. (1998) describes the development of arch bridges from stone heaps over small creeks. Interestingly, Bühler (2004) mentions the construction of natural bridges in the same way by Peruvian Indians.

Besides the mentioned first constructors of arch bridges, the old Greeks also knew the fault and arch structure. However, they did not pay much attention to it since they preferred strictly horizontal and vertical structural elements. The Greek vaults were only applied late (after 350 B.C.) and only used with small spans (less than 10 m) (Weber 1999). At that time, the Greeks were not the only one to build vaults. Also, the Nabatäer on the Arabian peninsula used vaults heavily for the covering of cisterns. Besides that, the Nabatäer are well-known for the mountain city of Petra in Jordan.

A first major step in the development of experienced arch bridges was during the time of the Etruscans. The Etruscans settled in the Northern middle of Italy before the time of the Roman Empire. The Etruscans have been seen as the inventor of the wedge stone arch. In wedge stone arches, every stone has a wedge-type shape, which allows a better shape of the arch compared to ashlar-shaped stones. However, the Etruscans still did not know mortar. And still, the placing of the stones in terms of adjustment of joints towards the circle centre was mainly done with low quality.

A second step in the development of arch bridges was done during the time of the Roman Empire. The Romans not only improved the quality of the placement of the stones significantly, but also invented the mortar and pentagonal-shaped stones to improve the link between the arch and the spandrel walls. This permitted a radical improvement from wide vaults built by the Etruscans to wide-spanned arch bridges with up to 36 m spans, such as the bridge in Alcántara in Spain. The bridge about the Teverone (nowadays Anio) close to Salario in Italy is one of the oldest stone arch bridges of the Romans. The time of construction has been dated to 600 B.C. The bridge was destroyed about 1,000 years after constructions by the East Goths (546 A.D.). The reconstruction was started in 569 A.D. The bridge had a span of 22 m and a rise of 11 m.

The ratio of ½ for rise to span already shows a major problem of the Roman arch bridges. Such high-rise bridges required steep ramps, which were difficult to cross by carriages. A second problem of the Roman bridges was wide piers. Here, the Romans developed countermeasures. Either the piers were constructed with further openings to permit an extended water flow in case of flooding or they simply positioned the superstructure of the bridge very high above the valley. An example of the first technique is the Fabricius Bridge in Rome. The bridge was probably constructed in 62 B.C. Interestingly, the Fabricius bridge is build on a complete circle: the upper part of the circle forms the arch of the bridge and the lower part of the circle forms an earth arch in the ground.

A further important Roman stone arch bridge is the bridge towards the Engelsburg (Leonhardt 1982). This bridge was constructed around 137 A.D. The bridge has long been considered one of the most beautiful bridges because when the water is smooth, the arch circle is closed by the mirror image in the water. The Romans strived for such expression of harmony.

Due to the excellent stonecutter work and the good foundations, many Roman arch bridges are still able to carry loads (von Wölfel 1999, Brown 1994, Zucker 1921, Jurecka 1979).

Gazzola (1963) published a catalogue of 293 known Roman arch bridge structures (von Wölfel 1999, Leliavsky 1982). According to Weber (1999), about 330 Roman arch bridges still exist. Most of these bridges are half-circular arch bridges, and some of them are already segmental circular arch bridges. In his book, O'Connor (1994) especially mentions the segmental arch Pont St. Martin in Northern Italy, built around 25 B.C. and reaching a span of 35.6 m. Most Roman arch bridges were built between 50 B.C. and 150 A.D. (Gaal 2004).

The high quality of the Roman bridges was possible because of the strong requirement of a sound Roman road system. Furthermore, the road system was the basis for the existence of the Roman Empire itself. According to Fletcher and Snow (1976), the length of the road system reached a length of 65,000 miles. The road system permitted a daily distance of about 85 km. However, the Roman courier service "Cursus Publicus" reached a daily distance of up to 335 km, with changing horses and couriers which was an incredible value for that time.

Under Emperor Trajan (98–117 A.D.), the Italian provinces of the Roman Empire had a road system of 16,000 km and about 8 million inhabitants. Therefore, in the core regions of the Roman Empire, a ratio of 2 km road per 1,000 inhabitants was reached (Von Wölfel 1999). In comparison, Germany currently has reached a value of 2.7 km per 1,000 inhabitants. Such high values could only be achieved by a strong economy. The Gross Domestic Product of the Roman Empire was at least 10% above the average Gross Domestic Product of the rest of the world at that time (Tables 1-2 and 1-3). The only region with a comparably strong economy was China.

Not only road bridges were of major importance in the Roman Empire, but viaducts—namely, another application of bridges—were also very important. Several of the best known Roman bridges are viaducts, such as the Pont du Gard (Garbrecht 1995). The viaduct of Segovia, probably constructed between 81 and 96 A.D., still reaches a distance of 818 m and the maximal height of 28 m (Garbrecht 1995). If one believes a painting by Zeno Diemer (Garbrecht 1995), then the intersection of Roman water supply pipelines such as the Aqua Claudia/Aqua Anio Novus with the Aqua Marica/Tepula/Julia was carried out at several levels and reminds one very much of modern highway intersections. The Roman water supply reached values comparable to modern water supply values (more than 140 litres per person per day).

Table 1-2. Per capita income in 1990 US dollars for different world regions (Streb 2003)

Land/region	Average per capita income in 1990 US dollar in the year			
	0	1000	1820	1998
Western Europe	450	400	1,232	17,921
USA, Canada	400	400	1,201	26,146
Japan	400	425	669	20,413
Latin America	400	400	665	6,795
Eastern Europe and USSR	400	400	667	4,354
Asia without Japan	450	450	575	2,936
Africa	444	440	418	1,368
World	444	435	667	5,709

Table 1-3. Average economical growth in percent for different world regions (Streb 2003)

Land/Region	Average economical growth in % in the years		
	0–1000	1001–1820	1821–1998
Western Europe	–0.01	0.14	1.51
USA, Canada	0.00	0.13	1.75
Japan	0.01	0.06	1.93
Latin America	0.00	0.06	1.22
Eastern Europe and USSR	0.00	0.06	1.06
Asia without Japan	0.00	0.03	0.92
Africa	0.00	0.00	0.67
World	0.00	0.05	1.21

Besides the Romans, the Persians also constructed arch bridges over a long period, presumably from 500 B.C. until 500 A.D. These arch bridges were often elements of piled up dams. Some authors assume that Roman constructions styles strongly influenced Persian bridge construction (von Wölfel 1999).

Only shortly after the massive introduction of stone arch bridges by the Romans arch bridges were also constructed in China. In contrast to the Romans, which without exception built strong and massive arch bridges, slender arch bridges were designed very early in China. The first bridge of this type was the bridge in Luoyang, probably constructed 282 A.D. Since then, many arch bridges were built in China, even arch bridges with several spans. At least one of these bridges, the Jewel Belt bridge in Suhou (built in 800 A.D.) is still in use. Furthermore, the Anji Bridge in China should be mentioned here. Based on several authors, this bridge was the first segmental bridge worldwide, designed and constructed in 605–607 A.D. The bridge reaches a span of 37 m (Brown 1994, Yi-Sheng 1978, Jurecka 1979, Graf 2005, Ding 1993 and Ding and Yongu 2001). Also, the Marco-Polo bridge in China, built in 1194, is a worldwide-known historical stone arch bridge in China (Brown 1994, Yi-Sheng 1978, Ding 1993 and Ding and Yongu 2001).

After the decline of the Roman Empire in the middle of the first millennium, the road system of the Romans degraded. This degradation also included the bridges. The economy experienced a strong depression (Table 1-3). This is a very impressive example of the relation between bridge construction and economical and social boundary condition (Heinrich 1983).

Only in the 11th and 12th centuries did a radical change reach Europe. This change was accompanied not only by new technologies in agriculture but also by the development of free bourgeoisie. The changes yielded to an improvement in economy, an improvement in supply of materials and goods, and a growth in trade. The number of cities at that time grew enormously. For example, the number of cities in Germany increased from about 100 at the beginning of the second millennium up to 3,000 until the 14th century (Heinrich 1983).

With the growth of cities and trade, the requirement and the possibility of constructing stone arch bridges also returned. A list of early medieval stone arch bridges is given in Table 1-4.

Table 1-4. List of medieval arch bridges in Europe (Velflík 1921, Heinrich 1983 and Mehlhorn and Hoshino 2007)

City	Bridge construction time	River
Toledo	Approximately 900	Tagus
Würzburg	Probably 1133–1146	Main
Regensburg	1135–1146	Danube
Prague	1158–1172	Moldavia
London	1176–1209	Thames
Avignon	1178–1188	Rhone
Dresden	1179–1260	Elbe

The origin of the bridge in Regensburg should be introduced furthermore. Regensburg was, at that time, one of the biggest cities in the German Empire and, besides Cologne, the second biggest trading centre. Several central Europe trade roads crossed in Regensburg. The relations to the South were of major importance since Venice and Genoa were the commercial and maritime super powers of the western Mediterranean Sea and had possessed excellent relations with the eastern Mediterranean Sea too. Because the German Empire at that time included major parts of Italy, Germany, and some parts of France, Regensburg was located very centrally. In contrast, Cologne was more important for trade with England (Ludwig and Schmidtchen 1992).

However, besides the central location of Regensburg in the German Empire, Regensburg had a difficult binding to the sea. Shipping at that time was of major importance, since transport on roads was the most expensive method. The cost of customs, escorts, damages on vehicles, overnight stay, and food for animals had to be provided. Transport on roads was about five times more expensive than transporting the goods on rivers and about ten times more expensive than transports on sea (Ganshof 1991).

Furthermore, not only the costs were important, but uncertainties about the time schedules were also considerable. In spring and in fall, fords were often impassable and traders had to wait weeks to pass the rivers. For example, Wilhelm the Conqueror had to wait three weeks in 1069 to pass the river Aire on his way to York (Harrison 2004). Sometimes major rivers were even seen as invincible barriers, such as the Yangtze in China (Yi-Sheng 1978).

Therefore, Regensburg had to improve the attraction of road transport. So, when in the year 1135 an unusual drought yielded to a very low river level, the construction of the stone arch bridge was launched. The bridge design was mainly based on historical Roman templates. However, in contrast to the original Roman bridges, the foundations were different, since Roman concrete was forgotten in medieval ages. Therefore, the piers had to be protected against water in a different way: islands were constructed around the piers. The arches had a span between 10.4 and 16.7 m, but the islands yielded to a greater constriction of water flow between the piers.

It has been assumed by historians that the experiences from the construction of the Regensburg arch bridge spread in Europe. For example, for the construction of the bridges in Prague and Dresden, knowledge from Regensburg was probably used. Furthermore, in many different cities bridge construction schools may have evolved after such important and successful constructions. For example in France, the friars of bridge construction were led by Saint Bénézet, the designer of the arch bridge in Avignon. Although the existence of such an order could never be proven, the construction of the arch bridge in Avignon was a milestone in the art of bridge construction. The bridge consisted not only of an incredible span of 33 m, but also on a very slender vertex. In direct comparison to the arch bridge in Regensburg, the arch bridge from Avignon looks much more slender and graceful, although the bridge was constructed only 40 years later. This excellent expression is also reached by the application of circular segments (Heinrich 1983, Brown 1994).

The application of circular segments actually became very popular only in the Renaissance. Besides the success of the segmental arch, the basket arch was applied more and more. Both yielded to more elegant views, wider spans, and lower bridge access roads than using a half-circular arch shape. A third shape, the ellipse shape, also became widely known, especially due to Dürers publication about the construction of ellipses. However, the ellipse could not assert itself and the basket arch was much more successful (Heinrich 1983).

During the 14th and 15th centuries, the basket arch and circular arch segments were widely used. For example, in Italy during that time, many famous bridges were constructed, such as the Ponte Vecchio in Florence. This bridge constructed in 1341–1345 reached a span of 32 m by a ratio of pier width to a span of only 1:6.5 (Heinrich 1983, Brown 1994). Only 200 years later, Ammanati designed another famous bridge in Florence: the Ponte Santa Trinità. The three-span bridge built from 1567 to 1569 reaches spans of 32 m. The shape of the arch also follows a basket arch, but the major clue of the bridge is the ratio of the pier width to the span of 1:7.

The bridge was so slender that the public could not believe the sufficient load-bearing behaviour of the bridge for some time. As a final example for the Italian bridge design art at that time, the Rialto Bridge in Venice, designed by Da Ponte and constructed in 1588–1591, is mentioned. The bridge has only a span of 28 m but the foundation of the bridge is extraordinary: about 12,000 wooden piles were used (Heinrich 1983, Brown 1994).

Only with the butcher bridge in Nuremberg, built in 1597–1602, could the German bridge designers reach the quality of the Italian bridge constructors again. The bridge reached a span of 34 m but shows many parallels to the Rialto bridge in Venice and was probably therefore strongly influenced by Italian bridge designers.

About 150 years later, under the French bridge designer Perronet, the construction of stone arch bridges reached its perfection. Perronet designed the bridge in Neuilly and the Concorde Bridge in Paris. The bridge in Neuilly built in 1768–1774 was a landmark in arch bridge design. The bridge consisted of five arches with a maximum span of nearly 38 m. The piers were extremely slender and had a width of only 4.2 m. This gives a ratio of span-to-pier width of 1:9.23.

The development of such excellent bridges did not come from anywhere. Just like the Romans, the central organized French state at that time had recognized the importance of a good road and bridge system. To provide that in 1716, a *Corps des Ingéniurs des Ponts et Chaussée* was founded. In 1747 in Paris, the school "Ecole des Ponts et Chaussées" was founded and Perronet was the head of this school for more than 47 years. Besides his teaching, he strongly improved the arch bridge design as examples have shown. He discovered that the arch piers, need to take vertical loads only or vertical loads and very limited horizontal loads if the adjacent arches are constructed at the same time. This would permit very slender piers, as constructed at the bridge in Neuilly. Furthermore, he heavily used slender basket arches and thus eliminated the bridge access road (Heinrich 1983, Brown 1994, Jesberg 1996).

Figure 1-38 summarizes the development of arch bridges over time. However, the last shapes already show the rise of new construction materials: steel and reinforced concrete.

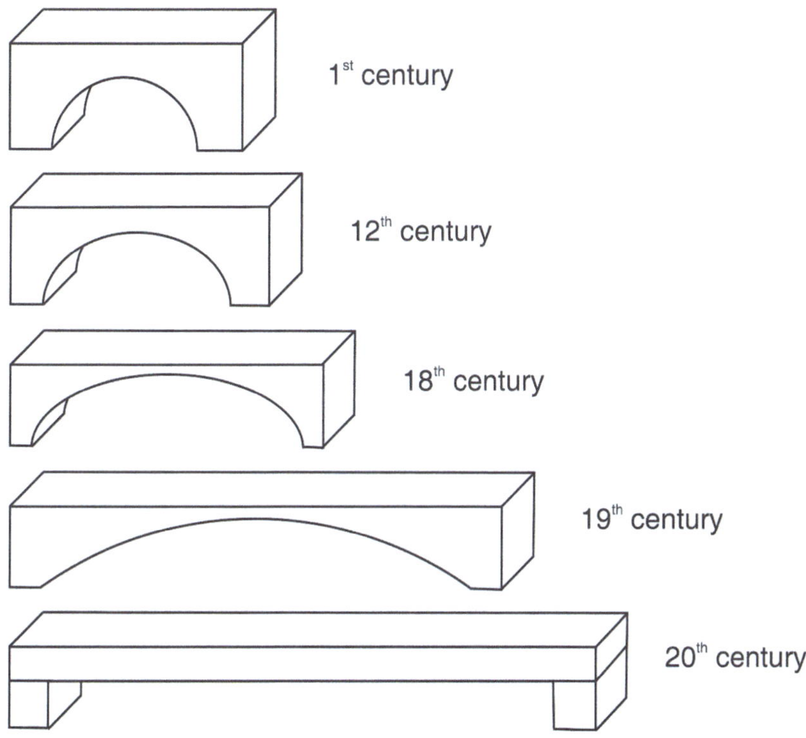

Fig. 1-38. Development of massive bridge shapes over time

Harrison (2004) also gives a summary about the historical development of arch bridges. He investigated the development of bridges over certain rivers in England from around 1540 to the middle of the 19th century (Tables 1-5 and 1-6). He furthermore tried to distinguish between arch bridges and other bridges. In contrast to the aforementioned authors, Harrison assumes that the first stone bridges in England were already there at the end of the 11th century. That would be around 50 years before the bridge in Regensburg. However, Harrison also mentions the problems in the definition of stone bridges. For example, stone bridges could have also been bridges with stone piers and wooden superstructures or with stone parapets. According to Harrison, the oldest (arch) bridge built in Oxford was ordered by Robert D'Oilliy at the end of the 11th century. A second stone bridge was built in Winchester and is perhaps even older. The estimation of the age and the construction type of the later bridge are, however, very uncertain. Around 1086 there should have been several stone bridges in England. At around 1100 the number of stone bridges increased and also the historical indications became more trustworthy (Harrison 2004).

Besides England, Harrison (2004) also states that stone bridges were built at that time in France. He assumes such bridges over the Loire in Blois and Tours before 1100. Further stone arch bridges were built in Saumur in 1162, in Orleans in 1176 and in Beaugency in 1160–1182 (Harrison 2004).

In general, stone arch bridges were very common in Great Britain between the late Middle Ages and the end of the 19th century. In the 16th century, stone arch bridges were the major part of the bridge stock. Some examples illustrate that: the old Exe Bridge in Exeter, which was found by excavations during 1960 and 1970, was probably built in the 12th century. The Framwellgate Bridge was built around 1400. The span of the bridge was exceeded only in the middle of the 19th century. A last example is the Warkworth Bridge erected around 1380 with a span of 20 m. Ruddock (1979) gives a good summary and description about constructions of arch bridges in Great Britain and Ireland in the 18th century. Therefore, the list of successful arch bridges in these countries could be further extended. Many of the historical arch bridges perform very well even today. And although age, increasing traffic, floods and storms attacking the bridges, still today the major cause of destruction of historical arch bridges is an insufficient road width of the bridges (Harrison 2004).

Table 1-5. Development of the number of bridges over English rivers over time (Harrison 2004)

River	1540	1765–75	1850
Avon (downstream from the Finford Bridge)	17	18	20
Great Ouse (from Claydon Brook to Ely)	17	24	36
Severn (from Montford Bridge)	10	10	16
Thames (from Lechlade)	17	23	36
Trent (from Stoke-on-Trent)	16	23	30
Ure und Ouse (from Bain Bridge)	10	12	16
Avon (Bristol) (downstream from Malmsbury)	13	18	21
Avon (Hants.) (from Salisbury)	7	10	11
Medway (from Ton Bridge)	8	10	12
Stour (Dorset) (from Blandford)	6	7	7
Tame (Staffs.) (from Water Orton)	6	9	9
Wear (from Stanhope)	9	12	15

Table 1-6. Construction material of bridges in England during 1540 (Harrison 2004)

River	Number of bridges	Number of stone bridges	Number of wooden bridges	Building material unknown
Avon (downstream from the Finford Bridge)	17	13	3	1
Great Ouse (from Claydon Brook to Ely)	17	8	2	7
Severn (from Montford Bridge)	10	9	1	0
Thames (from Lechlade)	17	7	9	1
Trent (from Stoke-on-Trent)	16	5	1	10
Ure und Ouse (from Bain Bridge)	10	6	2	2
Avon (Bristol) (downstream from Malmsbury)	13	9	0	4
Avon (Hants.) (from Salisbury)	7	6	0	1
Medway (from Ton Bridge)	8	7	0	1
Stour (Dorset) (from Blandford)	6	6	0	5
Tame (Staffs.) (from Water Orton)	6	4	0	2
Wear (from Stanhope)	9	6	0	3

The oldest historical arch bridges in the state of Saxony, Germany originate from the 13th century, such as the Elster Bridge in Kürbitz and a bridge in Plauen. The Hammer Bridge (1550–1576) and the Altväter Bridge (around 1570) on the Mulde were constructed in the 16th century. The state of Saxony experienced an economical boom period mainly due to the silver mining at that time. The end of the silver mining and the 30 Years War caused an economical decline and the end of the stone arch construction period. Only the inauguration of the Saxon elector and later Polish king Friedrich August I. again yielded to an economical prosperity period with further stone arch bridges. For example, in 1716–1719 in Grimma, and in 1715–1717, the Pöppelmann Bridge and a bridge in Nossen over the Mulde were constructed (Frenzel 2004).

During the origin of the Saxon railway net between 1846–1851, the world's biggest brick stone arch bridge, the Göltzschtal Bridge, was built. Furthermore, the Syratal Bridge in Plauen, constructed in 1905 with a span of 90 m, was a major landmark for stone arch bridges. It was (or still is) the masonry stone arch bridge with the longest span worldwide (Frenzel 2004).

However, already at that time, the boundary conditions for bridge construction changed increasingly. The Industrial Revolution during the 19th century, with the introduction of new means of transport such as the railway, yielded to an exponential growth of transported goods. This caused completely new requirements for bridge constructions. Furthermore, the scientific approach to solving engineering problems became more and more important. This can be clearly seen in the number of publications dealing with bridge design (BMA 2008). For example, in 1809, Wiebeking (1809) published material about a planned bridge. In 1847, Ardant, professor for construction art in Metz, published a book about the construction of bridges including material about arch bridges. He furthermore developed a device to test arch bridge models for bending and shear. Kurrer and Kahlow (1998) show the device from Ardant. Based on these results and Naviers' theory, Ardant was able to compute the deformation of the arch at the crown with

$$y = \frac{1}{2} \cdot \frac{K \cdot V}{E \cdot I} \cdot l \cdot h^2 \tag{1-2}$$

with y as deformation at crown, l as span, h as rise, V as sum of vertical loads, EI as bending stiffness, and K as factor for the consideration of the load distribution over the arch.

Schubert (1847/1848), the designer of the Göltzschtal Bridge, published his own vault theory in 1847 and 1848, and Scheffler published his vault theory in 1857 (Scheffler 1857).

After the 1920s, scientists started investigating deformations of model arch bridges in more detail. The major goal was the avoidance of the complicated numerical computation of static indeterminated arches. The tests included an arch model, a device for the application of the load and a microscope for the observation of the arch deformations. The devices were called deformers. There were different deformers developed, for example one by Beggs (1927), Bühler (1927), and Magnel and Schaechterle (Mörsch 1947). In Mörsch (1947), an illustration of the deformer by Schaechterle applied on a fixed arch is shown.

The last paragraphs gave the impression that statical computations arrived only very late in arch bridge design. In contrast, Fleckner (2003) assumes that already around the year 1200 simple static computations were carried out for vaults and piers in France. He proves that sufficient knowledge about mathematics and mechanics was provided at that time, especially in cloister schools. Fleckner (2003) further shows that independent from the different shapes and designs, all piers in churches experienced pressure under normal loading conditions. The avoidance of tensile stresses inside the piers is, according to Fleckner, satisfactory proof of the mentioned computation at the former times.

However, advanced statical computations were used for arch bridges using other materials such as reinforced concrete or steel.

1.8 Arch Bridges from Alternative Material

1.8.1 Steel Arch Bridges

The oldest historical cast iron arch bridge was constructed over the Severn Coalbrookdale in England in 1779 (Martínez 2004, Schaechterle et al. 1956, Brown 1994). The bridge consisted of five parallel arches with a span of 30.5 m. The designer was Thomas Pritchard. Twenty years later, Telford designed a cast iron arch bridge with a span of 40 m. In 1796 in Germany, the first iron road arch bridge was built over the Striegauer Water in Silesia. In the first half of the 19th century, several further such bridges were built in Germany and England. The maximum span until the middle of the 19th century was reached with 72 m in England (Martínez 2004).

In the middle of the 19th century, the material changed from cast iron to wrought iron, allowing much larger spans. In 1884, Gustav Eiffel designed the Garabit Viaduct in France with a span of 165 m. However, the application of wrought iron lasted a rather short time. Already since 1860, steel had became more and more popular. The Mississippi Bridge in St. Louis, built from 1867 to 1874, was probably the first large steel arch bridge worldwide with a span of 158.5 m. The bridge is a three-span bridge and was designed as combined road and railway bridge. The bridge is now named after the designer Eads (Martínez 2004).

In Germany, the Wuppertal Bridge in Müngsten reached 160 m in 1893, and the Rhine Bridge Engers-Neuwied reached 188 m in 1918 (Schaechterle et al. 1956).

Over the next decades, steel arch bridges experienced a major development. Using bending stiff slabs, arch bridges spans of up to 250 m were reached, with arch frameworks up to 500 m span possible. International examples are the Mälarseer Bridge in Stockholm (1935) with a 204 m span, the Henry Hudson Bridge in New York with a 244 m span, or the Sydney Harbour Bridge in Australia in 1930 with a 503 m span.

In 1977, the New Gorge River Bridge in Fayetteville was opened. The bridge has the span of 518 m. The current record is owned by the Lunpu arch bridge in Shanghai, China, with an incredible 550 m. The bridge was opened in 2003. Furthermore, the latest developments are so-called network arch bridges, which yield to extremely slender steel arch bridges (Tveit 2007 and Graße and Tveit 2007).

1.8.2 Wooden Arch Bridges

Due to the origin of wood, the application of curved structural elements made of wood represents a problem. However, with the skilled connection of single wood elements, wooden arch bridges become possible. One of the first visualizations and indications of a wooden arch bridge can be found at the Trajan pillar in Rome. The pillar was built in 110 A.D. and shows a wooden arch bridge over the river Danube, which was probably constructed during the Dacia wars (Steinbrecher 2006).

Leonardo da Vinci described a wooden arch bridge in the 15th century (Ceraldi and Ermolli 2004). In 1540, Andrea Palladio described a wooden arch bridge in Northern Italy (Fletcher and Snow 1976). In the 18th century, several wooden arch bridges were constructed in England, such as the Walton bridge over the river Thames (Ceraldi and Ermolli 2004). In the 18th century, also in France, the wooden arch Bridge Point de Choisy sur la Seine was built (Fletcher and Snow 1976). Also in Germany, wooden arch bridges were constructed (Holzer 2007).

Ivan Petrovic Kulibin designed a wooden arch bridge with a span of 300 m in the middle of the 18th century in Russia. The bridge was planned for Saint Petersburg to span the river Neva. In December 1776, a loading test on a model with scale 1:10 was carried out. Experts for evaluation of the model were, for example, Leonhard Euler and Daniel Bernoulli. Unfortunately, the bridge was never built (Bühler 2004).

In Japan, a historical wooden arch bridge has been considered as a most important historical bridge: the Kinai-Kyo Bridge (Bühler 2004, Troyano 2003). Yang et al. (2007) describe the development of wooden arch bridges in China.

The Colossus Bridge was the tallest wooden arch bridge ever built. It was designed by Lewis Wernwag over the river Schuylkill at Fairmont in Philadelphia in 1811. The bridge reached a span of 103.60 m. However, a fire destroyed the bridge in 1838 (Bühler 2004, Troyano 2003 and Fletcher and Snow 1976).

In Venice between 1933 and 1934, the wooden arch bridge Ponte dell' Accademia was constructed as a temporary structure. However, the bridge performed so well that the bridge remains functioning.

Modern wooden arch bridges show many advances compared to historical bridges. For example, parts of the structure are made of glued-laminated timber and the roadway is made of concrete as shown at the Crestawald bridge in Switzerland or the Wennerbridge in Austria (Bühler 2004). A last interesting example of a wooden arch bridge is the Leonardo Bridge in Norway (Goldberg 2006).

1.8.3 Concrete Arch Bridges

The first concrete bridge was constructed by Monier in 1875. However, it was only a pedestrian bridge. Only 30 years later, in 1904, Hennebique designed the Risorgimento Bridge in Rome with a span over 100 m (Martínez 2004). Until the end of the 19th century, concrete arch bridges were mainly built with a Spangenberg's boldness number up to 700. Spangenbergs number is computed by the span to the square divided by the rise. The Risorgimento Bridge in Rome reached a Spangenberg's number of 1,000, manifesting the change from the arch to the beam (Pauser 2002).

Maillard and Freyssinet further developed the reinforced concrete arch. For example, Freyssinet constructed an arch bridge in Villeneuve-sur-Lot with a span of 100 m in 1910. In 1925, he constructed an arch bridge in Plougastel consisting of three arches with a span of 180 m each. At that time, the formwork crystallized as problem.

From 1938 to 1942, the Sando Bridge (with a span of 264 m) was constructed. Until the finishing of the Arrabida Bridge in Porto, the Sando Bridge remained the largest arch bridge worldwide. The Arrabida Bridge was opened in 1963. The Arrabida arch reaches a span of 270 m. In the same year, the Gladesville Bridge in Australia was opened with a span of 305 m. In 1979 in Croatia, the Krk arch bridge with a span of 390 m, was finished. Currently, the largest concrete arch bridge is the Wanxian Bridge over the river Jangtse in China, with a span of 420 m. In Germany in recent years, the Kylltal Bridge was opened and, although it does not reach the span of the Wanxian Bridge, the bridge is a good example of a concrete arch bridge.

The tenders for the Millau Viaduct in France also included a reinforced concrete arch bridge with a span of 600 m. Kamisakoda et al. (2004) have also discussed concrete arch bridges with spans up to 600 m. Čandrlić et al. (2004) have published numerical investigations for a reactive powder concrete (RPC) arch bridge with a span of 1,000 m. Martínez (2004) suggests an economical usage of concrete arch bridges with a maximum span of 1,000 m and an normal span of 500 m.

Arch bridges built with modern construction materials such as steel and concrete are still very competitive bridges. New materials or construction technologies, such as concrete-filled steel tubular arch bridges, may even increase the applicability of arch bridges and the competitiveness such as recent examples in China show (Chen et al. 2004, Martínez 2004, Chen 2007).

1.9 Number of Arch Bridges

Since the history of stone arch bridges is strongly related to the development of new building materials, as stated before, the question arises how many stone arch bridges still exist about 150 years after the massive implementation of new building materials such as reinforced concrete or steel.

In Germany, currently about 120,000 bridges exist (Prüfingenieur 2004). Current estimations assume a value of 80 billion Euro for all the 120,000 bridges in Germany. Weber (1999) has estimated the value of all the German railway bridges with a span lower than 20 m with about 66 billion DM in 1993 prices. The reconstruction cost of the entire railway bridge stock would probably exceed 26 billion Euro according to Marx et al. (2006).

Although many of these bridges were constructed in recent decades, stone arch bridges still contribute significantly to this bridge stock. In some regions, they still represent the majority of bridges. For example, in the region of the road department of Zwickau (Saxony, Germany), about 1/3 of all bridges are historical stone/concrete arch bridges that were constructed at the end of the 19th century (Bothe et al. 2004). The overall number of road bridges in the German federal state of Saxony was estimated at the beginning of 1990 as approximately 4,000 (Bartuschka 1995). About 32% of this bridge stock was estimated as stone arch bridges (Bartuschka 1995). Purtak (2004) assumes several thousand stone arch bridges in Saxony.

Schmitt (2004) presumes that about 1/3 of all railway bridges with a span between 10 and 20 m are historical stone arch bridges in Germany. The German railway organization (DB AG) has estimated the overall number of stone arch bridges in the German railway system at about 35,000. That would represent about 40% of all bridges and culverts, and about 30% of bridges with a minimum span of 2 m (without the culverts) (Orbán 2004).

According to Marx et al. (2006), the German railway currently uses 29,200 railway bridges, 860 road bridges, and another 1,300 bridges for pipes, signal transfer, and other purposes. Most of the bridges have a span less than 30 m (96%). They assume that historical arch and vault bridges made up 28% of the entire bridge stock. About 25% of the bridges are rolling steel beams in concrete, 24% are steel bridges, 18% are reinforced concrete bridges, and 4% are prestressed bridges. According to them, most bridges were built between 1900 and 1920, and a second peak was between 1970 and 1995. The average bridge age is 70 years. This is only due to the fact that since the reunification of Germany more than 270 new bridges have been constructed. Before that, the average bridge age was 80 years old and bridges with an age of 170 years are still in use (Marx et al. 2006).

According to Weber (1999), stone or brick arch bridges competed with wooden bridges until approximately 1860. After the introduction of welding steel and the construction of rolling steel beams in concrete, the number of new stone arch bridges dropped (Fig. 1-39). Only after World War II, several arch bridges were rebuilt, however mainly from concrete. Since then, the construction of new stone arch bridges rests (Weber 1999).

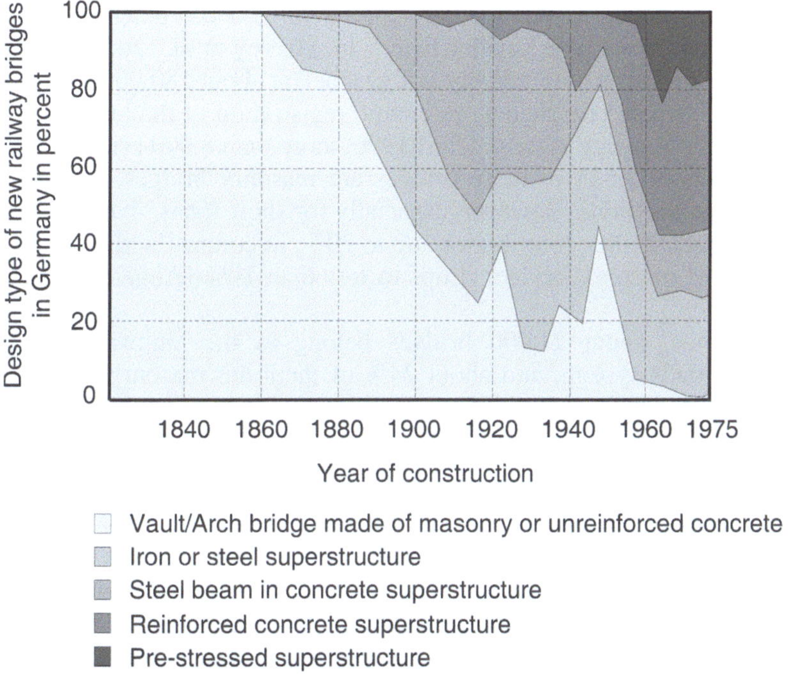

Fig. 1-39. Design-type railway bridges in Germany, according to Weber (1999), excluding wooden bridge structures. Wooden railway bridges were not permitted after 1865 based on the technical requirements of the association of German railway administration. Wooden railway bridges were built until 1860 as permanent structures with a span up to 40 m. In many other countries, wooden structures made a significant part of the bridge stock (Weber 1999).

In Great Britain, the overall number of bridges has been estimated at about 150,000 (Woodward et al. 1999). The number of stone arch bridges in the British railway system ranges in certain publications between 20,000 (UIC 2005, Orbán 2004, Melbourne et al. 2004,) and 40,000 (Choo et al. 1991). Murray (2004) assumes an overall number of about 40,000 arch bridges in Great Britain. Harvey et al. (2007) have estimated the overall British railway bridge stock at about 70,000. Smith (2003) estimates about 2,400 arch bridges, mainly brick and stone arch bridges owned and maintained by the British waterway administration. Woodward et al. (1999) have estimated that from 13,000 bridges belonging to highway and long-distance roads, about 0.7% are stone arch bridges.

Brencich and Colla (2002) estimate the number of stone arch bridges with a span greater than 8 m, and constructed in the second part of the 19 century in the Italian railway system, as at least 7,000. Cavicchi and Gambarotta (2004) number the quantity of masonry arch bridges in the Italian railway with a span greater than 2 m at 12,000, whereas 80% of this bridge stock has a span smaller than 5 m. Harvey et al. (2007) have estimated the overall Italian railway bridge stock at about 180,000.

In 1985, Spain launched a systematic registration of the bridge stock of the Spanish highway system. Until 1996, more than 8,000 bridges were recorded. About 2,824 of these bridges are masonry bridges, which represents about one third. However, especially for short spans, masonry bridges show a much higher contribution: up to 70%. In contrast with wider spans, the ratio of masonry bridges drops to less than 10% (Angeles Yáñez and Alonso 1996).

In France, about 21,000 bridges belong to the highway and long-distance roads system, and about 24% of them are masonry bridges. For comparison purposes, the France railway system includes 100,000 bridges (Harvey et al. 2007). In Spain, masonry bridges represent about 8%, and in Slovenia about 6%, of the highway bridge stock (Woodward et al. 1999). In Norway, masonry bridges make only a minor contribution to the highway bridge stock of about 1.3% (Woodward et al. 1999). The THC (2006) reports about stone arch bridges in Ireland.

Bién and Kamiński (2004) assume approximately 1,000 railway stone arch bridges and about 2,000 road stone arch bridges in Poland. However, they have counted only bridges with a minimum span of 3 m. If culverts are included, then the overall number increases to more than 20,000. Radomski (1996), in contrast, considers the contribution of masonry bridges to the Polish road bridge stock as negligible and does not record it statistically.

Karaveziroglou-Weber et al. (1998) estimate the number of stone arch bridges for central Greece as 300. Turer (2008) estimates the overall number of bridges in Turkey at about 5,000 and 100–200 are stone arch bridges. Dogangün and Ural (2007) also describe certain Anatolian stone arch bridges in some other regions of Turkey (Ural and Dogangün 2007).

The overall road bridge number in a Swiss Canton was given as 667 for 2,126 km of road distance. Masonry arch bridges make up 22%, and masonry-concrete arch bridges contribute 37% to the mentioned bridge stock (Brühwiler 2008).

The United States probably has a total bridge stock of more than 600,000 bridges (Gaal 2004, Dunker 1993). However, masonry arch bridges contribute only a minor stock with less than 1,000 (Boothby and Roise 1995). Perhaps, because this low number increases the efforts to retain the bridges, several descriptions and collections about stone arch bridges in the United States can be found (SABM 2006, Evans 2006, Forbes 2006, HBMW 2006). In Virginia, US, about 187 historical arch bridges built from concrete and masonry are recorded (Miller et al. 2000). Senker (2007) gives details about the number of stone arch bridges in Pennsylvania.

Generally, in the United States, good records about the bridges was introduced after the collapse of the Silver Bridge at Pleasant Point, West Virgina in 1967. The collapse yielded to the introduction of the Federal Highway Act in 1968, which requires the recording and documentation of bridges. However, as mentioned before, the number of arch bridges as part of the highway system is extremely low. Even reinforced concrete arch bridges contribute only minor to the bridge stock with about 0.2% (Gaal 2004, Dunker 1993).

All types of arch bridges contribute according to unofficial statistics, with 70% to the overall bridge stock of China (Dawen and Jinxiang 2004). Ou and Chen (2007) describe stone arch bridges in a region of China. Even in recent times, stone arch bridges have been constructed in China such as the New Danhe Bridge that opened in 2000 (Chen 2007). The overall number of road bridges in China is 533,618 for 3,457,000 km road (Yan and Shao 2008). The first stone arch bridge in Korea was erected in 413, the latest still existing stone arch bridge originates from the year 760 (Hong et al. 2008). Hong et al. (2007) describe certain properties of historical Korean stone arch bridges.

The Indian railway has 119,724 bridges–of which 20,967 are arch bridges according to Gupta (2008). Damage on Indian stone arch bridges is also reported (Bridge 2006). Mathur et al. (2006) describe the assessment and strengthening of arch bridges in the Indian railway system.

Shrestha and Chen (2007) mention stone arch bridges in Nepal. Examples of arch bridges in Russia, for example in Saint Petersburg, are given in The Heritage Council (THC 2006). Seiler (2006) reports about masonry railway bridges in Eritrea (Africa). Paredes et al. (2007) mention stone arch bridges in Colombia, South America.

In general, it is difficult in all the statistics to figure out the meaning of masonry bridges, arch bridges, and stone arch bridges, and so on. This, to a certain extent, limits the comparability of all these numbers.

Therefore, a systematic analysis of the arch/vault bridge stock in the international railway system was carried out by the International Union of Railways. Thirteen railway organizations contributed and over 200,000 railway arch bridges were counted (UIC 2005). Harvey has estimated the number of arch bridges in Europe even much higher. Based on the UIC (2005) report, stone arch bridges contribute up to 60% of the bridge stock of the railway organizations joining the study. A more detailed summary of the results of the investigation is shown in Table 1-7 and Figs. 1-40 and 1-41.

It is interesting to compare the data with a former study published by Weber (1999). He also has investigated the amount of stone arch bridges in the European railway stock. The results are summarized in Table 1-8.

Table 1-7. Number of stone arch bridges in different European railway organizations (UIC 2005)

Railway organization	Number of stone arch bridges and culverts	Number of stone arch bridges	Ratio of stone arch bridges on all bridges and culverts	Ratio of stone arch bridges on all bridges without culverts
French (SNCF)	78,000[1]	18,060	76.8	43.5
Italian (RFI)	56,888		94.5	
British (NR)	17,867	16,500[1]	46.9	
Portuguese (REFER)	11,746	874	89.8	39.6
German (DB)	35,000[1]	8,653	38.9	27.5
Spanish (RENFE)		3,144		49.3
Czech (CD)	4,858	2,391	18.9	35.8

[1] Estimated number

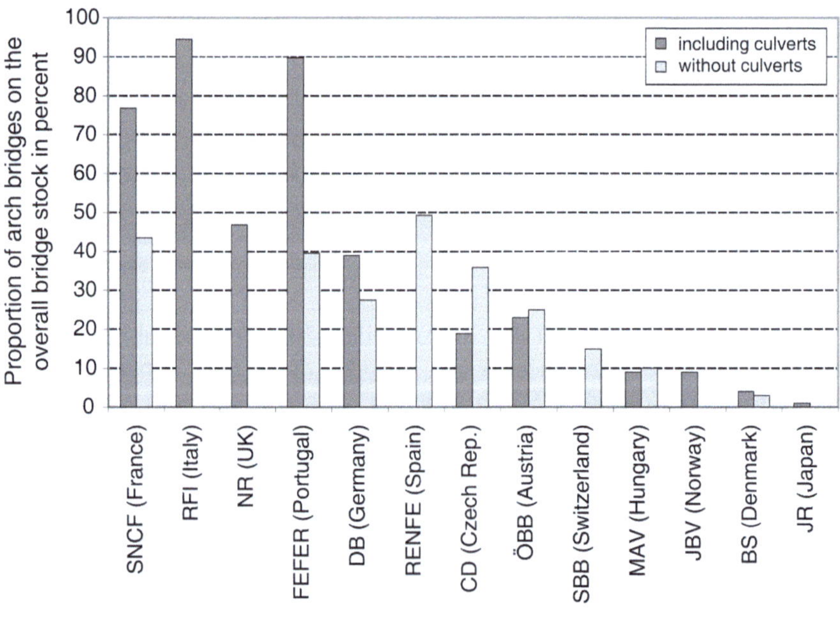

Fig. 1-40. Proportion of arch bridges on the bridge stock of several European railway organizations (UIC 2005)

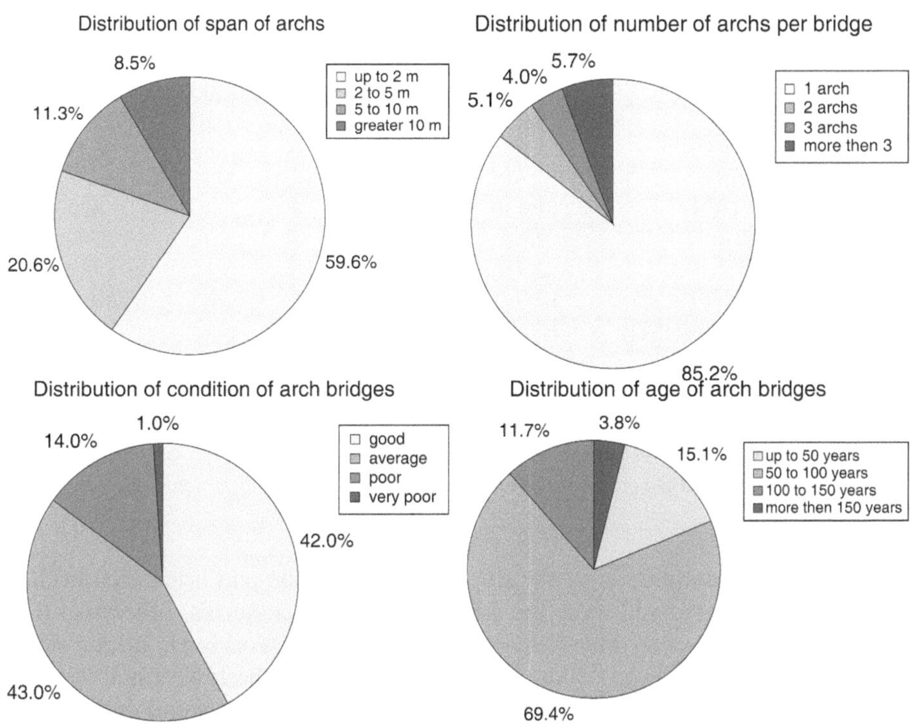

Fig. 1-41. Distribution of certain properties of the historical stone arch bridges in the European railway bridge stock (UIC 2005)

Table 1-8. Number of stone arch bridges in different European railway organizations according to Weber (1999)

Railway organization	Overall railway distance in km	Number of railway bridges[1]	Bridge density[2]	Number of arch bridges	Ratio in percent[3]	Oldest arch bridge from
Belgium (SNCB/NMBS)	3,432	3,400	10	600	18	1845
British (BR)	16,528	26,240	16	13,000	50	1825
Bulgarian (BDŽ)	4,299	982	2	62	6	1867
Danish (DSB)	2,344	1,500	6	135	9	1853
German (DB)	4,087	32,017	8	9,146	29	1837
Finnish (VR)	5,874	1,905	3	60	3	1861
French (SNCF)	32,731	28,259	9	13,167	47	1840
Greek (CH, OSE)	2,484	21,000	8	710	34	1883
Italian (FS)	16,112	59,473	37	37,400	63	1850
Irish (CIE)	1,944	2,752	14	1,484	54	1839
Yugoslavian (JŽ)		2,770		619	22	1874
Luxemburg (CFL)	275	282	10	149	53	1859

Dutch (NL)	2,753	2,790	10	50	2	1842
Norwegian (NSB)	4,027	2,700	7	311	12	1888
Austrian (ÖBB)	5,605	5,048	9	1,200	24	1838
Polish (PKP)	25,254	8,500	3	1,020	12	1842
Portuguese (CP)	3,054	1,928	6	883	46	1875
Rhaetian (RhB)	375	489	13	931		1888
Rumanian (CFR)	11,430	4,067	4	240	6	1859
Swedish (SJ)	9,846	3,500	4	100	3	1857
Swiss (SBB)*	2,985	5,267	18	914	17	1847
Spanish (RENFE)	13,041	6,371	5	3,205	50	1860
Czechoslovakia (ČSD)	13,100	9,411	7	3,213	34	1845
Hungarian (MAV)	7,605	2,375	3	278	12	1845

[1] Crossed by trains

[2] Number of bridges per 10 km railway distance

[3] Ratio of arch bridges on the overall bridge stock

* The value of 1,000 arch bridges at the Swiss Railway is also mentioned by Berset (2005)

All these numbers give the impression that stone and brick arch bridges performed very well over the last decades and centuries, otherwise they would not still contribute heavily to European and worldwide bridge stock. According to Jackson (2004), arch bridges require the lowest maintenance costs of all bridge types. The high maintenance costs currently are mainly caused by the high age of the bridges. The low lifetime costs may be one reason why new brick and stone arch bridges were constructed in Great Britain recently (Wallsgrove 1995). A new stone arch bridge has also been erected recently in Portugal (Arêde et al. 2007).

However, besides the maintenance costs and limitations of the historic bridges in fulfilling modern serviceability, historic bridges also have to bear heavily on the change in live loads over the last century and the last few decades. Therefore, it is of great interest not only to understand the modelling of live loads as discussed in detail in Chapter 2, but also how to limit and control the live load.

References

Angeles Yáñez M & Alonso AJ (1996) The actual state of the bridge management system in the state national highway network of Spain. Recent Advances in Bridge Engineering. Evaluation, Management and Repair. Proceedings of the US-Europe Workshop on Bridge Engineering, organized by the Technical University of Catalonia and the Iowa State University, JR Casas, FW Klaiber & AR Marí (Eds). Barcelona, 15–17 July 1996, International Center for Numerical Methods in Engineering CIMNE, pp 99–114

Ardant P-J (1847) Theoretisch-praktische Abhandlung über Anordnung und Construction der Sprengwerke von grosser Spannweite mit besonderer Beziehung auf Dach- und Brücken-Constructionen aus geraden Theilen, aus Bögen, oder aus der Verbindung beider; für praktische Baumeister, so wie für Vorträge über Ingenieur-Mechanik; mit einem Atlas von 28 Tafeln und in den Text gedruckten Holzschnitten. Auf Befehl des Französischen Kriegsministeriums gedruckte Abhandlung. Deutsch hrsg. von Aug[ust] von Kaven. Hannover (Hahn)

Arêde A, Costa P, Costa A, Costa C & Noites L (2007) Monitoring and testing of a new stone arch bridge in Vila Fria, Portugal. Proceedings of the Arch 07 – 5th International Conference on Arch Bridges, PB Lourenço, D Oliveira & V Portela (Eds), 12–14 September 2007, Madeira, pp 323–330

Bartuschka P (1995) Instandsetzungsmaßnahmen an hochwertiger Bausubstanz von Gewölbebrücken. Diplomarbeit. Technische Universität Dresden, Institut für Tragwerke und Baustoffe

Becqué RF (1983) De brug in Sociaal en historisch perspectief. In: Over Bruggen, H de Jong (Ed), Delft University Press, The Netherland, pp 221–272

Beggs, GE (1927) Der Gebrauch von Modellen bei der Lösung von statisch unbestimmten Systemen. Beton und Eisen, Heft 16, pp 300–304

Bennet D (1999) The Creation of Bridges – From Vision to Reality – The Ultimate Challenge of Architecture, Design and Distance. Arum Press Ltd., London

Berset T (2005) Plastische Berechnung von Naturstein-Gewölbebrücken. Kolloquium am Institut für Baustatik und Konstruktionen, ETH Zürich, 28. Juni 2005

Bién J & Kamiński T (2004) Masonry Arch Bridges in Poland. Arch Bridges IV – Advances in Assessment, Structural Design and Construction, P Roca & C Molins (Eds), CIMNE, Barcelona, pp 183–191

Birkhoff GD (1933) Aesthetic Measure. Harvard University Press, Cambridge, MA

BMA (2008) Bibliotheca Mechanico-Architectonica. www.bma.arch.unige.it

Boothby TE & Roise CK (1995) An overview of masonry arch bridges in the USA. Arch Bridges, C. Melbourne (Ed), Proceedings of the 1st International Conference on Arch Bridges, Bolton 1995, Thomas Telford, London, pp 11–18

Bothe E, Henning J, Curbach M, Bösche T & Proske D (2004) Nichtlineare Berechnung alter Bogenbrücken auf der Grundlage neuer Vorschriften. Beton- und Stahlbetonbau 99, Heft 4, pp 289–294

Brencich A & Colla C (2002) The influence of construction technology on the mechanics of masonry railway bridges, Railway Engineering 2002, 5th International Conference, 3–4 July 2002, London

Bridges (2006) http://www.iitk.ac.in/nicee/EQ_Reports/Bhuj/bridge1.htm

Brown DJ (1994) Brücken – Kühne Konstruktionen über Flüsse, Täler, Meere. Callwey Verlag München

Brühwiler E (2008) Rational approach for the management of medium size bridge stock. Bridge Maintenance, Safety, Management, Health Monitoring and Informatics, H Koh & DM Frangopol (Eds), Taylor & Francis Group, London, pp 2642–2649

Bühler A (1927) Zur Anwendung des Modellverfahrens von Professor Beggs. Beton und Eisen, Heft 16, pp 304–306

Bühler D (2004) Brückenbau. Deutsche Verlags-Anstalt GmbH, München

Čandrlić V, Radić J & Gukov I (2004) Research of concrete arch bridges up to 1000 m span. Arch Bridges IV – Advances in Assessment, Structural Design and Construction, P Roca & C Molings (Eds), CIMNE, Barcelona, pp 538–547

Cavicchi A & Gambarotta L (2004) Upper bound limit analysis of multispan masonry bridges including arch-fill interaction. Arch Bridges IV – Advances in Assessment, Structural Design and Construction, P Roca & C Molins (Eds), Barcelona, pp 302–311

Ceraldi C & Ermolli ER (2004) Timber arch bridges: a design by Leonardo. Arch Bridges IV – Advances in Assessment, Structural Design and Construction, P Roca & C Molings (Eds), CIMNE, Barcelona, pp 69–78

Chandler HW & Chandler CM (1995) The analysis of skew arches using shell theory. Arch Bridges, C Melbourne (Ed)_, Proceedings of the 1st International Conference on Arch Bridges, Bolton 1995, Thomas Telford, London 1995, pp 195–204

Chen B (2007) An overview of concrete and CFST arch bridges in China. Arch'07 – 5th International Conference on Arch Bridges. Madeira, pp 29–44

Chen B (2008) Arch bridges. College of Civil Engineering, Fuzhou University China, www.arch-bridges.cn

Chen B, Chen Y & Hikosaka H (2004) The application of concrete filled steel tubular arch bridges and study on ultimate load-carrying capacity. Arch Bridges IV – Advances in Assessment, Structural Design and Construction, P Roca & C Molings (Eds), CIMNE, Barcelona, pp 38–52

Choo BS & Gong NG (1995) Effect of skew on the strength of masonry arch bridges. Arch Bridges, C Melbourne (Ed), Proceedings of the 1st International Conference on Arch Bridges, Bolton 1995, Thomas Telford, London 1995, pp 205–214

Choo BS, Coutie MG & Gong NG (1991) Finite-element analysis of masonry arch bridges using tapered elements. Proceedings of the Institution of Civil Engineers, Part 2, December 1991, pp 755–770

Corradi M (1998) Empirical methods for the construction of masonry arch bridges in the 19th century. Arch Bridges: history, analysis, assessment, maintenance and repair. Proceedings of the second international arch bridge conference, A Sinopoli (Ed), Venice 6–9 October 1998, Balkema Rotterdam, pp 25–36

Czechowski A (2001) Structural Reliability Assurance – A Process Approach. Safety, Risk, Reliability – Trends in Engineering, Malta

Dawen P & Jinxiang H (2004) Development, actuality and strengthening of double curved arch bridges in China. Arch Bridges IV – Advances in Assessment, Structural Design and Construction, P Roca & C Molins (Eds.), Seite 444–450

Diamantidis D (2001) Assessment of Existing Structures. Joint Committee of Structural Safety JCSS, Rilem Publications S.A.R.L., The publishing Company of Rilem

Ding D & Yongu L (2001) Erfolge des Bogenbrückenbaus in China. Bautechnik 78, pp 63–66

Ding D (1993) Antike und moderne chinesische Brücken. Beton- und Stahlbetonbau 88, pp 289–296

Dogangün A & Ural A (2007) Characteristics of Anatolian stone arch bridges and a case study for Malabadi Bridge. Proceedings of the Arch 07–5th International Conference on Arch Bridges, PB Lourenço, D Oliveira & V Portela (Eds), 12–14 September 2007, Madeira, pp 179–186

Drei A & Fontana A (2001) Load carrying capacity of multiple-leaf masonry arches. Historical Constructions, PB Lourenço & P Roca (Eds), Guimarães, 2001, pp 547–555

Dunker KF (1993) Why America's bridges are crumbling, Scientific American, 266, No. 3, March, pp18–25

Ebeling W & Schweitzer F (2002) Zwischen Ordnung und Chaos. Komplexität und Ästhetik aus physikalischer Sicht. In: Gegenworte, Zeitschrift für den Disput über Wissen. Berlin-Brandenburgische Akademie der Wissenschaften, Heft "Wissenschaft und Kunst", pp 46–49

Evans GR (2006) The Clements Stone Arch Bridge – A Preservation Challenge. http://home.comcast.net/~g.k.evans/

Fanning PJ, Salomini VA & Sloan DT (2003) Nonlinear modelling of a multi-span arch bridge under service load conditions. Extending the Life of Bridges, Concrete and Composites Buildings, Masonry and Civil Structures, MC Forde (Ed), The Commonwealth Institute, London, 1st–3rd July 2003

FAZ (2005) Frankfurter Allgemeine Zeitung. Kartographie der Sehnsucht. 14th August, Nr 32, p V 1

Fleckner S (2003) Gotische Kathedralen – Statische Berechnungen. Langfassung eines Artikels im Bauingenieur 1/2003

Fletcher R & Snow JP (1976) A history of the development of wooden bridges. Paper No. 1864, American Society of Civil Engineers, Transactions. ASCE Historical Publications Nr. 4, New York

Forbes A (2006) Rehabilitation of the Keeseville Stone Arch Bridge, http://www.dot.state.ny.us/design/css/files/2001r07.pdf

Frenzel M (2004) Untersuchung zur Quertragfähigkeit von Bogenbrücken aus Natursteinmauerwerk. Diplomarbeit Technische Universität Dresden

Gaal GCM (2004) Prediction of Deterioration of Concrete Bridges. Dissertation, TU Delft

Ganshof, FL (1991) Das Hochmittelalter. Propyläen Weltgeschichte, Band 5, Propyläen Verlag, Berlin

Garbrecht G (1995) Meisterwerke antiker Hydrotechnik. Einblicke in die Wissenschaft. B.G. Teubner Verlagsgesellschaft. Stuttgart

Gazzola P (1963) Ponti romani. Contributo ad un indice sistematico con studio critico bibliografico, Ponti romani 2, Florence

Gilbert M & Melbourne C (1995) Analysis of multi-ring brickwork arch bridges. Arch Bridges, C Melbourne (Ed), Proceedings of the 1st International Conference on Arch Bridges, Bolton 1995, Thomas Telford, London 1995, pp 225–238

Gilbert M (1998) On the analysis of multi-ring brickwork arch bridges. Arch Bridges: History, Analysis, Assessment, Maintenance And Repair. Proceedings of the Second International Arch Bridge Conference, A Sinopoli (Ed), Venice 6.–9. October, Balkema Rotterdam, pp 109–118

Goldberg D (2006) Spanning the Globe – A user's guide to the new golden age of bridges. http://www.wired.com/wired/archive/13.01/bridge.html

Goyet J (2001) Basic Considerations on Risk-Based Inspection for Offshore Structures. Safety, Risk, Reliability – Trends in Engineering, Malta

Graf B (2005) Bridges that Changed the World, Prestel Publishing, Munich

Graße W & Tveit P (2007) Netzwerkbogenbrücken – Geschichte, Tragverhalten und ausgeführte Beispiele. 17. Dresdner Brückenbausymposium, 13. März 2007, Dresden, pp 183–198

Griefe NF (2006) Bridge Case Study: Forteviot Bridge over the River Earn. http://www.trp.dundee.ac.uk/research/glossary/bridge.html

Gupta RK (2008) Judging suitability of arch bridges for higher axle loading by load testing. Bridge Maintenance, Safety, Management, Health Monitoring and Informatics, H Koh & DM Frangopol (Eds), Taylor & Francis Group, London, pp 468–475

Harrison D (2004) The Bridges in Medieval England – Transport and Society 1400–1800. Clarendon Press, Oxford

Harvey B (2006) Fun Arches http//billharvey.typepad.com/photos/fun_arches/index.html

Harvey B, Ross K & Orban Z (2007) A simple, first filter assessment for arch bridges. Proceedings of the Arch 07 – 5th International Conference on Arch Bridges, PB Lourenço, D Oliveira & V Portela (Eds), 12–14 September 2007, Madeira. pp 275–279

Haser H & Kaschner R (1994) Spezielle Probleme bei Brückenbauwerken in den neuen Bundesländern. Berichte der Bundesanstalt für Straßenwesen, Verlag für neue Wissenschaft, Bergisch Gladbach

HBMW (2006) Historic Bridges of the Midwest http://bridges.midwestplaces.com/ar/pulaski/lakeshore/

Heinrich B (1983) Brücken – Vom Balken zum Bogen. Rowohlt Taschenbuch Verlag GmbH, Hamburg

Hodgson J (1996) The Behaviour of Skewed Masonry Arch Bridges. PhD Thesis, University of Salford, 1996

Holzer SM (2007) Der Bogen im Dach. Bautechnik 84, Heft 2, pp 130–146

Hong NK, Koh H-M & Hong SG (2008) Structural safety of historical stone arch bridges in Korea. Bridge Maintenance, Safety, Management, Health Monitoring and Informatics, H Koh & DM Frangopol (Eds), Taylor & Francis Group, London, pp 757–764

Hong NK, Koh HM, Hong SG, Bae BS & Yoon WK (2007) Characteristics of historical arch bridges in Korea. Proceedings of the Arch 07 – 5th International Conference on Arch Bridges, PB Lourenço, D Oliveira & V Portela (Eds), 12–14 September 2007, Madeira, pp 187–192

Huges TG & Blackler MJ (1997) A review of the UK masonry arch assessment methods. Proceedings of Institution Civil Engineering Structures and Buildings, 122, August, pp 305–315

ICOMOS (2001) Recommendations for the analysis, conservation and structural restoration of architectural heritage. International Scientific Committee for Analysis and Restoration of Structures of Architectural Heritage, Draft Version No 11, September 2001

Jackson P (2004) Highways Agency BD on New Masonary Arch Bridges. 27th February 2004. http://www.smart.salford.ac.uk/February_2004_Meeting.php

Janberg N (2008) International Database and Gallery of Structures. N Janberg Internet content services, www.structurae.net

Jensen JS, Plos M, Casas JR, Cremona C, Karoumi R & Melbourne C (2008) Guideline for load and resistance assessment of existing European railway bridges. Bridge Maintenance, Safety, Management, Health Monitoring and Informatics, H Koh & DM Frangopol (Eds), Taylor & Francis Group, London, pp 3658–3665

Jesberg P (1996) Die Geschichte der Ingenieurbaukunst. Deutsche Verlagsanstalt, Stuttgart

Jurecka C (1979) Brücken – Historische Entwicklung, Faszination der Technik. Verlag Anton Schrool & Co, Wien

Jurecka C (1979) Brücken. Historische Entwicklung, Faszination der Technik. Verlag Anton Schroll Co., Wien, Wien-München

Kamisakoda K, Nakamura H & Nakamura A (2004) Studies on long-span concrete arch bridge for construction at Ikarajima in Japan. Arch Bridges IV – Advances in Assessment, Structural Design and Construction, P Roca & C Molings (Eds), CIMNE, Barcelona, pp 548–558

Karaveziroglou-Weber M, Stavrakakis E & Karayianni E (1998) Damage of existing stone bridges in Greece. Arch Bridges: History, Analysis, Assessment, Maintenance and Repair. Proceedings of the Second International Arch Bridge Conference, A Sinopoli (Ed), Venice 6.–9. October 1998, Balkema Rotterdam, pp 361–367

Klein W (2008) Ist Schönheit messbar? 12. Berliner Kolloquium der Gottlieb Daimler- und Karl Benz Stiftung, 7. Mai 2008, Konrad Adenauer Stiftung, Berlin

Koch W (1998) Baustilkunde – Das Standardwerk der europäischen Baukunst von der Antike bis zur Gegenwart. Band 1 und 2. Faktum Lexikon Institut – Bassermann Verlag, Bertelsmann Lexikon Verlag, Gütersloh

Kurrer K-E & Kahlow A (1998) Arch and vault from 1800 to 1864. Arch Bridges: History, Analysis, Assessment, Maintenance and Repair. Proceedings of the Second International Arch Bridge Conference, A Sinopoli (Ed), Venice 6.–9. October, Balkema Rotterdam, pp 37–42

Kurrer K-E (2002) Geschichte der Baustatik. Ernst und Sohn Verlag für Architektur und technische Wissenschaften GmbH, Berlin

Leliavsky S (1982) Arches and short span bridges. Design Textbook in Civil Engineering: Volume VII, Chapman and Hall, London

Leonhardt F (1982) Brücken. Deutsche Verlagsanstalt. Stuttgart

Lourenço PB (2002) Computations on historic masonry structures, Structure Engineering Material, 4, pp 301–319

Ludwig K-H & Schmidtchen V (1992) Metalle und Macht – 1000 bis 1600 – Propyläen Technikgeschichte, Band 2, Propyläen Verlag Berlin

Martínez SPF (2004) Arches: evolution and future trends. Arch Bridges IV – Advances in Assessment, Structural Design and Construction, P Roca & C Molings (Eds), CIMNE, Barcelona, pp 11–25

Mathur M, Kumar SV & Singh K (2006) Assessment and retrofitting of arch bridges

Mehlhorn G & Hoshino M (2007) Handbuch Brücken, Entwerfen, Konstruieren, Berechnen, Bauen und Erhalten, G Mehlhorn (Ed), Springer, Heidelberg, Berlin

Melbourne C & Tao H (1995) The behaviour of open spandrel masonry arch bridges. Arch Bridges, C Melbourne (Ed), Proceedings of the 1st International Conference on Arch Bridges, Bolton, Thomas Telford, London, pp 239–244

Melbourne C & Tao H (1998) The behaviour of open spandrel brickwork arch bridges. Arch Bridges: History, Analysis, Assessment, Maintenance and Repair. Proceedings of the Second International Arch Bridge Conference, A Sinopoli (Ed), Venice 6.–9. October 1998, Balkema Rotterdam, pp 263–269

Melbourne C & Tao H (2004) The behaviour of open spandrel masonry arch bridges. Arch Bridges IV – Advances in Assessment, Structural Design and Construction, P Roca & C Molins (Eds), Barcelona, pp 125–133

Melbourne C (1998) The collapse behaviour of a multi-span skewed brickwork arch bridge. Arch Bridges: History, Analysis, Assessment, Maintenance and Repair. Proceedings of the Second International Arch Bridge Conference, A Sinopoli (Ed), Venice 6.–9. October, Balkema Rotterdam, pp 289–294

Melbourne C, Tomor AK & Wang J (2004) Cyclic load capacity and endurance limit of multi-ring masonry arches. Arch Bridges IV – Advances in Assessment, Structural Design and Construction, P Roca & C Molins (Eds), CIMNE, Barcelona, pp 375–384

Miller AB, Clark KM & Grimes MC (2000) Final Report: A Survey of Masonry and Concrete Arch Bridges in Virginia, Virginia Department of Transportation

Molins C & Roca P (1998) Load capacity of multi-arch masonry bridges. Arch Bridges: History, Analysis, Assessment, Maintenance And Repair. Proceedings of the Second International Arch Bridge Conference, A Sinopoli (Ed), Venice 6.–9. October, Balkema Rotterdam, pp 213–222

Mori Y & Kato T (2003) Practical method of reliability-based condition assessment of existing structures. Applications of Statistics and Probability in Civil Engineering, Der Kiureghian, Madanat & Pestana (Eds), Millpress, Rotterdam, pp 613–620

Mori Y & Nonaka M (2001) LRFD for assessment of deteriorating existing structures, Structure Safety, 23, pp 297–313

Mörsch E (1947) Die Statik der Gewölbe und Rahmen, Teil B, Verlag Konrad Wittwer Stuttgart

Mörsch E (1999) Gewölbte Brücken. Betonkalender 2000, Part 2, Verlag Ernst und Sohn, Berlin, pp 1–53

Murray M (2004) CIRIA Research Project RP692 Masonry and Brick Arch Bridges. Condition appraisal and remedial treatment. http://www.smart.salford.ac.uk/February_2004_Meeting.php

NN (2005) Minzhu Bridge: http://bridge.tongji.edu.cn/chinabridge/arch/133-2.HTM

O'Connor C (1994) Roman Bridges, Cambridge University Press, Cambridge

Orbán Z (2004) Assessment, reliability and maintenance of masonry arch railway bridges in Europe. Arch Bridges IV – Advances in Assessment, Structural Design and Construction, P Roca & C Molins (Eds), Barcelona 2004, pp 152–161

Ou Z & Chen B (2007) Stone arch bridges in Fujian, China. Proceedings of the Arch 07 – 5th International Conference on Arch Bridges, PB Lourenço, D Oliveira & V Portela (Eds), 12–14 September 2007, Madeira, pp 267–274

Paredes JA, Galindo JA & Mora DF (2007) Brick arch bridges in the high cauca region of Colombia (1718–1919). Proceedings of the Arch 07 – 5th International Conference on Arch Bridges, PB Lourenço, D Oliveira & V Portela (Eds), 12–14 September 2007, Madeira. pp 193–202

Pauser A (2002) Massivbrücken ganzheitlich betrachtet. Geschichte – Konstruktionen – Herstellung – Gestaltung. Österreichische Zementindustrie. Wien

Pauser A (2003) Konstruktions- und Gestaltungskonzepte im Brückenbau. Betonkalender 2004, Part 1, Ernst und Sohn Verlag für Architektur und technische Wissenschaften, Berlin, pp 27–96

Petersen C (1990) Statik und Stabilität der Baukonstruktionen. 2., durchgesehene Auflage, Vieweg Verlag

Petryna Y (2004) Schädigung, Versagen und Zuverlässigkeit von Tragwerken des Konstruktiven Ingenieurbaus. Habilitationsschrift, Ruhr Universität Bochum – Bauingenieurwesen, Schriftenreihe des Instituts für Konstruktiven Ingenieurbau, Heft 2–2004, Shaker Verlag

Piecha A (1999) Die Begründbarkeit ästhetischer Werturteile. Fachbereich: Kultur- u. Geowissenschaften, Universität Osnabrück, Dissertation

Prüfingenieur (2004) Der Prüfingenieur, April 2004, Nachrichten, p 7

Purtak F (2004) Zur nichtlineare Berechnung von Bogenbrücken aus Natursteinmauerwerk. 22. CAD-FEM Users Meeting 2004 – International Congress on FEM Technology with ANSYS CFX & ICEM CFD Conference, 10.–12. November 2004, Dresden

Radomski W (1996) Application of external prestressing for strengthening of bridge structures in Poland. Recent Advances in Bridge Engineering. Evaluation, management and repair, JR Casas, FW Klaiber & AR Marí (Eds). Proceedings of the US-Europe Workshop on Bridge Engineering, organized by the Technical University of Catalonia and the Iowa State University, Barcelona, 15.–17. July 1996, International Center for Numerical Methods in Engineering CIMNE, pp 597–612

REHABCON (2000) REHABCON IPS 2000–00063: Annex B: UK procedures for bridge inspections, assessment and strengthening, http://www.cbi.se/rehabcon/Annex_B.pdf, 2006

Rücker W, Hille F & Rohrmann R (2006) Guideline for the Assessment of Existing Structures. SAMCO Final Report 2006

Ruddock T (1979) Arch Bridges and Their Builders 1735–1835. Cambridge University Press

SABM (2006) The Stone Arch Bridge at Minneapolis, Minnesota. http://voteview.com/stonearch.htm

Schaechterle K (1937) Fortschritte im Gewölbebrückenbau, Die Strasse, p 368

Schaechterle K (1942) Ueber gewoelbte Eisenbahnbruecken aus Natur- und Kunststeinmauerwerk, Die Bautechnik, p 145

Schaechterle K, Hammacher R, Brunkow W, Neuffer W & Seitz H (1956) Brückenbau. Hütte - Des Ingenieurs Taschenbruch – Bautechnik. 28. Auflage. Verlag von Wilhelm Ernst & Sohn, Berlin

Scheffler H (1857) Theorie der Gewölbe, Futtermauern und eisernen Brücken sowol zum wissenschaftlichen Studium, als ganz besonders für den praktischen Gebrauch der Ingenieure. Braunschweig, Schulbuchhandlung

Schlaich J (2003) Brücken: Entwurf und Konstruktion. Betonkalender 2004, Teil 1, Ernst und Sohn Verlag für Architektur und technische Wissenschaften, Berlin, pp 1–26

Schmitt V (2004) Eisenbahnbrücken kleiner und mittlerer Stützweite im Stahlverbund. 14. Dresdner Brückenbausymposium. 9. März, pp 115–121

Schneider J (1996) Sicherheit und Zuverlässigkeit im Bauwesen – Grundwissen für Ingenieure. Vdf Hochschulverlag an der ETH Zürich und B.G. Teubner, Stuttgart

Schubert JA (1847/1848) Theorie der Construction steinerner Bogenbrücken. Theil 1: Dresden/Leipzig (Arnold) 1847; Theil 2: Dresden/Leipzig (Arnold) 1848; Figurentafeln: Dresden/Leipzig (Arnold) 1848

Schueremans L, Van Rickstall F, Ignoul S, Brosens K, Van Balen K & Van Gemert D (2003) Continuous Assessment of Historic Structures – A State of the Art of applied Research and Practice in Belgium. KULeuven, Department of Civil Engineering

Seiler B (2006) Eritrea: Mit Mallets durchs Gebirge. http://www.farrail.com/seiten/touren/eritrea-2006.html

Senker JH (2007) Pennsylvania stone arch bridges: evaluation and assessment. Proceedings of the Arch 07 – 5th International Conference on Arch Bridges, PB Lourenço, D Oliveira & V Portela (Eds), 12–14 September 2007, Madeira, pp 257–265

Shrestha KM & Chen B (2007) An overview on "Arch bridges in Nepal". Proceedings of the Arch 07 – 5th International Conference on Arch Bridges, PB Lourenço, D Oliveira & V Portela (Eds), 12–14 September 2007, Madeira, pp 331–339

SIA 269 (2007) Grundlage der Erhaltung von Tragwerken. Draft Swiss code, March 2007

Smith F (2003) The Reconstruction of brick arch bridges on the Chesterfield canal. Structural faults + repair 2003. Abstracts of the tenth International Conference of "Extending the life of bridges Concrete + Composites Buildings, Masonry + Civil Structures". The Commonwealth Institute, London, p 65

Staudek T (1999) On Birkhoff's Aesthetic Measure of Vases. Faculty of Informatics, Masaryk University, Brno, Czech Republic

Steinbrecher D (2006) Geschichte des Holzbaus am Beispiel des Holzbrückenbaues in Amerika und Europa. BTU Cottbus. http://www.tu-cottbus.de/Stahlbau/lehre/holzbau

Straub H (1992) Die Geschichte der Bauingenieurkunst. Ein Überblick von der Antike bis in die Neuzeit. Birkhäuser Verlag Basel

Streb J (2003) Problemorientierte Einführung in die Volkswirtschaftslehre. Vorlesungsbegleiter Teil III: Wirtschaftsgeschichte, Wintersemester 2003/04

THC (2006) The Heritage Council: Bridge Strengthening Procedures, Artificial Roosts and Roost Construction Within Bridges: Bridge Strengthening Procedures, Ireland, http://www.heritagecouncil.ie/publications/bats/ch6.html

Troyano LF (2003) Bridge Engineering – A Global Perspective. Thomas Telford

Turer A (2008) The need, challenges, and opportunities for research and application of bridge health monitoring, a Turkish experience. Bridge Maintenance, Safety, Management, Health Monitoring and Informatics, H Koh & DM Frangopol (Eds), Taylor & Francis Group, London, pp 3063–3068

Tveit P (2007) Visit to the Steinkjer network arch 44 years later. Proceedings of the Arch 07 – 5th International Conference on Arch Bridges, PB Lourenço, D Oliveira & V Portela (Eds), 12–14 September 2007, Madeira, pp 305–314

UIC (2005) International Union of Railways: Improving Assessment, Optimization of Maintenance and Development of Database for Masonry Arch Bridges, http://orisoft.pmmf.hu/masonry/

Ural A & Dogangün (2007) Arch bridges in East Black sea region of Turkey and effects of infill material on a sample bridge. Proceedings of the Arch 07 – 5th International Conference on Arch Bridges, PB Lourenço, D Oliveira & V Portela (Eds), 12–14 September 2007, Madeira, pp 543–550

Van der Vlist AA, Bakker MM, Heijnen WJ, Roelofs HJJ & Oosterhoff J (1998) Bruggen in Nederland 1800–1940. Bruggen van beton, steen en hout. Nedelandse Bruggen Stichting – Uitgeverij Matrijs

Velflík AV (1921) About four medieval stone bridges in Bohemia founded in 1169–1357. Technický obzor 18, pp 91–96

Vockrodt HJ (2005) Instandsetzung historischer Bogenbrücken im Spannungsfeld von Denkmalschutz und modernen infrastrukturellen Anforderungen. 15. Dresdner Brückenbausymposium, 15. März 2005, Dresden, pp 221–241

Vockrodt HJ, Feistel D & Stubbe J (2003) Handbuch Instandsetzung von Massivbrücken. Untersuchungsmethoden und Instandsetzungsverfahren. Basel, Boston, Berlin: Birkhäuser Verlag

von Hänsel-Hohenhausen M (2005) Vom Antlitz in der Welt. Nachwort im Buch: Meine Lieder werden leben, Pohl, I. Frankfurter Verlagsgruppe, Frankfurt/M.

von Wölfel W (1999) Kanäle, Brücken und Zisternen - Berichte aus den antiken Kulturen Ägyptens, Mesopotamiens, Südwestarabiens, Persiens und Chinas, Bautechnik Spezial, Verlag Ernst & Sohn, Berlin

Wallsgrove J (1995) The Aesthetics of Load bearing Masonry Arch Bridges. Arch Bridges, C Melbourne (Ed), Proceedings of the 1st International Conference on Arch Bridges, Bolton 1995, Thomas Telford, London, pp 3–9

Weber WK (1999) Die gewölbte Eisenbahnbrücke mit einer Öffnung. Begriffserklärungen, analytische Fassung der Umrisslinien und ein erweitertes Hybridverfahren zur Berechnung der oberen Schranke ihrer Grenztragfähigkeit, validiert durch einen Großversuch. Dissertation, Lehrstuhl für Massivbau der Technischen Universität München

Wiebeking C-F (1809) Beyträge zur Brückenbaukunde, worin auch die neue Bauconstruction wohlfeiler und dauerhafter Bogenbrücken, nach welcher mehrere große Brücken vom Verfasser angegeben und ausgeführt sind, dargestellt ist; welche als eine Fortsetzung des Perronet'schen Werkes betrachtet werden können. München, Selbstverlag

Woodward RJ, Kaschner R, Cremona C & Cullington D (1999) Review of current procedures for assessing load carrying capacity – Status C, BRIME PL97-2220. January 1999

Yan BF & Shao XD (2008) Application of China bridge management system in Qinyuan city. Bridge Maintenance, Safety, Management, Health Monitoring and Informatics, H Koh & DM Frangopol (Eds), Taylor & Francis Group, London, pp 2675–2682

Yang Y, Chen B & Gao J (2007) Timber arch bridges in China. Proceedings of the Arch 07 – 5th International Conference on Arch Bridges, PB Lourenço, D Oliveira & V Portela (Eds), 12–14 September 2007, Madeira, pp 171–178

Yeomans D (2006) The safety of historic structures. The Structural Engineer, 21st March, pp 18–22

Yi-Sheng M (1978) Bridges in China, old and new. From the Ancient Chaochow Bridge to the Modern Nanking Bridge over the Yangtze. Foreign Languages Press, Peking

Žnidarič A & Moses F (1997) Structural safety of existing road bridges. 7th International Conference on Structural Safety and Reliability (ICOSSAR 97), Kyoto, pp 1843–1850

Zucker P (1921) Die Brücke – Typologie und Geschichte ihrer künstlerischen Gestaltung. Ernst Wasmuth a. G., Berlin

2 Loads

2.1 Introduction

As mentioned in the first chapter, bridges are designed to enable the transport of goods over the shortest distance compared to the original geographical shape. Therefore, the bridge structure is exposed to certain loads, not only the loads passing over the bridge, but also other types of loading that are related to the inherent properties of bridges. Such loads include temperature caused by changing weather conditions, the sun or snow, wind loads, or simply the dead load of the bridge.

Historical arch bridges, in contrast to newly constructed bridges, require further considerations of loading. For example, a historical bridge may be unable to bear a modern traffic load, but the road can be weight restricted to enable its continued use. In this chapter, loads are mainly considered in reference to the viewpoint of historical arch bridges.

2.2 Road Traffic Loads

In the 1880s, Benz, Daimler, and Maybach more or less developed the petrol driven car in parallel. This invention is the basis for the most modern used road vehicle. However, that does not mean that only heavy trucks have been used since this time. On the 11th of July 1893, the "Blue Wonder Bridge" was opened in the German city of Dresden. The bridge was first tested by many street car wagons loaded with rocks and anchors, including several road rollers powered by steam engines, a street sprayer pulled by horses, and one military company.

Neglecting earlier road traffic, the introduction of the petrol driven cars can be seen as the beginning of an extremely successful growth of traffic in many countries worldwide. This development, as shown in Figs. 2-1 and 2-2, began in Germany approximately 100 years ago with a marginal number of motorcars, by the 1950s already more than 2.5 million cars used in Germany and nowadays about 44 million cars are reported (KBR 2001). The

D. Proske, P. van Gelder, *Safety of Historical Stone Arch Bridges*, DOI 10.1007/978-3-540-77618-5_2,
© Springer-Verlag Berlin Heidelberg 2009

maximum load for cars amounts to 44 tonnes. Lorries with higher loads are permitted by special agreement. The number of allowances for such heavy cars has increased in the last few years in Germany exponentially (Naumann 2002). On highways in Germany and Austria, lorries currently travel with a weight of up to 100 tonnes (Hannawald et al. 2003, Pircher et al. 2009).

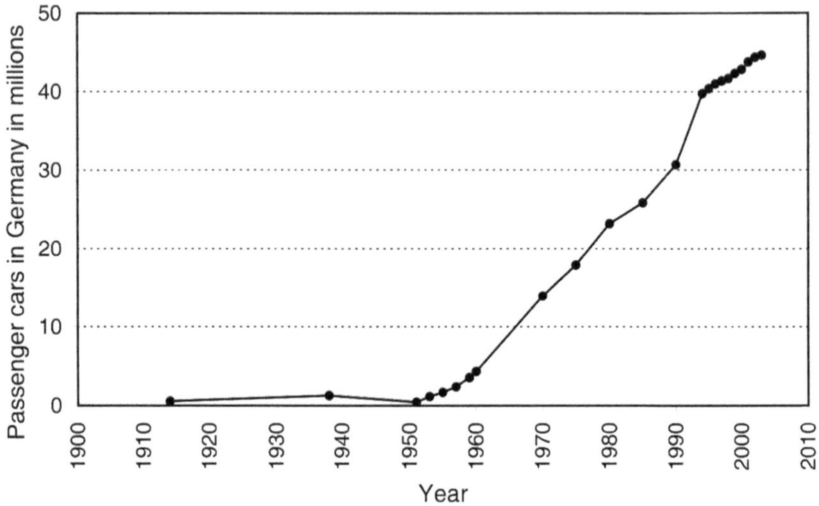

Fig. 2-1. Development of the number of motorcars in Germany mainly based on KBA (2001)

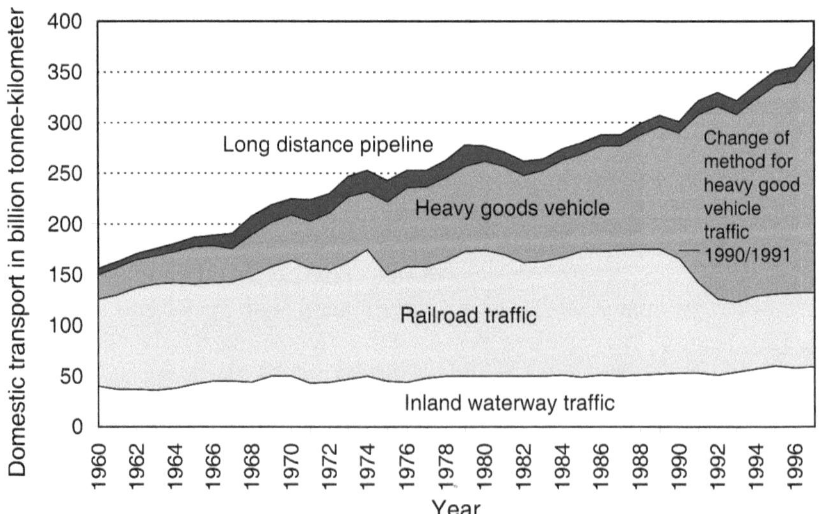

Fig. 2-2. Development of different means of transport

During the planning and design of road bridges, the engineer has to forecast road traffic for decades to come. To simplify this task, different traffic load models are included in codes of practice. Load models attempt to model traffic loads on bridges on the one hand precisely, and on the other hand, with a limited amount of work for the engineer. Good models fulfill both of these requirements at the same time. The last requirement is often questioned especially when Eurocode 1 traffic load models with many different load combinations are considered. Independent from this criticism, the intensive scientific work behind the models is appreciated. Here only the exemplarily works by Merzenich and Sedlacek (1995) are mentioned.

The road traffic model of the German DIN-reports 101 is heavily based on the Eurocode 1 model, especially ENV 1991-3. This road traffic model is valid for bridges with a maximum overall span and a maximum width of 42 m. A dynamic load factor is already considered in the characteristic loads, if necessary. In contrast to the ENV 1991-3, the German DIN-report 101 considers only three road traffic load models: load model 1 with twin axle and uniform distributed loads, load model 2 with a single axle for short structural elements, and load model 4 to describe a human scrum. A further load model in the ENV 1991-3 that considers special lorries has not been considered in the German DIN-report 101.

Table 2-1 gives characteristic traffic load values for load model 1 and Table 2-2 shows the distribution of the loads on a roadway according to the DIN-report 101 and the historical DIN 1072.

Table 2-1. Characteristic loads for load model 1 according to the DIN-report 101

Load position	Twin axle		Uniform distributed load		
	Axle load Q_{ik} in kN	Reduction factor α_{Qk}	$Q_{ik} \times \alpha_{Qk}$	q_{ik} in kN/m^2	Reduction factor α_{qk}
Lane 1	300	0.8	240	9,0	1.0
Lane 2	200	0.8	160	2.5	1.0
Lane 3	100	0.0	0	2.5	1.0
Further lands	0	-	0	2.5	1.0
Remaining area	0	-	0	2.5	1.0

Table 2-2. Characteristic loads for load model 1 according to the DIN-report 101 and DIN 1072

DIN-report 101	DIN 1072
Characteristic loads for load model 1 in the area of the twin axle	Characteristic loads for load model 1 in the area of the SLW (Heavy Load Vehicle)

DIN-report 101 (area of the twin axle):

$P_1 = 3 \times 100$ kN

$P_1 \times \varphi$ $P_1 \times \varphi$ $P_2 = 3 \times 150$ kN

P_2 P_2 $p_2 = 3$ kN/m²

DIN 1072 (area of the SLW):

$Q_1 = 2 \times 120$ kN $Q_2 = 2 \times 80$ kN

Q_1 Q_1 Q_2 Q_2

$q_1 = 9$ kN/m² $q_2 = 2.5$ kN/m²

Characteristic loads for load model 1 outside the area of the twin axle	Characteristic loads for load model 1 outside the area of the SLW (Heavy Load Vehicle)

DIN-report 101 (outside the area of the twin axle):

$p_1 \times \varphi$ $p_2 = 3$ kN/m²

DIN 1072 (outside the area of the SLW):

$q_1 = 9$ kN/m² $q_2 = 2.5$ kN/m²

3.0 | 3.0

3.0 | 3.0

Geometry of twin axle	Geometry of SLW (Heavy Load Vehicle)

Direction

0.4

0.4

2.0

1.2

1.5 | 1.5 | 1.5 | 1.5

0.2   0.2   0.2

2.0 3.0

6.0

The road traffic model of the DIN 1072 includes two bridge standard classes: the bridge class 60/30, which is used for new motorways, highways, city roads (most roads); and the bridge class 30/30, which is used for secondary roads. The bridge class 60/30 includes a main lane, a secondary lane, and remaining areas, as does the load model 1. The main lane is exposed to a uniformly distributed load of 5 kN/m^2 and six single loads of the SLW 60 (Heavy Load Vehicle). Furthermore, in the DIN 1072, the loads are dependant on the span and the coverage increases by a dynamic load factor. The secondary lane is exposed to a uniformly distributed load of 3 kN/m^2 and six single axes of the SLW 30. No dynamic load factor is applied to the secondary lane and to the remaining areas. Also, the remaining areas are exposed to a uniformly distributed load of 3 kN/m^2. In contrast to the Eurocode or DIN-report 101 load model, the uniformly distributed loads continue in the area of the SLW. Table 2-2 shows the load patterns according to the DIN-report 101 and the DIN 1072.

Besides the two standard bridge classes, the DIN 1072 has introduced bridge classes (BK 16/16, BK 12/12, BK 9/9, BK 6/6, BK 3/3) for checking or recalibration (Table 2-3). Further historical load models for standard 20 and 8 tonnes lorries can be found in Leliavsky (1982).

Table 2-3. Characteristic loads for recalibration classes of DIN 1072

Bridge class		16/16	12/12	9/9	6/6	3/3
Overall load in kN for the lorry		160.00	120.00	90.00	60.00	30.00
Front wheels	Wheel load in kN	30.00	20.00	15.00	10.00	5.00
	Contact width in m	0.26	0.20	0.18	0.14	0.14
Back wheels	Wheel load in kN	50.00	40.00	30.00	20.00	10.00
	Contact width in m	0.40	0.30	0.26	0.20	0.20
Single axle	Wheel load in kN	110.00	110.00	90.00	60.00	30.00
	Contact width in m	0.40	0.40	0.30	0.26	0.20
Uniform distributed load p_1 in kN/m^2		5.00	4.00	4.00	4.00	3.00
Uniform distributed load p_2 in kN/m^2		3.00	3.00	3.00	2.00	2.00

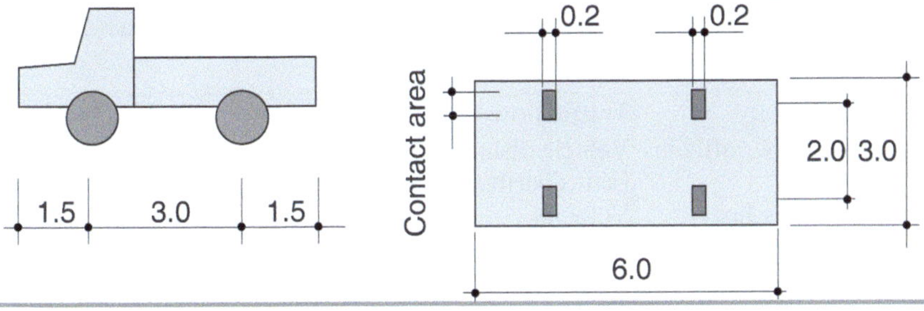

The DIN 1072 offers a wide range of different characteristic road traffic loads and is therefore permitted a fine gradation for the usage of historical bridges. This gradation cannot be found either in the Eurocode or in the German DIN-report 101. If the codes of practice no longer offer special load patterns for weight-restricted historical bridges, it would be helpful to develop different characteristic road traffic loads for such weight-restricted historical bridges.

There are many different theoretical scientific works about road traffic models that can be used as a basis for such a development. Some works in the German-speaking area were carried out by König and Gerhardt (1985), Spaethe (1977), Schütz (1991), Krämer and Pohl (1984), Pohl (1993), Puche and Gerhardt (1986), Bogath (1997), Bogath and Bergmeister (1999), Ablinger (1996), Crespo-Minguillón and Casas (1997), and O'Connor and O'Brien (2005).

Besides the Eurocode load model, different road models are used in other countries, for example the HA-KEL and HB load model in Great Britain, the HB-17 load model in the United States, or the load models T 44 and L 44 in Australia. The great diversity of load models is partially caused by the high number of influencing parameters for traffic load prediction models.

The development of traffic load models is, of course, strongly related to the essential properties of road traffic. Road traffic is and will be (for an indeterminate time) the most important means of traffic: it offers high speed, high usability by the masses, and all uses omnipresent. On roads, every non rail-tied vehicle can reach every road-realized goal at all times, and the number of road-developed goals is immense compared to all other means of traffic. This advantage of roads causes major drawbacks for the road traffic models, since the numbers of influencing parameters is extremely high. To develop models that can be used by engineers under practical conditions, the number of input parameters has to be strongly restricted. Table 2-4 shows some load-influencing factors classified in four groups.

Table 2-4. Load-influencing factors for the estimation of the characteristic road traffic loads according to Schütz (1991)

Traffic intensity	Traffic flow	Vehicle group	Single vehicle
Average daily traffic intensity	Vehicle distance	Frequency of	Number of axles
	Lane distribution	single vehicle	Axle load
Average daily heavy traffic intensity	Velocity	types	Axle distance
			Vibration properties
Maximum hourly traffic intensity			

Additional to the traffic parameters, further parameters describing the structural conditions have to be considered. Such parameters are the static system of a bridge or the quality of the roadway. Such a quality of the roadway can be considered in terms of local, regular, and irregular bumpiness. A classification of the quality of the roadway in terms of the road class is given in Table 2-5. A more detailed work about the road roughness can be found in Bogsjö (2007).

Table 2-5. Roadway quality based on road classes (Merzenich and Sedlacek 1995)

Road class	Roadway quality
Highway	Excellent
Federal highway	Good until very good
State road	Good
Country road	Average

After the identification of significant input parameters, realistic values for these parameters have to be found. These values are usually identified by traffic measurement. However, although many measurement stations on highways exist, the number on country roads associated with historical arch bridges is rather limited. In Germany in 1991, about 300 measurement stations were placed on highways and federal highways, but only 15 measurement stations could be found on country roads (Loos 2005).

Also, on the European level, the major amount of traffic measurements are carried out on highways, especially regarding the development of the international European traffic load model, which focuses heavily on high lorry traffic density measurements of long-distance traffic. Based on these measurements, classes of lorries were identified. Merzenich and Sedlacek (1995) have proposed four different lorry classes (Table 2-6).

Table 2-6. Suggestion for lorry classes according to Merzenich and Sedlacek (1995)

Class	Description of lorry	Representative lorry	
Class 1	Lorry with two axes	Two-axle vehicle	
Class 2	Lorry with more than two axes	Three-axle vehicle	
Class 3	Semi-trailer track	Two axle track with three axle semi-trailer	
Class 4	Tractive units	Three-axle vehicle with two axle with trailer	

Within the classes, a bimodal random distribution of the vehicle masses is observed as shown in Fig. 2-3 (Geißler 1995, Quan 2004). This distribution consists of a distribution for the weights of both the unloaded and the loaded lorries. Pohl (1993), for example, has considered in his model the distributions of the weight of the lorry without load and the distribution of the load itself, whereas other authors have considered axle loads as randomly distributed variables (Fig. 2-4). An overview of these different models can be found in Geißler (1995).

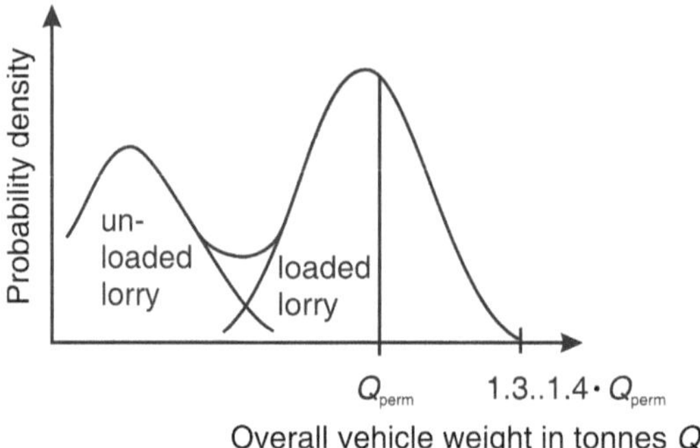

Fig. 2-3. General bimodal random distribution of the overall vehicle weight (Geißler 1995, Quan 2004)

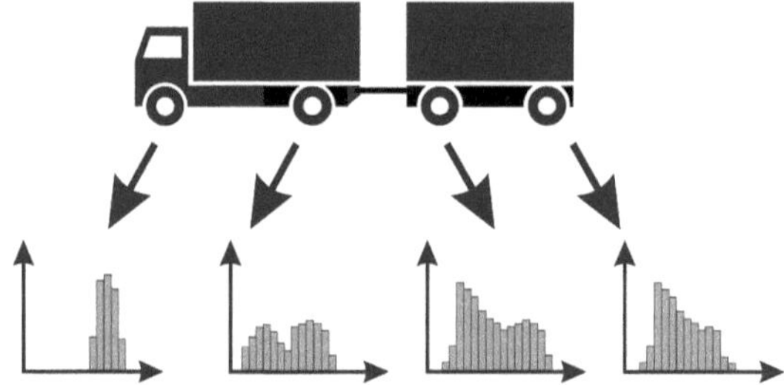

Fig. 2-4. General bimodal random distribution of the axle load (Geißler 1995)

Based on the limitation of the current Eurocode or DIN-report, which neglects models for weight-restricted bridges, an extension of the Eurocode road traffic model is recommended. However, the general procedure outlined in the Eurocode for the development of traffic models will be used. The goal of the following research is the development of α-factors that can be applied to the Eurocode traffic model 1 to permit the re-computation of load-restricted bridges.

To develop such a model, as previously suggested, requires measurement data. Such traffic measurements were carried out on the Dresden weight-restricted (15 tonnes) Blue Wonder Bridge (Fig. 2-5). Software was used to process the axle weight measurement, including identification of vehicle types. Such identification is mainly based on axle distances, and also on the correlation between axle loads, as suggested by Cooper (2002). Results of the measurements of heavy weight vehicles are shown in Fig. 2-6.

Fig. 2-5. Blue Wonder Bridge in Dresden. Although the bridge is not a historical arch bridge, it is a historical weight-restricted bridge and therefore it is assumed that the traffic properties are comparable to historical arch bridges

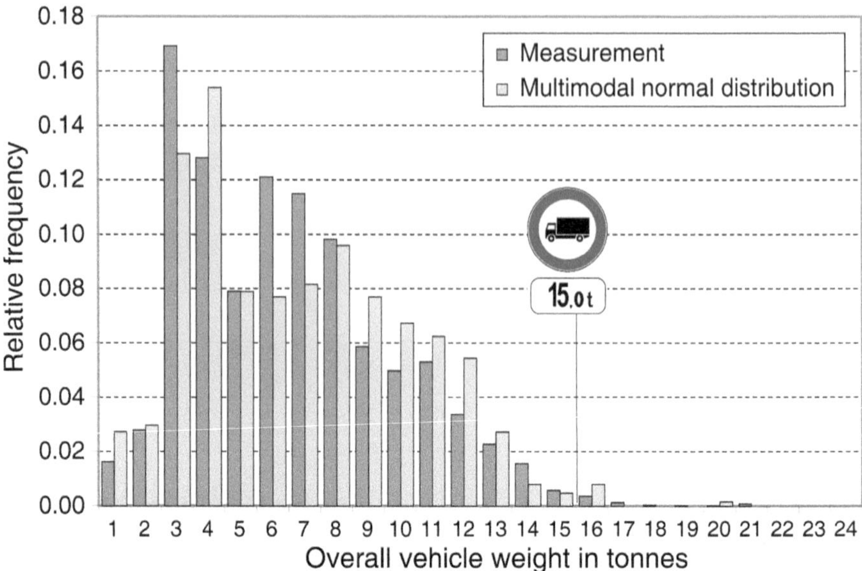

Fig. 2-6. Relative frequency of measured overall vehicle weight and adjusted multimodal normal distribution of heavy weight vehicles in October 2001 at the Blue Wonder Bridge in Dresden

The factors that will be developed should deliver traffic models that are comparable to the recalibration classes of the DIN 1072. The number inside the class gives the weight restriction for the lane, such as bridge class 30/30, 16/16 and 12/12.

Besides the input data obtained from measurement, a Monte Carlo simulation is required. The input for the simulation is the decomposition of the heavy vehicle-measured data classified into four lorry types (standard lorry or truck, truck with trailer, semi-trailer, and busses). Seeing that the measurement data from the Blue Wonder does not include all relevant information, further data were taken from Merzenich and Sedlacek (1995). The approximation of the axle load distribution was done by bimodal distribution.

In general, the α-factors will be determined in the following steps:

1. Simulation of the vehicle type based on the traffic contribution of the Blue Wonder Bridge data.
2. Simulation of the overall vehicle weight based on the available Blue Wonder Bridge data on vehicle weights.
3. Simulation of the axle load contributions to the overall vehicle weight based on the work by Merzenich and Sedlacek (1995).

4. Computation of the axle loads based on the overall vehicle weight and the axle-load contribution.
5. Simulation of the axle distances based on Merzenich and Sedlacek (1995).
6. Computation of the maximum bending moment for a single-span beam.

There are some simplifications involved in the simulation process. For example, five-axle vehicles are simplified so that the fifth axle has the same weight as the fourth axle. Also, the decisive load pattern was identified by iteration. The simulation itself was repeated 5,000 times for one length of a single-span beam. Of course, the length of the single-span beam was varied. For the maximum bending moment of a single-span beam with a certain length, the simulation yielded to a frequency distribution. A log-normal distribution has been applied parallel to the approximation of the frequency data by a normal distribution. The characteristic traffic load value is assumed as a value with a 1,000-year return period of this distribution (Merzenich and Sedlacek 1995).

After the computation of the maximum bending moment by the simulation process, the α-factor was computed by the required adaptation of the standard traffic model 1 according to the Eurocode 1 and the DIN-report 101. The standard traffic model 1, including the α-factor, should give comparable results in terms of moments as the simulated computation.

Together with flowing traffic conditions, traffic jam conditions must also be considered. Traffic jam conditions are mainly relevant for long-span conditions and have to be considered in the computation of the factors.

The described procedure was applied for the bridge class 16/16. However, for the bridge classes 12/12 and 30/30, the procedure had to be slightly changed, since measurements were only based on the bridge class 16/16.

For the bridge class 12/12, the mean value of the measurement data from the Blue Wonder Bridge was multiplied with $12/16 = 0.75$. The standard deviation and the contribution of the different vehicle types to the traffic were kept constant. For the adaptation of the bridge class 30/30, this procedure was again extended. Based on measurements from the Blue Wonder and Auxerre-traffic (Fig. 2-7), a new overall traffic weight distribution was constructed by changing mean values, standard deviation, and contribution of different vehicle types to the traffic from the Blue Wonder Bridge traffic (Fig. 2-8). The simulation was then repeated.

To prove the suggested approach, the α-factor of 1.0 for unified traffic load in the traffic model 1 in the Eurocode was verified for the Auxerre-traffic (Fig. 2-9). For the axle load, a value of 0.9 was found. However, this slight difference can be interpreted as an additional safety element. The computed α-factors are summarized in Table 2-7.

Table 2-7. Reduction factor α for different bridge classes for the new

Bridge class	Roadway quality	Lane 1		Lane 2	
		α_{Q1}	α_{q1}	α_{Q2}	α_{q2}
3/3*	Average	0.1	0.22		
6/6*	Average	0.2	0.24		
9/9*	Average	0.25	0.26		
12/12	Good	0.30	0.28	0.20	1.00
	Average	0.30	0.30	0.25	1.00
16/16	Good	0.35	0.30	0.35	1.00
	Average	0.35	0.40	0.45	1.00
30/30	Good	0.55	0.70	0.50	1.00
	Average	0.60	0.70	0.80	1.00
Simulation "Auxerre-traffic"	Good	1.00	0.90	1.00	1.00
Load model 1 DIN-report 101		0.80	1.00	0.80	1.00

*First drafts

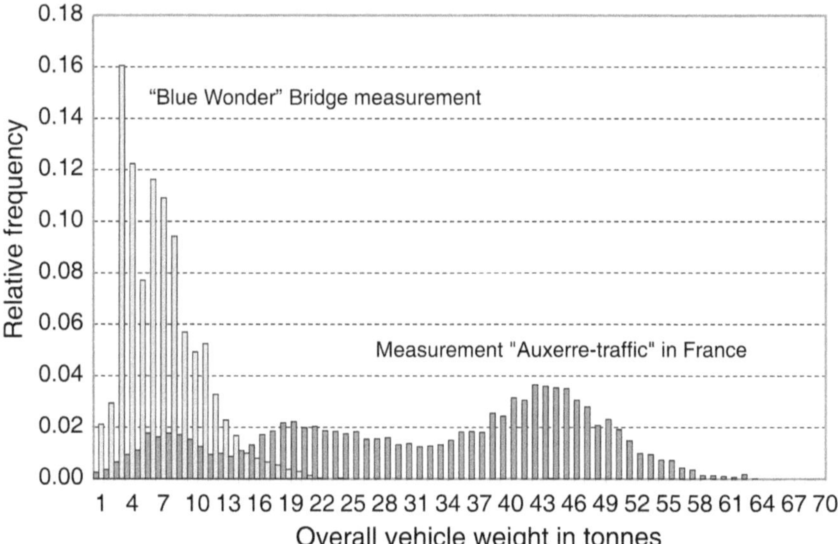

Fig. 2-7. Comparison of overall vehicle weights measured at the Blue Wonder Bridge in Dresden and at the Auxerre-traffic in France. The latter was mainly used for the development of the Eurocode traffic model 1

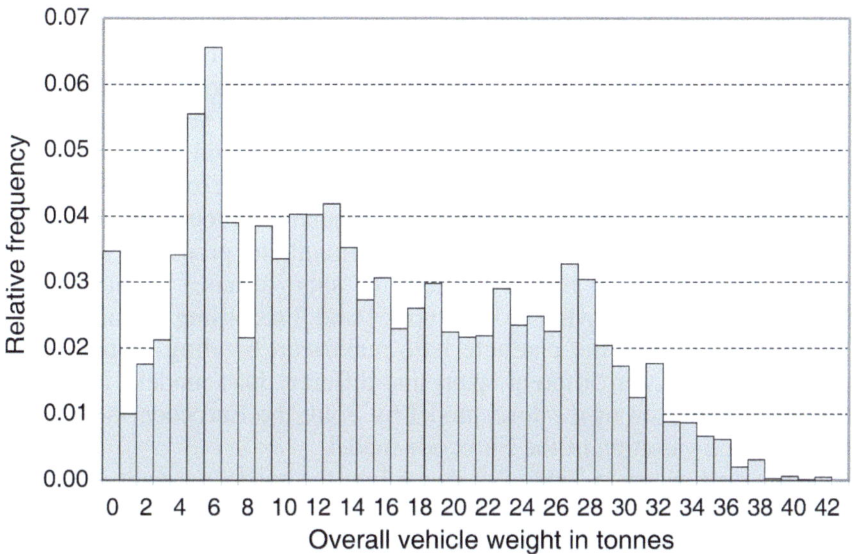

Fig. 2-8. Development of a synthetic traffic distribution for the bridge class 30/30 based on measurements from the Blue Wonder in Dresden and from the Auxerre-traffic in France based on Loos (2005)

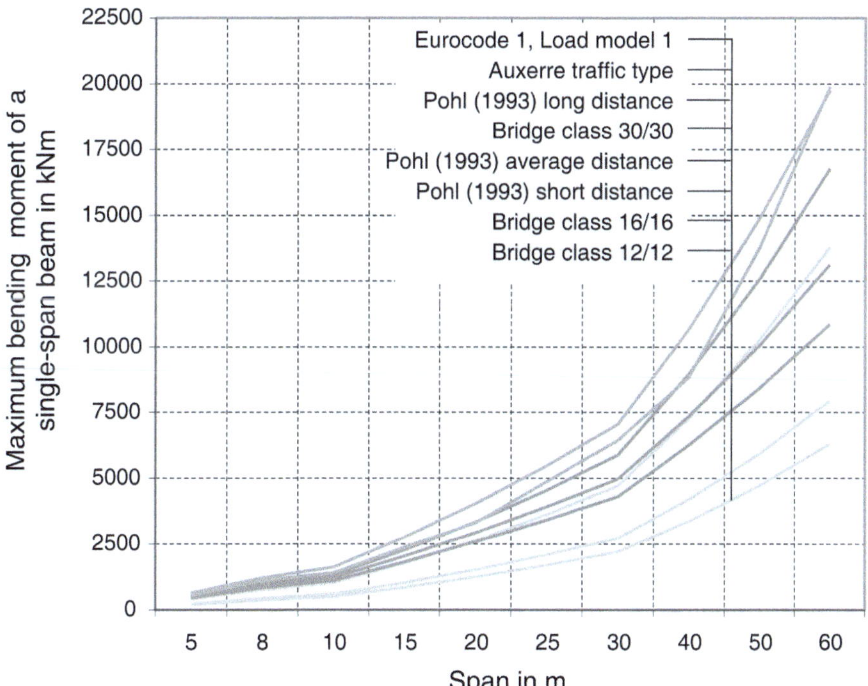

Fig. 2-9. Maximum bending moments caused by characteristic traffic loads, including dynamic load factor according to Loos (2005)

To provide further control of the suggested factors, the results have also been compared with the road traffic model by Pohl (1993). Pohl distinguishes between long-distance, average-distance and short-distance traffic. Long-distance traffic represents more or less heavy vehicle traffic on German highways. Average-distance traffic can be found on federal highways and on country roads, and short-distance traffic can be found on weight-restricted routes. Therefore, the simulation procedure using the model of Pohl is slightly different than the simulation procedure explained above. Furthermore, in our simulation, the short-distance traffic is separated into two types (types 1 to 4 or types 1 and 2 according to Table 2-6).

Figure 2-9 shows the characteristic maximum bending moment of a single-span beam for different spans and different load models. It permits a direct comparison of the load model by Pohl, the Eurocode model, and the suggested variation of the Eurocode model.

The factors given in Table 2-7 depend on the roadway quality. Usually this property is not given in codes, however here it is assumed that for country roads lower roadway quality can be found that has significant impacts on the chosen α-factor due to the dynamic properties. Here, the model from Merzenich and Sedlacek (1995) has been applied. In general, the different roadway qualities do not have such a strong effect on the main lane, however the second lane is influenced more by this property.

It should be stated here that the factors found are a much more appropriate method for the recomputation of historical bridges than the sometimes found simple diminishing factors of the Eurocode load model 1 of 0.9, 0.8, 0.7 and so on, as has been suggested in Vockrodt (2005). Applying these factors of 0.9, 0.8 or 0.7 to the Eurocode model 1 violates general assumptions in the statistical properties of model 1.

Further adaptations of the Eurocode model 1 are known, or different models have been proposed besides the presented schematic. Such an additional adaptation has been presented by Novák et al. (2007).

A different proposal for the consideration of local traffic conditions for local traffic loads has been offered by Bailey and Hirt (1996). This method shows the corresponding stochastic basis much stronger than the Eurocode model 1. The adaptation of traffic load to local traffic according to Bailey and Hirt (1996) considers

- Maximum overall lorry weights in q_{max} in kN/m
- Mean overall lorry weights μ_Q in kN/m
- Standard deviation of the overall lorry weights σ_Q in kN/m
- Traffic volume N

- Proportion of heavy truck traffic *HV* on traffic volume
- Proportion of the freely flowing traffic *F* on traffic volume

Based on these data, six coefficients are computed:

$$c_1 = \frac{q_{max}}{73} \cdot 0{,}2 + 0{,}8 \quad (40 \le q_{max} \text{ in kN/m} \le 80) \tag{2-1}$$

$$c_2 = \frac{1}{\dfrac{\mu_Q}{14{,}5} \cdot 0{,}65 + 0{,}35} \quad (6 \le \mu_Q \text{ in kN/m} \le 20) \tag{2-2}$$

$$c_3 = \frac{1}{\dfrac{\sigma_Q}{6{,}0} \cdot 0{,}6 + 0{,}4} \quad (2 \le \sigma_Q \text{ in kN/m} \le 8) \tag{2-3}$$

$$c_4 = \frac{1}{\dfrac{HV}{0{,}25} \cdot 0{,}7 + 0{,}3} \quad (0{,}1 \le HV \le 0{,}4) \tag{2-4}$$

$$c_5 = \frac{1}{\log(N) \cdot 0{,}08 + 0{,}33} \quad (10^5 \le N \le 10^9) \tag{2-5}$$

$$c_6 = \frac{F}{94} \cdot 0{,}2 + 0{,}8 \quad (40 \le F \le 100) \tag{2-6}$$

The coefficients then permit the adaptation of the traffic load:

$$\alpha_Q = \frac{6 \cdot c_1 \cdot c_2 \cdot c_3 \cdot c_4 \cdot c_5 \cdot c_6}{(c_1 + c_2 + c_3 + c_4 + c_5 + c_6)} \tag{2-7}$$

The adaptation is then carried out as

$$Q = \frac{Q_d}{\alpha_Q} \tag{2-8}$$

The example in Table 2-8 is taken from Bailey and Hirt (1996). The coefficients are then $c_1 = 0.99$; $c_2 = 1.02$; $c_3 = 1.0$; $c_4 = 1.72$; $c_5 = 1.09$; $c_6 = 1.01$, and finally $\alpha_Q = 1.68$.

Table 2-8. Computation example from Bailey and Hirt (1996)

Local traffic properties	Value
Maximum overall weight of lorries in q_{max} in kN/m	70
Mean overall weight of lorries μ_Q in kN/m	14
Standard deviation of the overall weight of lorries σ_Q (kN/m)	6
Traffic volume N	20×10^6 in 10 years
Proportion of heavy truck traffic HV on the traffic	0.05 rounded up to 0.1
Proportion of the freely flowing traffic F on the traffic volume	1% standing 2% with 40 km/h 500 vehicles per hour

The major problem of this schematic is probably the capture of the data. Many publications concerned with traffic loads can be found. Exemplary publications are by Casas and Crespo-Minguillon (1996), Hannawald et al. (2003), COST-345 (2004), and Allaix et al. (2007). Further traffic load models can be found in COST-345 (2004), Vrouwenvelder and Waarts (1993), and Prat (2001).

2.3 Railroad Traffic Load

The first railway was probably introduced by Richard Trevithick in 1804. He simply put a steam engine on wheels. At first, railways were used for goods traffic, but have also been used since 1829 for passenger traffic, when Robert Stephenson won the race. In the following 100 years, railroads experienced incredible growth (Table 2-9).

Table 2-9. Development of railroad length in Germany according to Mann (1991)

Year	Railroad length in Germany in km
1840	549
1850	6044
1870	19575
1910	61148

This incredible growth yielded also to an enormous demand for bridge construction. Therefore, between 1845 and 1890 especially, many railway bridges were constructed as arch or vault bridges. Later, most superstructures were built from steel.

Because of the rapidly growing demand for railway passenger and goods transport, locomotives were permanently improved. This improvement was related to an increased overall weight. For example, from 1835 to 1922, the weight of locomotives expanded from 10 to 175 tonnes. This corresponds to an increase in uniform load from 2.5 to 13.67 tonnes per metre (Beyer 2001). As an example, the growth of railway loads in Spain is shown in Table 2-10.

Table 2-10. Growth of railway loads in Spain according to Vega (1996)

Year	Uniform load in kg/m
1877	4,900
1902	6,230
1925	10,780
1956	13,400
1975	12,000

According to Weber (1999), in the beginning of railway technology in Germany (1845–1876), the railway load design was a train with real axle loads. A locomotive with an overall length of 7.0 m and wheel distances of 2.4 and 1.4 m was used. The wheel load was 5.0 and 4.5 tonnes, respectively. The maximum speed was 40 km/h. Only in 1877, the Wurttemberg railway company introduced a railway load pattern.

The use of the railway load pattern UIC 71 for new bridges was developed deterministically based on the sum of load patterns of real railway loads concurrently in Germany and many other European countries. It was introduced in 1971. The introduction of a new load pattern for railways called IM 2000 has so far failed. In addition to the general railway loads for heavy railways, the so-called, heavy load patterns SW/0 and SW/2 are used. For consideration of local conditions, the UIC 71 can be adapted by factor α. The factor can reach from 0.75 to 1.33 (see also Fig. 2-10). This would permit an adaptation of the load for special conditions on historical bridges.

If this adaptation is insufficient, then there exists a further opportunity by using special static standard railway load models. These special railway models are related to lower railway track classifications. Figure 2-11 includes examples of such lower classifications and the standard load patterns. However, the application of such load patterns requires a strong collaboration with the railway company (in Germany the DB AG, O'Connor et al. 2008).

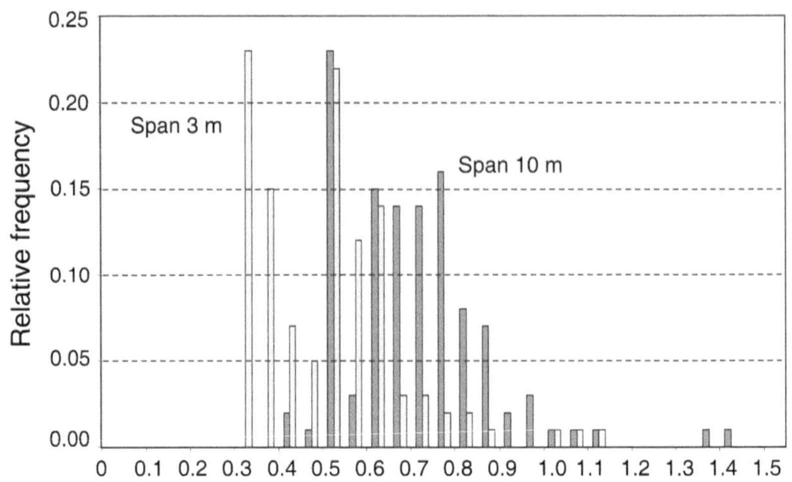

Fig. 2-10. Observed moments on bridges in comparison with the assumed moments based on the UIC loads (Lieberwirth 2004)

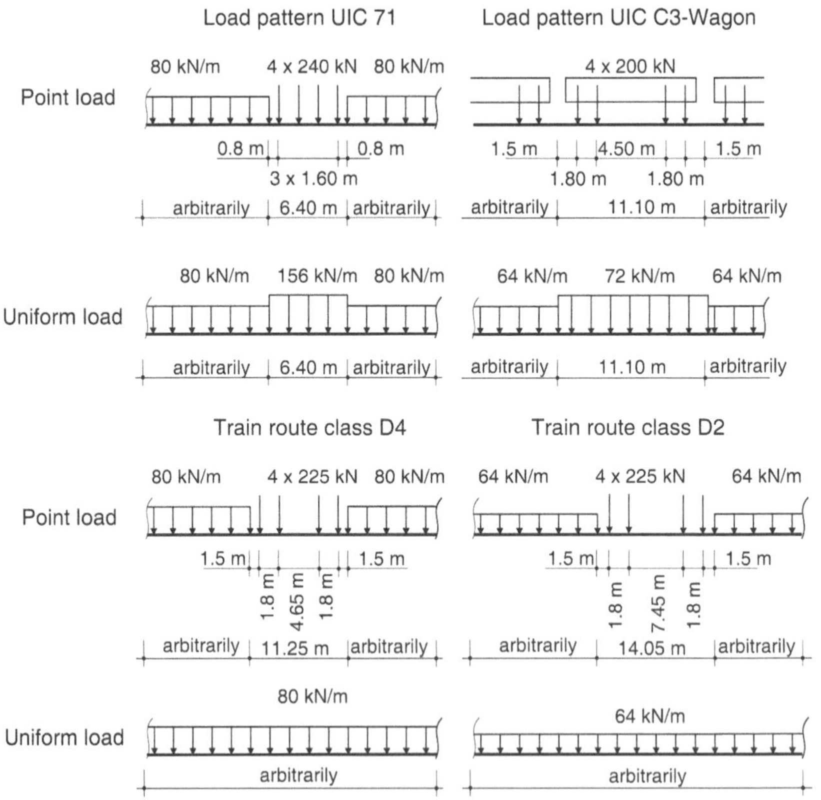

Fig. 2-11. Continued

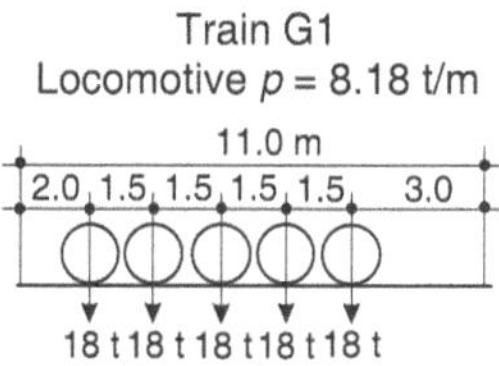

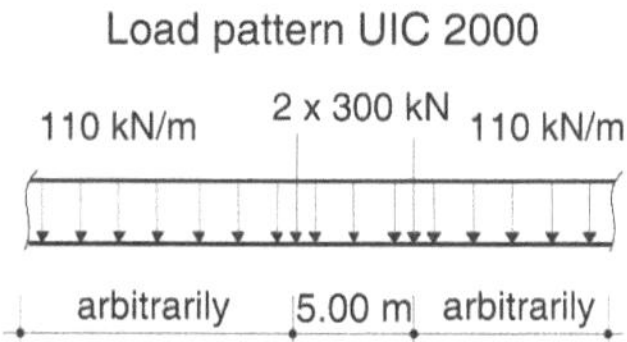

Fig. 2-11. Railway load patterns for the UIC 71 and the recomputation railway patterns C3 and D4 according to the DB AG. Furthermore, the possible future load patterns IM-2000 (Weber 1999) and the historical load pattern G1 are shown

Besides consideration about the weight of the locomotives, which was mainly driven by the first steam locomotives, the heavy goods railway transport has significantly changed. Locomotives are much stronger compared to the old days, which yields not only to higher starting tensile forces but also to heavier railway wagons. Nowadays, the axle loads of railway wagons exceed the axle loads of locomotives and show much higher deviations caused, for example, by overload or humidity of the bulk freights.

An international load border system for goods railway wagons gives the maximum load class, up to which wagons on certain railway routes can be loaded (Tables 2-11 and 2-12).

Table 2-11. Load border system for goods railway wagons (t = tonnes)

A/B1	B2/C2	C3/C4	D2	D3	D4
38.0 t	56.0 t	65.0 t	56.0 t	67.0 t	77.0 t
Maximum speed 120 km/h					

Table 2-12. Interpretation of the load border system in Germany (DB AG)

	Classification Uniform vehicle load in tonne/m	Wheel set load			
		A 16 tonnes	B 18 tonnes	C 20 tonnes	D 22.5 tonnes
1	5.0	A	B1		
2	6.4		B2	C2	D2
3	7.2			C3	D3
4	8.0			C4	D4

The load border system influences the railway traffic on a certain route in different ways—for example, the speed can be limited. This influences the dynamic load factor and therefore decreases the load on structures on this route. Additionally, railway goods trains can only be loaded to a certain value. Usually, the wheel set load is declared on the goods wagon and the wagons are only then loaded up to the permitted value corresponding to the load border system.

Since the route classification often depends on the weakest link, which is usually a bridge, it becomes clear that great interest from railway companies exists to utilize all load-bearing capabilities of historical arch bridges in order to reach an acceptable route class. If the historical arch bridge remains an impairment for the route classification, then sooner or later the bridge will be replaced. However, this problem provides an opportunity to apply modern concepts for both the load description and the structural resistance description.

Background information about the European railway loads can be found in some ERRI reports (1993, 1994a, 1994b and 1998), UIC-report 702 (1974), Tobias et al. (1996), or Lieberwirth (2004).

2.4 Initial Drive Forces

Initial drive forces from railway trains and road trucks can introduce high main spin direction parallel horizontal forces. However, this depends on the structural components of the arch superstructure. In Germany, single-span arch bridges with sufficient coverage have rather low average span distances, continuous rail initial drive forces, and breaking forces. This is based on the Ril 805 (1999) and Ril 804 (2003) for the recomputation of historical railway bridges.

Weber (1999) assumes that by using continuous rails, the initial drive forces and the breaking forces are directly transferred to the connected railway. Therefore, these forces are not transferred by the arch bridges and are not measurable in the arch itself. A historical draft of the Eurocode 1, Sect. 3.4, had assumed that initial drive forces and breaking forces do not have to be considered for historical vault and arch bridges.

Ril 804 considers historical bridges by reducing the maximum initial drive force of 1 MN by a factor ξ.

$$F_x = 33,3 \cdot l \cdot \xi \tag{2-9}$$

The loading length is limited to 30 m. For single arches, this is $\xi = 0.5$. For long arch lines, special considerations are required since the loading length can easily exceed the above-mentioned length. Here, for example, the type of construction of the arch and the piers has to be considered. It should be noted that even when a lower load pattern than UIC 71 is used for the estimation of vertical loads, the initial drive force must be considered. This is understandable if one considers that the strongest locomotives in Germany are used in the train route class D4. Currently, the strongest locomotive of the DB AG (Germany) is the modernized BR 241 (Worowschilowgrad). This locomotive is mainly used for heavy goods trains and reaches a maximum pull force of 450 kN over 20.82 m.

2.5 Breaking Forces

The normative railway breaking force according to Ril 804 is evaluated based on the same function as the initial drive force:

$$F_x = f_{x,Br} \cdot l \cdot \xi \quad \text{in kN.} \tag{2-10}$$

For every track, $f_{x,Br} = 20$ kN/m has to be used for passenger and good trains. However, for heavy goods transport, an increased value of 35 kN/m has to be applied since the heavier wagons can cause greater breaking forces.

For tall and slender piers, a load combination of relatively low vertical traffic load (low route class) with rather high initial drive or breaking forces and side wind is often relevant for design. To fulfill the stability requirements for long bridges, different pier types are often used: these are called group piers, which are thicker than the normal piers and have the function to transfer high horizontal forces to the foundations. The result is that only one group pier has to take the horizontal forces between the group piers. The coupling of the different bridge parts needs to then be proven independently. For the coupling, the backfill concrete on the arches is often considered. Post-applied precast concrete way elements are usually not of considerable strength for the horizontal force transfer due to the joints.

Breaking forces for road bridges were originally computed based on the DIN 1072 (1985) and supplementary sheets. Values range from approximately 10 to 900 kN. Currently, the Eurocode 1 or the DIN-report is the basis for the estimation of the breaking forces. Figure 2-12 shows breaking force values subject to the bridge length and different vehicle types.

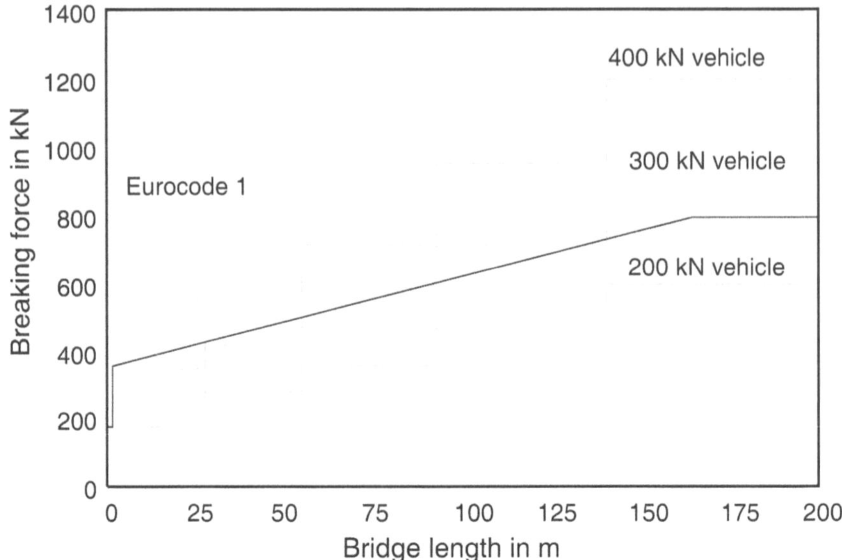

Fig. 2-12. Breaking forces on bridges subject to the bridge length and the type of design vehicle according to Merzenich and Sedlacek (1995)

As with railway bridges, the computation of the flow of the breaking forces into the foundation requires models that consider the stiffness of the single piers. Further work on this topic can be found in Merzenich and Sedlacek (1995), Pfohl (1983), and Weihermüller and Knöppler (1980).

2.6 Wind Loading

Arch bridges are exposed not only to horizontal forces parallel to the main spin of the bridge but also to horizontal forces rectangular to the main spin of the bridge. Such forces are impact forces or wind forces. Wind forces indeed act on historical arch bridges, however in most cases the wind forces are not relevant due to the high dead load of arch bridges. As mentioned before, sometimes wind forces have to be considered in combination with other horizontal forces, mainly for tall piers.

Codes of practice for wind forces on bridges are again the Ril 804 and Ril 805 for railway bridges in Germany, and the DIN-report 101. It should be mentioned that often wind forces with and without traffic on bridges have to be considered separately.

There are many background documents about wind force evaluation including Lieberwirth (2003).

2.7 Impact Forces

Horizontal impact forces rectangular to the main spin of bridges can be caused by technical and natural means. Technical means are mainly different means of transport, such as impacts from ships, cars, railways, or airplanes and helicopters. Natural processes are mainly gravitational mass movements such as avalanches, debris flows, flash floods, landslides, or ice pressure.

Many historical bridges have been destroyed by flash floods or other impacts. Such forces can easily reach several Mega Newton. Background for the computation of impact forces by means of transport can be found in Proske (2003), and for natural processes in Proske and Hübl (2007) and Proske et al. (2008).

2.8 Settlements

The static integrity of the arch bridge is strongly related to the low deformation of the foundation and the abutment. This is the reason why arch bridges are usually not built in regions with soft ground, such as gravel or sand. Therefore, settlements of the foundation of arch bridges can cause substantial damage to the bridge. However, for historical arch bridges, settlements usually should not increase further and if the bridge has survived a number of decades or centuries, settlement should no longer be of concern.

The consequences of settlements are discussed, for example, in Jagfeld (1998, 2000) and Ochsendorf (2002). Ochsendorf gives an all-inclusive number for the decrease of the ultimate load-bearing behaviour of arch bridges of 15%. Furthermore, settlements can change the location of the hinges inside the arch and can therefore completely change the failure mechanisms. Goldschneider (2003) discusses the combinations of ground and historical foundations.

2.9 Temperature Loading

The consideration of temperature loading on historical arches is still disputed due to a number of different opinions.

Based on the UIC-report, a temperature gradient over the vault cross section has to be considered. The constant temperature change has to be assumed with ± 15 K. A comparable method for arch bridges was used in a former Eurocode draft.

For temperature loading, Pietsch (1961) recommends either a computation with consideration of the plasticity of masonry or an artificial decrease of the stiffness E_0 of the masonry, depending on the temperature change. He proposes for the

$$\text{daily temperature change, } E_t = 1,0 \cdot E_0 , \tag{2-11}$$

$$\text{average temperature change, } E_m = 0,5 \cdot E_0 , \tag{2-12}$$

$$\text{yearly temperature change, } E_j = 0,2 \cdot E_0 . \tag{2-13}$$

This yields to rather low internal forces inside the arch due to temperature loading. The factor of 0.2 fits very well with a publication by Thürmer (1995). He calculated a loss of stress caused by temperature inside a historical arch bridge by a nonlinear computation of 70% compared to the linear-elastic computation. This would represent a factor of 0.3.

Bothe et al. (2004) presents the computation of a historical arch bridge. He calculates that a linear-elastic computation without temperature loading results in a maximum live load of 56 kN/m and consideration of temperature loading results in a 14 kN/m maximum live load. However, if a nonlinear computation based on maximum strain and overall stability is carried out, the maximum live load becomes independent from the temperature loading. If a certain characteristic concrete strength is assumed for the resistance, then the maximum live load without temperature loading is 102 kN/m, compared to 100 kN/m with temperature loading. Figure 2-13 shows the formulae for the computation of the characteristic concrete strength, including further additional formulas. Besides the three nonlinear computations with and without temperature, the linear-elastic computation results are also included in the figure. The diagram shows clearly that the consideration of temperature loadings on historical arch bridges does not result in a significant decrease of the ultimate load-bearing capacity, if nonlinear computations are carried out.

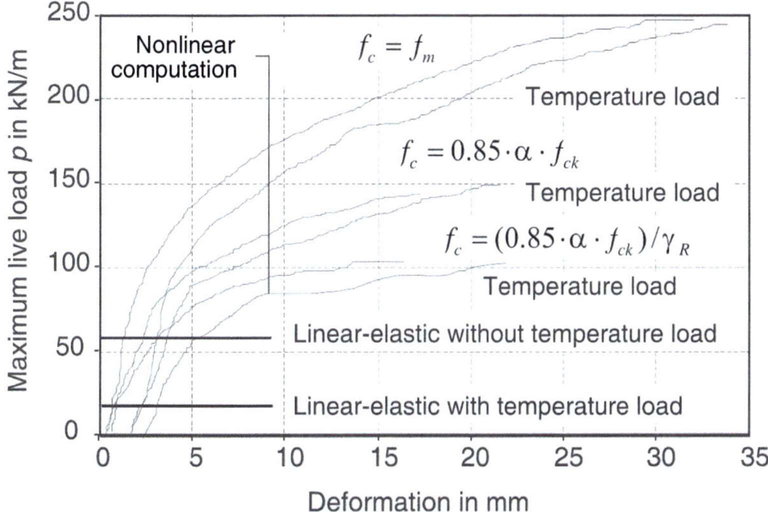

Fig. 2-13. Computation of maximum permitted live loads on a historical arch considering four load cases. First, a linear-elastic computation is carried out considering temperature and no temperature. Furthermore, three nonlinear computations with and without temperature are shown. These computations differ in the assumption of the concrete compression strength f_c as shown. As we can see, α is the long-term loading factor (0.85), f_{ck} is the characteristic concrete compression strength, and γ_R is the partial safety factor (1.8)

This statement fits very well with the regulation of the historical DIN 1072 (1952), whereon masonry historical arch bridges with backfill, functional arch without hinges, a maximum span of 20.0 m, and with a certain pier ratio temperature loading had not to be considered.

In contrast to this, there are some known works where temperature loads caused significant effects on historical arch bridges. Patzschke (1996) describes an example where temperatures in the summer of 1992 caused the appearance of a three-hinge arch on a railway multi span brick arch bridge. Since the development of hinges in the arch caused major deformations and limited the use of this railway route, structural hinges were installed to limit the deformations. Schlegel et al. (2003) also mention that temperature caused major deformations on the Göltzschtal Bridge in Saxony.

Further investigations into the influence of temperature loadings on spandrel walls and parapets are given by Robinson et al. (1998). They use ABACUS to simulate climatic effects on arch bridges. The temperatures considered range from 0° to 31°C. Measurements were also considered in this study.

2.10 Snow Loading

Snow loads on bridges can usually be neglected. This is understandable since no traffic can pass the bridge under high snow loads. However, the traffic weight is usually much higher than the snow load and therefore dominates design. Further information about snow load can be found in Lieberwirth (2003), JCSS (2004), Soukhov (1998), and Mehlhorn (1997).

2.11 Dead Load

All loads discussed so far have been live loads. They change direction and presence. However, the most important load on arch bridges is the dead load or self-weight. In contrast to many other structures, the dead load only allows the function of the arch. If the self-weight is missing, then the stability of the arch itself is not given.

Therefore, the correct computation of the dead load is of utmost importance. For these computations, the density and volume of certain structural elements and materials are required. The volume is mainly computed on historical or up-to-date documentation about the structure. Although the computation of the volumes and weight is rather simple, one should carefully investigate documents or the structure about the assembly and the geometry. In some cases, drilling or other further investigation technologies may be required to find cavities in the structure. Such cavities can be blasting chambers or vaults inside the structure.

Once the geometry is known, either based on measurements or based on codes of practice like the DIN 1055-1, the density of the material can be chosen. As an example, for the bridge shown in Fig. 2-14, the following densities were used:

Density Porphyr g_P =28 kN/m^3 (DIN 1055-1, 2002, Tabelle 6, Zeile 12),
Density Sandstone g_S =27 kN/m^3 (DIN 1055-1, 2002, Tabelle 6, Zeile 15),
Density Concrete g_B =24 kN/m^3 (DIN 1055-1, 2002, Tabelle 1, Zeile 21).

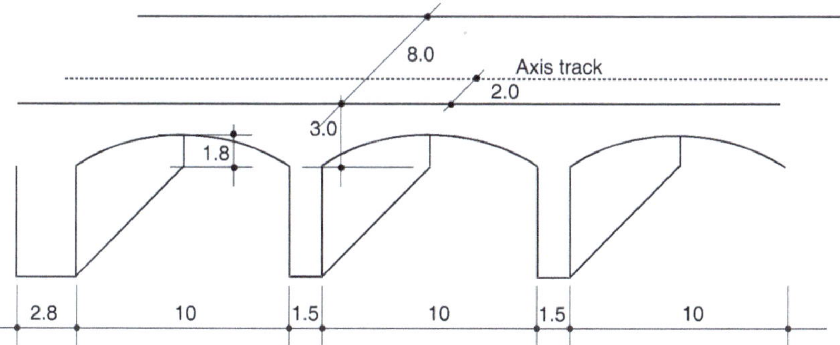

Fig. 2-14. Example geometry draft of a bridge (no internal volumes are shown)

If the densities cannot be taken from the codes, mean values of measurement data can be taken. The mean value as characteristic values for densities and volumes correspond with Eurocode 1, if the coefficient of variation is less than 0.05. If the deviation is higher, then 95% fractile values can be chosen. However, this is rather unusual, since slight deviations are covered by a partial safety factor of 0.95 and 1.05.

Although partial safety factors are discussed later in Chapter 7, some remarks should be given here, since the dead load on arch bridges is a very specific case. This is caused by the fact that the dead load is increasing the ultimate load-bearing capacity of arches up to a certain value. Over a certain value, the dead load becomes like the live load, and decreases the ultimate load-bearing behaviour.

Whereas in the Ril 805 for the German railway bridges the partial safety factor for the dead load is $\gamma_G = 1.20$, in the Eurocode 1 or in the German DIN 1055-100, $\gamma_G = 1.35$. Works by Wiese et al. (2005) have shown that this value is too high and considers not only uncertainties from the dead load but also from the modelling. A former draft of the Eurocode gives some special considerations to historical arch bridges:

- The partial safety factor for the self-weight of the arch can be decreased to 1.1 for unfavourable effects and 1.0 for favourable effects. Herrbruch et al. (2005) have suggested 1.35 and 0.9, respectively, according to the current codes.
- The partial safety factor for the self-weight of further elements of the superstructure (like coverings) can be chosen as 1.2 for unfavourable and 0.9 for favourable effects.

References

Ablinger W (1996) Einwirkungen auf Brückentragwerke – Achslastmessungen und stochastische Modelle, Diplomarbeit, Institut für konstruktiven Ingenieurbau, Universität für Bodenkultur

Allaix DL, Vrouwenvelder ACWM & Courage, WMG (2007) Traffic load effects in prestressed concrete bridges. 5th International Probabilistic Workshop, L Taerwe & D Proske (Eds), Ghent, Belgium, pp 97–110

Bailey SF & Hirt AH (1996) Site specific models of traffic action effects for bridge evaluation. Recent Advances in Bridge Engineering. Evaluation, Management and Repair, JR Casas, FW Klaiber & AR Marí (Eds), Proceedings of the US-Europe Workshop on Bridge Engineering, organized by the Technical University of Catalonia and the Iowa State University, Barcelona, 15–17 July 1996, International Center for Numerical Methods in Engineering CIMNE, pp 405–425

Beyer P (2001) 150 Jahre Göltzschtal- und Elstertalbrücke im sächsischen Vogtland 1851–2001. Vogtland Verlag Plauen

Bogath J & Bergmeister K (1999) Neues Lastmodell für Straßenbrücken. Bauingenieur 6, pp 270–277

Bogath J (1997) Verkehrslastmodelle für Straßenbrücken. Dissertation, Institut für konstruktiven Ingenieurbau, Universität für Bodenkultur

Bogsjö K (2007) Evaluation of stochastic models of parallel road tracks. Probabilistic Engineering Mechanics, 22, No. 4, October 2007, pp 362–370

Bothe E, Henning J, Curbach M, Bösche T & Proske D (2004) Nichtlineare Berechnung alter Bogenbrücken auf der Grundlage neuer Vorschriften. Beton- und Stahlbetonbau 99, Heft 4, pp 289–294

Casas JR & Crespo-Minguillon C (1996) Serviceability and fatigue evaluation of existing bridges. Recent Advances in Bridge Engineering. Evaluation, Management and Repair, JR Casas, FW Klaiber & AR Marí (Eds), Proceedings of the US-Europe Workshop on Bridge Engineering, organized by the Technical University of Catalonia and the Iowa State University, Barcelona, 15–17 July 1996, International Center for Numerical Methods in Engineering CIMNE, pp 441–464

Cooper D (2002) Traffic and moving loads on bridges. Dynamic Loading and Design of Structures, AJ Kappos (Ed), Chapter 8, Spon Press, London, pp 307–322

COST 345 (2004) European Commission Directorate General Transport and Energy: Procedures Required for the Assessment of Highway Structures: Numerical Techniques for Safety and Serviceability Assessment – Report of Working Groups 4 and 5

Crespo-Minguillón C & Casas JR (1997) A comprehensive traffic load model for bridge safety checking, Structural Safety, 19 (4), pp 339–359

DIN-report 101: Einwirkungen auf Brücken. Beuth Verlag, Berlin, 2001

ENV 1991-3 Eurocode 1: Grundlagen der Tragwerksplanung und Einwirkungen auf Tragwerke, Teil 3: Verkehrslasten auf Brücken, 1995

ERRI-report (1993) D 192/RP 1: Theoretische Grundlagen zur Überprüfung des bestehenden Lastbildes UIC 71, Utrecht

ERRI-report (1994a) D 192/RP 2: Vergleich der Einwirkungen des derzeitigen und künftigen Schienenverkehrs auf den internationalen Strecken mit denen aus Lastbild UIC 71 auf deterministischer Grundlage, Utrecht

ERRI-report (1994b) D 192/RP 3: Lastbild für die Berechnung der Tragwerke der internationalen Strecken, Utrecht

ERRI-report (1998) D 221/RP 1: Vertikale Eisenbahnverkehrslasten: Kombinationsbeiwerte – Hintergrundinformationen zum Thema "Eisenbahnverkehrslasten", Utrecht

Geißler K (1995) Beitrag zur probabilistischen Berechnung der Restnutzungsdauer stählerner Brücken. Dissertation, Heft 2 der Schriftenreiche des Institutes für Tragwerke und Baustoffe an der Technischen Universität Dresden

Goldschneider M (2003) Baugrund und historische Gründungen. Untersuchen, Bewerten und Instandsetzen. Empfehlungen für die Praxis, Fritz Wenzel und Joachim Kleinmanns (Hrsg) "Erhalten historisch bedeutsamer Bauwerke". Karlsruhe

Hannawald F, Reintjes K-H & Graße W (2003) Messwertgestützte Beurteilung des Gebrauchsverhaltens einer Stahlverbund-Autobahnbrücke. Stahlbau 72 (7), pp 507–516

Herrbruch J, Groß JP & Wapenhans W (2005) Wirklichkeitsnähere Tragwerksanalyse von Gewölbebrücken. Mauerwerk 9, Heft 3, pp 89–92

Jagfeld M (1998) Tragverhalten gemauerter Wölbkonstruktionen bei großen Auflagerverschiebungen – Vergleich verschiedener FE Modelle. 16. CAD-FEM User's Meeting, 7.–9. Oktober 1998 in Bad Neuenahr

Jagfeld M (2000) Tragverhalten und statische Berechnung gemauerter Gewölbe bei großen Auflagerverschiebungen – Untersuchungen mit der Finit-Element-Methode. Schriftenreihe des Lehrstuhls für Hochbaustatik und Tragwerksplanung, Technische Universität München, Aachen

JCSS (2004) Joint Committee of Structural Safety (2004). Probabilistic Modelcode. www.jcss.ethz.ch

KBA (2001) Kraftfahrzeugbundesamt: Festschrift 50. Jahre Kraftfahrt-Bundesamt 1951–2001. August 2001

König G & Gerhardt HC (1985) Verkehrslastmodell für Straßenbrücken. Bauingenieur 60, pp 405–409

Krämer W & Pohl S (1984) Der Ermüdungsnachweis in dem Standard TGL 13460/01. Ausgabe 1984 – Grundlagen und Erläuterungen. Die Straße 24, Heft 9, pp 257–263

Leliavsky S (1982) Arches and short span bridges. Design Textbook in Civil Engineering: Volume VII, Chapman and Hall, London

Lieberwirth P (2003) Ein Beitrag zur Wind- und Schneelastmodellierung. 1. Dresdner Probabilistik-Symposium – Sicherheit und Risiko im Bauwesen, D Proske (Ed), Technische Universität Dresden, Dresden, pp 123–138

Lieberwirth P (2004) Ausgewählte Aspekte zur statistischen Beschreibung von Bahnverkehrslasten. 2. Dresdner Probabilistik Symposium, Dresden

Loos S (2005) Entwicklung von Lastmodellen auf Grundlage des DIN-Fachberichtes 101 für gewichtsbeschränkt beschilderte Straßenbrücken. Diplomarbeit. Technische Universität Dresden, Institut für Massivbau

Mann G (1991) (Ed) Propyläen Weltgeschichte – Eine Universalgeschichte. Achter Band. Propyläen Verlag Berlin – Frankfurt am Main

Mehlhorn G (1997) Der Ingenieurbau, Band 8 – Tragwerkszuverlässigkeit und Einwirkungen, Verlag Ernst & Sohn, Berlin

Merzenich G & Sedlacek G (1995) Hintergrundbericht zum Eurocode 1 – Teil 3.2: Verkehrslasten auf Straßenbrücken. Forschung Straßenbau und Straßenverkehrstechnik. Bundesministerium für Verkehr, Heft 711

Naumann (2002) Aktuelle Schwerpunkte im Brückenbau. 12. Dresdner Brückenbausymposium. Lehrstuhl für Massivbau und Freunde des Bauingenieurwesens e.V., pp 43–52

Novák B, Brosge B, Barthel K & Pfisterer W (2007) Anpassung des Verkehrslastmodells des DIN FB-101 für kommunale Brücken, Beton- und Stahlbetonbau 102, Heft 5, pp 271–279

O'Connor A & O'Brien EJ (2005) Traffic load modelling and factors influencing the accuracy of predicted extremes. Canadian Journal of Civil Engineering 32, pp 270–278

O'Connor A, Pedersen C, Enevoldsen I, Gustavsson L & Hammarbäck J (2008) Probability Based Assessment of a Large Riveted Truss Railway Bridge. Bridge Maintenance, Safety, Management, Health Monitoring and Informatics, H Koh & DM Frangopol (Eds), Taylor & Francis Group, London, pp 1812–1819

Ochsendorf JA (2002) Collapse of Masonry Structures. University of Cambridge, Dissertation, Cambridge

Patzschke F (1996) Der Wahrener Viadukt in Leipzig – Rekonstruktion einer Gewölbebrücke der DB AG. 6. Dresdner Brückenbausymposium, Fakultät Bauingenieurwesen, Technische Universität Dresden, pp 135–148

Pfohl H (1983) Reaktionskraft am Festpunkt von Brücken aus Bremslast und Bewegungswiderständen der Lager. Bauingenieur 58, pp 453–457

Pietsch H (1961) Zur Frage der Beanspruchung aus Temperatur in massiven Gewölbebrücken am Beispiel des voll eingespannten Bogens mit Übermauerung. Dissertation Hochschule für Verkehrswesen, Dresden

Pircher M, Lechner B, Mariani O & Kammersberger A (2009) Schädigung einer schlaff bewehrten Betonbrücke durch Verkehrsbelastung. Beton- und Stahlbetonbau, 104(3), pp 154–163

Pohl S (1993) Definition von charakteristischen Werten für Straßenverkehrsmodelle auf der Basis der Fahrzeuge sowie Fraktilwerte der Lasten des Eurocode 1-Modells; Forschungsbericht Bundesanstalt für Straßenwesen

Prat M (2001) Traffic load models for bridge design: recent developments and research. Progress in Structural Engineering and Materials 3, pp 326–334

Proske D & Hübl J (2007) Historical arch bridges under horizontal loads. Proceedings of the 7th Arch Conference, Madeira

Proske D (2003) Ein Beitrag zur Risikobeurteilung alter Brücken unter Schiffsanprall. Dissertation, Technische Universität Dresden

Proske D, Kaitna R, Suda J & Hübl J (2008) Uncertainty in debris flow impact force assessment. Proceedings of the 6th International Probabilistic Workshop, CA Graubner, H Schmid & D Proske (Eds), Darmstadt

Puche M & Gerhardt HC (1986) Absicherung eines Verkehrslastmodells durch Messungen von Spannstahlspannungen. Bauingenieur 61, pp 79–81

Quan Q (2004) Modeling current traffic load, safety evaluation and SFE analysis of four bridges in China. Risk Analysis IV, CA Brebbia (Ed) Wessex Institute of Technology Press, Southampton

Ril 804 (2003) Eisenbahnbrücken (und sonstige Ingenieurbauwerke) planen, bauen und instand halten. Deutsche Bahn AG

Ril 805 (1999) Tragsicherheit bestehender Eisenbahnbrücken

Robinson JI, Prentice DJ & Ponniah D (1998) Thermal effects on a masonry arch bridge investigated using ABAQUS. Arch Bridges: History, Analysis, Assessment, Maintenance and Repair. Proceedings of the Second International Arch Bridge Conference, A Sinopoli (Ed), Balkema, Rotterdam 1998, pp 431–438

Schlegel R, Will J & Popp J (2003) Neue Wege zur realistischen Standsicherheitsbewertung gemauerter Brückenviadukte und Gewölbe unter Berücksichtigung der Normung mit ANSYS. 21. CAD-FEM User's Meeting 2003, Potsdam

Schütz KG (1991) Verkehrslasten für die Bemessung von Straßenbrücken, Bauingenieur 66, pp 363–373

Soukhov D (1998) The Probability Distribution Function for Snow Load in Germany. LACER No. 3, 1998, pp 263–274

Spaethe G (1977) Beanspruchungskollektive von Straßenbrücken, Die Straße 17, Heft 4, pp 241–246

Thürmer E (1995) Verbreiterung einer Bogenreihenbrücke im Zuge der Autobahn A4 bei Hainichen. 5. Dresdner Brückenbausymposium, Fakultät Bauingenieurwesen, Technische Universität Dresden, pp 37–54

Tobias DH, Foutch DA & Choros J (1996) Loading Spectra for railway bridges under current operation conditions. Journal of Bridge Engineering, 1 (4), pp 127–134

UIC-report 702 (1974) Lastbild für die Berechnung der Tragwerke der internationalen Strecken, Internationaler Eisenbahnverband UIC

Vega T (1996) Bridge management at RENFE. Recent Advances in Bridge Engineering. Evaluation, Management and Repair, JR Casas, FW Klaiber & AR Marí (Eds), Proceedings of the US-Europe Workshop on Bridge Engineering, organized by the Technical University of Catalonia and the Iowa State University, Barcelona, 15–17 July 1996, International Center for Numerical Methods in Engineering CIMNE, pp 115–123

Vockrodt H-J (2005) Instandsetzung historischer Bogenbrücken im Spannungsfeld von Denkmalschutz und modernen historischen Anforderungen. 15. Dresdner Brückenbausymposium, Technische Universität Dresden, pp 221–241

Vrouwenvelder ACWM & Waarts PH (1993) Traffic loads on bridges. Structural Engineering International, 3, pp 169–177

Weber WK (1999) Die gewölbte Eisenbahnbrücke mit einer Öffnung. Begriffserklärungen, analytische Fassung der Umrisslinien und ein erweitertes Hybridverfahren zur Berechnung der oberen Schranke ihrer Grenztragfähigkeit, validiert durch einen Großversuch. Dissertation, Lehrstuhl für Massivbau der Technischen Universität München

Weihermüller H & Knöppler K (1980) Lagerreibung beim Stabilitätsnachweis von Brückenpfeilern, Bauingenieur, 55, pp 285–288
Wiese H, Curbach M, Al-Jamous A, Eckfeldt L & Proske D (2005) Vergleich des ETV Beton und DIN 1045-1, Beton- und Stahlbetonbau, 100, pp 784–794

3 Computation of Historical Arch Bridges

"An arch bridge can bear everything, except a static computation."

Unknown author (Weber 1999)

3.1 Introduction

The described loadings on bridges in the preceding chapter cause a certain reaction of the arch bridge if applied to the bridge. This reaction is either admissible or inadmissible. Inadmissible reactions of bridges include damages or failure of the bridge. To prevent such effects, the reaction of bridges exposed to loads is usually studied in advance.

Such studies or structural analysis requires the development of an appropriate numerical model.

These models usually reflect the knowledge of humans about the structure. This fact becomes especially visible for structure types with a long history, such as arch bridges. Here, completely arbitrary models are models from Von Leibbrand (1897), Haase (1899), Gilbrin (1913), Fain (1953), Wolf (1989), Lachmann (1990), and Falter (1998). Of course, this chapter gives a much more detailed and structured list about certain types of models. However, a simplified rule for choosing an appropriate model cannot be given. Even very simple empirical rules have shown to be a solid basis for bridges with ages of centuries and millenniums.

The choice of the model or model type depends on the respective question and the provided resources. During the COST-345 (2004) report for the EU commission, different description classes for structures and their applications were discussed. Table 3-1 lists different levels of assessment. This chapter starts with a discussion of simple empirical rules and advances to the latest numerical models in terms of finite element and discrete element models. Many further recommendations about the different levels of bridge assessment can be found in literature (Schueremans et al. 2003, Diamantidis 2001, Rücker et al. 2006, Enevoldsen 2008, Brühwiler

D. Proske, P. van Gelder, *Safety of Historical Stone Arch Bridges*, DOI 10.1007/978-3-540-77618-5_3,
© Springer-Verlag Berlin Heidelberg 2009

2008, O'Connor et al. 2008, Jensen et al. 2008, Ruiz et al. 2008, and SIA 269 2008).

Table 3-1. Analysis methods recommended for each level of assessment (COST-345 2004)

Structure type	Subtype	Description level 1	2	3	4	5
Bridges	Arch bridges	Empirical or two-dimensional-model, linear-elastic arch frame	Two- or three-dimensional, linear-elastic or nonlinear, elastic or plastic, allowing for cracking	Two- or three-dimensional, linear or nonlinear, elastic or plastic, grillage or FEM (upstand model if necessary), allowing for soil-structure interaction, cracking, site- specific loading and material properties	FEM analysis of specific details of the structure being assessed not considered in the previous levels	Reliability analysis based on probabilistic models

3.2 Empirical Rules

3.2.1 Historical Rules

In 1717, Gauter listed the following five tasks during the design process of natural stone arch bridges (taken from Heyman 1998):

- Choice of the shape of the arch
- Choice of the arch thickness at the key
- Choice of the thickness of the foundation and abutment
- Choice of the thickness of the piers depending on the design of the arch
- Choice of the thickness of the wing walls

Over the building time of arch bridges with approximately 2000 the design of arch bridges has been mainly carried out using empirical models.

Empirical methods are here understood as methods that describe the ultimate load-bearing behaviour of arch bridges based on simple geometrical rules for the design of the arch elements. Such a simple rule is shown in

Fig. 3-1 based on suggestions by Alberti. Further borders of the load-bearing behaviour of arch bridges are given by Corradi (1998) based already on some theoretical considerations (Table 3-2). Such empirical rules were also introduced for the abutment, as the Blondel rule (Straub 1992) proves. However, the foundations should not be of interest here. The major interest here is the description of arches using empirical rules.

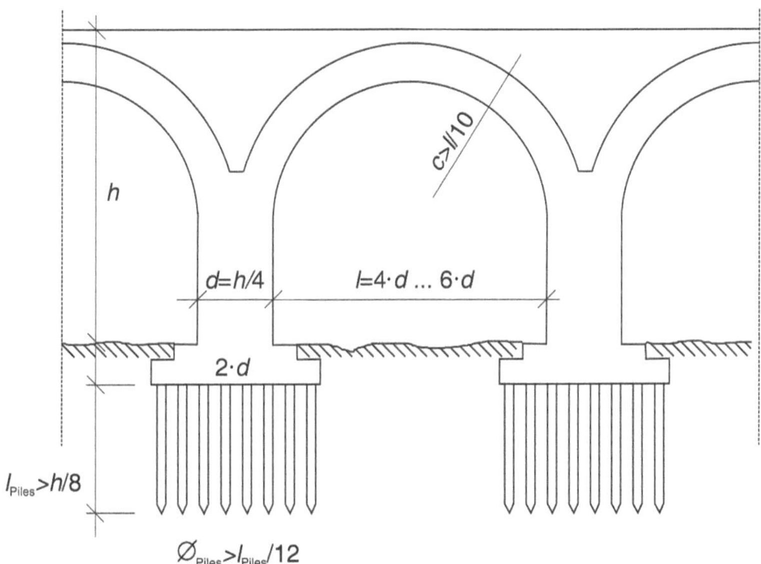

Fig. 3-1. Empirical rules for the design of arch bridges from Alberti around 1,450 (taken from Heinrich 1983)

Table 3-2. Types of failure for circular arches (Corradi and Filemio 2004)

Types of failure	Without backfill	Declining backfill	Horizontal backfill
		β=45°	
	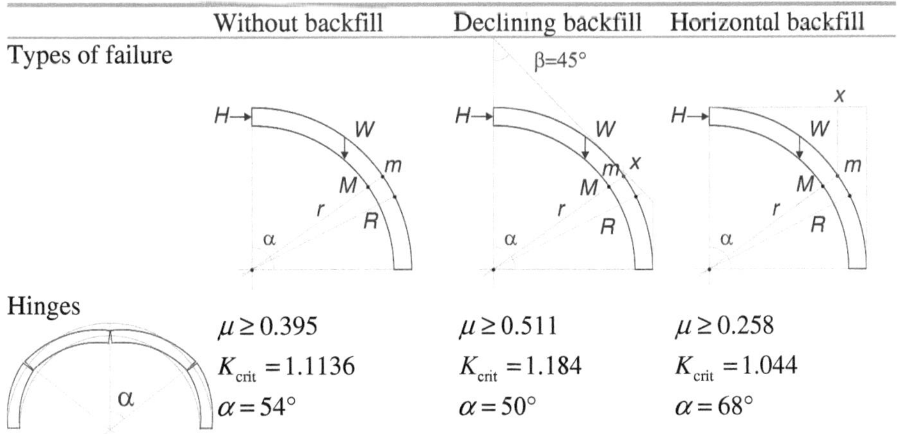		
Hinges	$\mu \geq 0.395$	$\mu \geq 0.511$	$\mu \geq 0.258$
	$K_{crit} = 1.1136$	$K_{crit} = 1.184$	$K_{crit} = 1.044$
	$\alpha = 54°$	$\alpha = 50°$	$\alpha = 68°$

	Without backfill	Declining backfill	Horizontal backfill
Hinges and sliding	$0.309 \leq \mu < 0.395$	$0.406 \leq \mu < 0.511$	$0.236 \leq \mu < 0.258$
	$1.2205 \leq K_{crit} < 1.1136$	$1.264 \leq K_{crit} < 1.184$	$1.1138 \leq K_{crit} < 1.044$
	$29° \leq \alpha < 54°$	$20° \leq \alpha < 50°$	$44° \leq \alpha < 68°$
Sliding	$\mu < 0.309$	$\mu < 0.406$	$\mu < 0.236$
	$K_{crit} = 1.2205$	$K_{crit} = 1.264$	$K_{crit} = 1.1138$
	$\alpha = 29°$	$\alpha = 20°$	$\alpha = 44°$

R = Radius extrados, r = Radius intrados, μ = Coefficient of friction, α = Angle collapse point, H = Normal force at crown, W = Self weight

$$K_{crit} = \frac{R}{r}$$

3.2.1.1 Rules for the shape of the arch

Major parameters for the description of the shape of the arch are the span l and the rise f. They both form a ratio of f/l, which is heavily used for a first characterisation of the arch shape. Purtak (2004) has published a statistic about this ratio (Fig. 3-2).

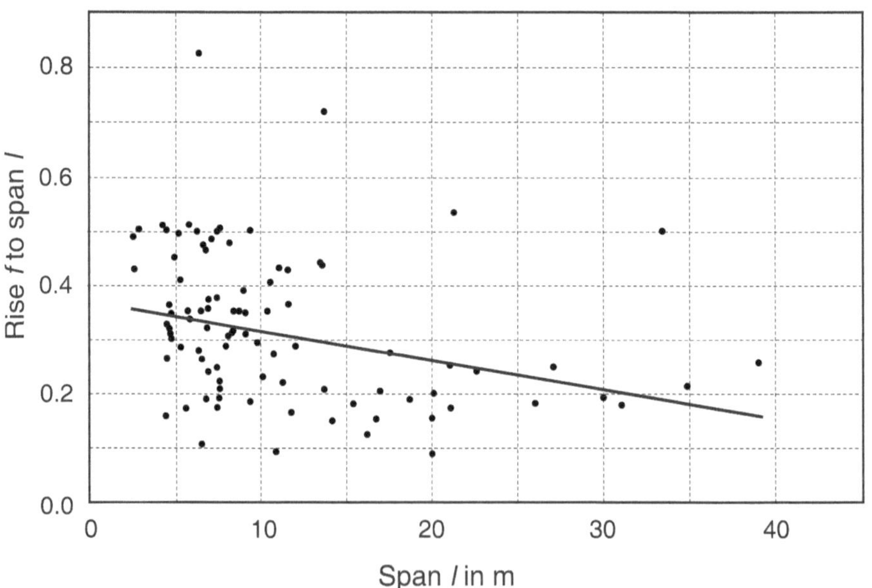

Fig. 3-2. Ratio of arch rise f to arch span l related to the span l (Purtak 2004)

Furthermore, the ratio of the rise f to the span l can be used as

$$s = \frac{f}{l} = \frac{f}{2 \cdot a} \tag{3-1}$$

Full circular arches have an s of 1/2. For segmental arches, s reached values between 1/6 and 1/9. Sometimes additionally a minimum angle of 60° at the springing was required that corresponded to an s of 1/7.5. At the end of the 18th century, the basket arch became more and more popular. At the Neuilly Bridge, Perronet used 11 circle segments to shape the arch. However, often lower numbers of circles were used such as three, five, or seven. Perronet's Bridge reached $s = 1/4$. Also, for the latter, ellipses s-values of 1/4 were used, or in general greater than 1/5 were used. However, in contrast to such rules, bridges with much lower s values were also constructed, such as the Nemours Bridge by Perronet in 1792 with $s = 1/10$ or $s = 1/15$ (Souppes) (Corradi and Filemio 2004, Corradi 1998).

Additionally to the choice of the s value, the number of circle elements in an arch was ruled for basket arches. For example, if the s was equal to 1/3 and the span of the arch was higher than 10 m, then three circular elements were suggested. Between 10 and 40 m span, five circular elements were recommended, and for a span greater than 40 m, about seven segments should be chosen. For $s = 1/4$, the number of circular segments should then be five, seven, and nine. For even greater s values, a circle with a radius r of

$$r = \frac{l^2 + 4 \cdot f}{8 \cdot f} \tag{3-2}$$

was suggested (Corradi and Filemio 2004).

However, the low number of circular segments in basket arches yielded to an aesthetical problem of the transition of the different segments. Therefore, later recommendations gave a higher number of segments. Also, parallel to the basket arches as mentioned before, alternative shape functions were chosen: Sejourne recommended ellipses and Viviani used a cycloid for the arch shape of a bridge over the river Arzana. At two bridges between Pistoia and Modena, a circular evolvente was used for the arch shape. Additionally, during the medieval ages, often ogival designs with parabolic functions were used. And in the 18th century, additional catenaries were applied. For example, bridges in Boucicault, Orelans, and Avignon used this function (Corradi and Filemio 2004, Corradi 1998).

Tolkmitt recommended the shape function subject to the loading. In general, Tolkmitt chose the following function

$$p = \sum_{k=0}^{n} a_k \cdot z^{2k} \tag{3-3}$$

with a_k as constant and z as longitudinal axis. The shape should then follow the function

$$\frac{d^2 y}{dz^2} = -\frac{p}{H} . \tag{3-4}$$

The solution of this differential equation is as follows:

$$y = \frac{1}{H} \sum_{k=0}^{n} \frac{a_k}{(2 \cdot k + 1) \cdot (2 \cdot k + 2)} \cdot k^{2 \cdot k + 2} . \tag{3-5}$$

Usually the series is interrupted after the second term. However, Freyssinet has used $n=4$ for the Bernard Bridge over the Balbigny–Regny railway line. Tables for the computation of the shape were among others provided by Kögler (Corradi and Filemio 2004).

A further arch shape function is given with

$$y = \frac{1}{a} \log(\cos(a \cdot z)) . \tag{3-6}$$

Lebert's parabolic functions were

$$y = k \cdot \log(\cos(a \cdot z)) \tag{3-7}$$

and

$$y = m \cdot \log(\cosh(a^2 \cdot z)) , \tag{3-8}$$

with a, k, and m as coefficients (Corradi and Filemio 2004).

Further information about the arch shapes can be found in Chapter 1.

3.2.1.2 Arch thickness d at crown

In 1669, Fabri published a geometrical arch model that corresponds with the typical three-hinge model nowadays. Based on this model, the thickness of the arch e subject to the extrados radius R could be estimated (Kurrer 2002)

$$e_1 = 2 \cdot R \cdot (3 - 2 \cdot \sqrt{2}) = 0.343 \cdot R \tag{3-9}$$

and the radius of the intrados r was given with

$$r = R \cdot (4 \cdot \sqrt{2} - 5) = 0.657 \cdot R \,. \tag{3-10}$$

The line of thrust is therefore completely inside the arch. However, the arch thickness can further be computed for a failure angle of the arch. For example, if 45° are assumed, then the thickness becomes

$$e_{o.Fabri} = 0,5 \cdot R \cdot (2 - \sqrt{2}) = 0.293 \cdot R \,. \tag{3-11}$$

According to Fabri, arches with this thickness are always stable. A lower thickness can also be stable, but that depends on the special conditions (Kurrer 2002).

In 1730 Couplet introduced a thickness value based on the following observations:

$$e_{u.Couplet} = 0,096 \cdot R \,. \tag{3-12}$$

Arches with lower thickness are always unstable (Kurrer 2002).

As mentioned before, the thickness depends on the assumed failure angle in the arch. Whereas Fabri and Couplet assumed 45°, Heyman estimated the angle with 58.8° and computed the required thickness (Kurrer 2002) as follows:

$$e_{u,Heyman} = 0,101 \cdot R \,. \tag{3-13}$$

Alberti suggested the following minimum thickness of the arch at the crown (Kurrer 2002):

$$e_{u,Alberti} = \frac{2 \cdot r}{10} = 0,131 \cdot R \,. \tag{3-14}$$

Croizette-Desnoyer's formulae is another empirical rule for thickness (Corradi and Filemio 2004, Martín-Caro and Martínez 2004, Modena et al. 2004)

$$e = a + b \cdot \sqrt{2 \cdot r} \tag{3-15}$$

with a and b as coefficients subject to the type of loading (railway or road traffic). Genio Civile suggested the following formulae (Corradi and Filemio 2004):

$$e = 0.05 \cdot h + 0.40 \cdot l + 2 \cdot (10 + l) \cdot \frac{l}{100 \cdot f} \,. \tag{3-16}$$

Sejourne has also suggested a formulae (Martín-Caro and Martínez 2004). Busch and Zumpe (1995) have investigated historical arch bridges and have given the following rule for the arch thickness at the crown:

$$e_S = 0.37 + 0.028 \cdot l > 0.5 \text{ m}. \tag{3-17}$$

This value can additionally be adjusted between 0.3 and 1.9 depending on the masonry quality, the coverage and the loading type. The minimum coverage thickness can also be computed by

$$h = 0.5 + 0.015 \cdot l. \tag{3-18}$$

The axial force N in the arch at the crown is given by

$$N = H = \frac{q \cdot l^2}{8 \cdot f} \tag{3-19}$$

where q represents the loading.

The axial force can be used to compute the stress inside the arch, which can then be compared to the compress strength of the arch. A simplified rule for stress evaluation at an arch crown was given by Dischinger (1949):

$$\sigma = \gamma_i \cdot \frac{l^2}{8 \cdot f}, \tag{3-20}$$

where γ_i is the characteristic weight of the bridge. For long span, low-pitched arches, γ_i amounts to 30 kN/m^3 and for slender, tall bridges, the value reached 40 kN/m^3.

How the rules for the thickness of the arch key or arch crown were used in practice is shown by some statistical investigations from Purtak (2004) and Busch and Zumpe (1995). Figures 3-3 and 3-4 show the results of these investigations in terms of key thickness versus span.

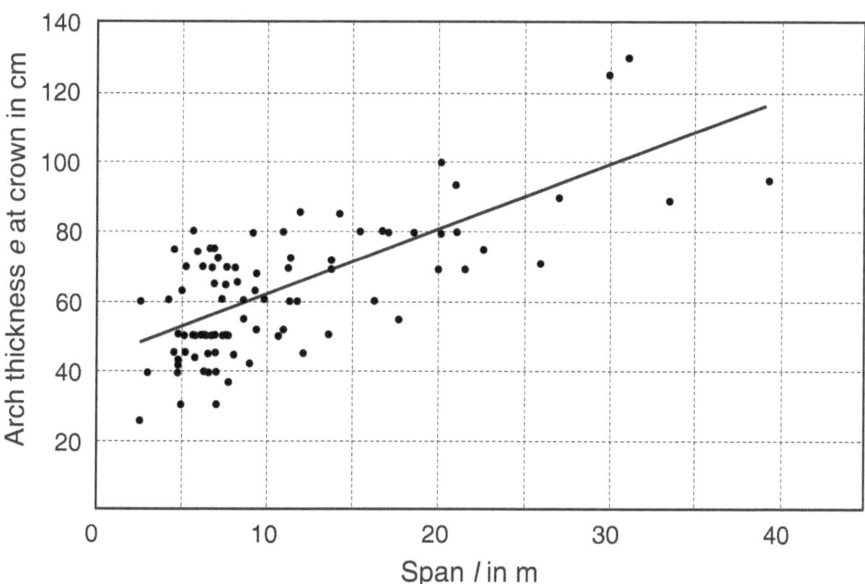

Fig. 3-3. Ratio of arch thickness at crown to span according to Purtak (2004)

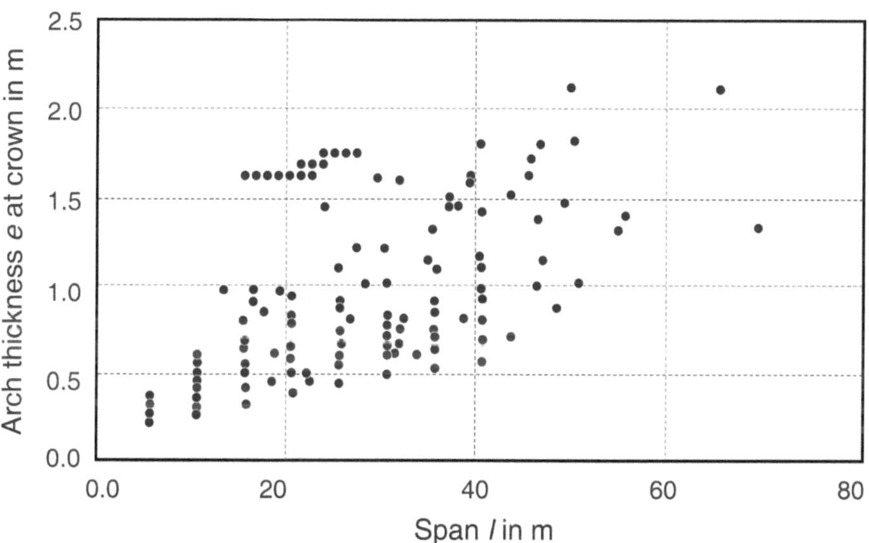

Fig. 3-4. Ratio of arch thickness at crown to span according to Busch and Zumpe (1995)

Besides the thickness of the arch at the crown, in many cases, simply the key stone thickness e versus the span l was given. Alberti gave $e/l = 1/15$,

Palladio 1/12, and Serlio 1/17. These values were mainly based on observations of Roman bridges. For example, the Pont de Gard reached $e/l = 1/15$.

Gautier recommended

$$e = \frac{1 \cdot l^2}{18} \tag{3-21}$$

at the beginning of the 18th century for bridges made of strong stones and with a span greater than 10 m. For bridges with softer stones, a minimum thickness of 0.32 m was required.

Many further rules were established, most of them in the 19th century, and are listed in Table 3-3 and visualised in Fig. 3-5 (Corradi and Filemio 2004, Corradi 1998).

Table 3-3. Different empirical arch thickness formulas

Dupuit for segmental arches	$e = 0.15 \cdot l^{0.5}$
Dupuit for semi-circular arches	$e = 0.20 \cdot l^{0.5}$
Rankine	$e = 0.191 \cdot R^{0.5}$
Gautier for semi-circular arches	$e = 0.32 + 1/15 \cdot l$
Perronet for semi-circular arches	$e = 0.325 + (1/24 - 1/144) \cdot l$
Lesguillier for semi-circular arches	$e = 0.10 + 0.20$
Dejardin for semi-circular arches	$e = 0.30 + 0.045 \cdot l$
Dejardin for circular arches	$e = 0.30 + 0.025 \cdot l$
Dejardin for elliptical arches	$e = 0.30 + 0.014 \cdot l$
Further equation	$e = 0.2 \cdot l$
German and Russian engineers for segmental arches	$e = 0.43 + 0.1 \cdot \rho$
Perronet	$e = 0.325 + 0.0694 \cdot \rho$
Perronet for semi-circular arches	$e = 0.325 + 0.035 \cdot l$
Lesguillier for segmental arches	$e = 0.10 + 0.20 \cdot l^{0.5}$
L'Eveillé for segmental arches	$e = 0.33 + 0.033 \cdot l$
German and Russian engineers for semi-circular arches	$e = 0.43 + 0.05 \cdot l$
Gauthey for semi-circular arches	$e = 0.33,\ l < 2$ m
Gauthey for semi-circular arches	$e = 0.33 + 0.020833 \cdot l,$ 2 m $< l < 16$ m
Gauthey for semi-circular arches	$e = 0.0416 \cdot l,$ 16 m $< l < 32$ m
Gauthey for semi-circular arches	$e = 1.33 + 0.020833 \cdot (l - 32\text{m}),$ $l > 32$ m
E. Roy for semi-circular arches	$e = 0.30 + 0.04 \cdot l$
Michon for semi-circular arches	$e = 0.40 + 0.04 \cdot l$

R – radius of the circle passing through the crown joint and the intrados springing in metre, l – span in metre, e – key stone thickness in metre and ρ as curvature radius, for segmental arch bridges, $\rho = \dfrac{l}{2}$,

for full circular arch bridges, $\rho = \dfrac{l^2}{8 \cdot f} + \dfrac{f}{2}$

and for elliptical arch bridges, $\rho = \dfrac{l^2}{4 \cdot f}$

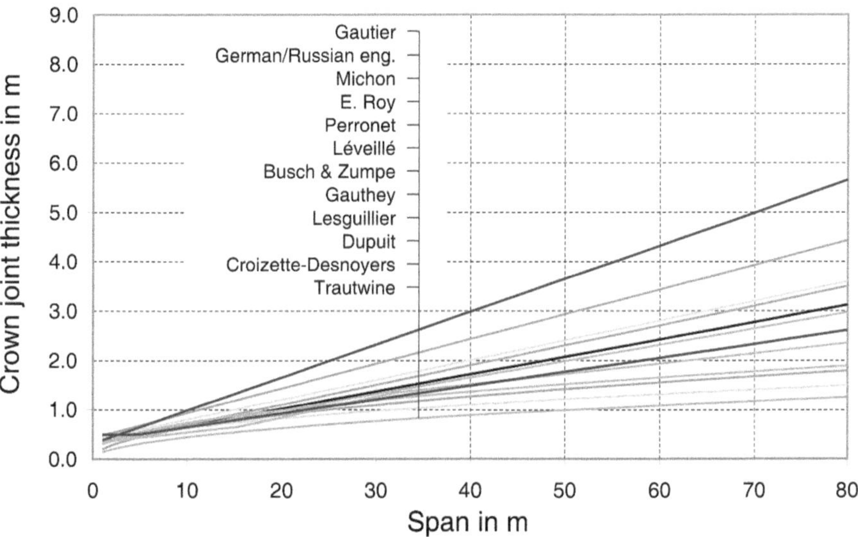

Fig. 3-5. Crown joint thickness e versus the span l according to different functions (Huerta 2004, Corrodi 1998)

Castigliano detected that the equations from circular arches and segmental arches can be converted by considering a limited span of the circular arch in terms of $l'=0.866 \cdot l$. Using this equation, the factor for Dupuits formula changes from 0.2 to 0.14. The conversion was also valid for L'Eveillé's equation (Coraddi 1998).

Kaven's equation was given as

$$e = 0.25 \text{ m} + l \cdot \left(0.025 + 0.00333 \cdot \frac{l}{f} \right). \tag{3-22}$$

The equation was valid for spans smaller than 12 m and arches made of very strong stone material. Furthermore, the height of the superstructure above the crown was limited to 1.5 m. Greater fillings required a specific adaptation

$$\text{Road bridges: } \sqrt{1+0.214\cdot h}\,, \tag{3-23}$$

$$\text{Railway bridges: } \sqrt{1+0.14\cdot h}\,.$$

Additionally, if the bridge was made up of brick stone, the arch thickness should be further increased by a factor of 1.5.

Huste's equation was given as

$$e = \alpha\cdot\sqrt{\rho} \tag{3-24}$$

with
$\alpha = 0.165$ for arches built of strong stone material,
$\alpha = 0.220$ for arches built of brick stones,
$\alpha = 0.247$ for arches built of soft stone material,
According to Corradi (1998), Italian engineers used

$$e = 0.20\ \text{m} + \frac{l}{40} + \frac{20+l}{1000} + \frac{l}{f}\,. \tag{3-25}$$

Résal's equation is given as (Coraddi 1998)

$$e = 0.15\ \text{m} + 0.20\cdot\frac{l}{2\cdot\sqrt{f}} \tag{3-26}$$

Croizette-Desnoyers's equation is given as (Corradi 1998)

$$e = k_1 + k_2\cdot\sqrt{\rho} \tag{3-27}$$

with ρ as curvature of the arch. The factors depend on the shape and the type of the bridge, and are given in Table 3-4 (Corradi 1998).

Table 3-4. Croizette-Desnoyers's factors (Corradi 1998)

	Circular arch			
	Road bridge		Railway bridge	
f/l	k_1	k_2	k_1	k_2
1/2	0.15	0.15	0.20	0.17
	Segmental arch			
	Road bridge		Railway bridge	
f/l	k_1	k_2	k_1	k_2
1/4	0.15	0.15	0.20	0.17
1/6	0.15	0.14	0.20	0.16

| | Segmental arch | | | |
	Road bridge		Railway bridge	
1/8	0.15	0.13	0.20	0.15
1/10	0.15	0.12	0.20	0.14
1/12	0.15	0.12	0.20	0.13

For skewed bridges, Résal recommended increasing the thickness according to

$$e' = \frac{e}{\sqrt{\sin \alpha}} \, ,$$

(3-28)

where α is the skewness of the bridge in grad.

Further equations are given in Leliavsky (1982). Taken from that reference, the equation of Heinzerling is given as

$$e = k_1 + k_2 \cdot r \, .$$

(3-29)

The factors are given in Table 3-5.

Table 3-5. Heinzerlings factors (Leliavsky 1982)

k_1	k_2	
0.4	0.028	Brick stone masonry
0.42	0.032	Ashlar masonry

The equation of Schwarz reads as follows:

$$e = 0.2 + \frac{1}{21} \cdot \frac{l \cdot Q}{zul \, \sigma} \, .$$

(3-30)

Further equations from Melan, Landberg and Mehrtens should only be mentioned (Leliavsky 1982).

3.2.1.3 Arch thickness at springing

In 1845, Dejardin suggested the following formula for the thickness of arches:

$$e_1 = \frac{e}{\cos \varphi}$$

(3-31)

The angle is measured from the crown (Corradi 1998). If this formula is applied, then the extrados follows a Nicomede conchoids or Pericycloid if the intrados follows a circular shape. However, the suggested formula can not be applied for half-circle arches because the thickness of the stones at springing becomes infinite. Therefore, the thickness of the arch stones is often not altered for an angle of 60°.

The maximum thickness of the arch stones then becomes

$$e_1 = 1.4 \cdot e .$$
(3-32)

This equation corresponds with a suggestion by Tavernier in 1907. However, in contrast to the description of the change of the stones, he furthermore suggests (Corradi 1998)

$$e_1 = \frac{e}{\sqrt{\cos \varphi}} .$$
(3-33)

Another equation is given as

$$1.125 \cdot (\rho + e) \leq R < 1.25 \cdot (\rho + e) .$$
(3-34)

The results of the later equation give values between Dejardin's and Tavernier's suggestions. Furthermore, Croizette-Desnoyers's rules were widely applied:

$$s = \frac{f}{l} = \frac{f}{2 \cdot a} .$$
(3-35)

The parameters for the formula are given in Table 3-6.

Table 3-6. Parameters for arch stone thickness

s	1/3	1/4	1/5	1/6	1/8	1/10	1/12
Half circular	-	1.80	-	1.40	1.25	1.15	1.10
Elliptic	1.80	1.60	1.40	-	-	-	-

Weber (1999) describes the change of the vault thickness as quadratic or bilinear.

3.2.1.4 Thickness of foundation

Besides the design of the arch shape, the thickness of the arch itself at the crown and at the springing of the abutment has to be chosen carefully. Many empirical relationships were published concerning this question. Design variables were S_s as the thickness of the abutment, r as the rise of the arch, L as width, e as the thickness of the crown joint, h as the distance from the springing line to the foundation base, and h_1 as the height of the backing. The parameter H is defined as

$$H = h + f + e + 0.60 , \quad h_1 < 0.60 \text{ m} ,$$
(3-36)

$$H = h + f + e + h_1 , \quad h_1 > 0.60 \text{ m} .$$

Many different formulas exist for the computation of thickness S_s of the abutment (Table 3-7).

Table 3-7. List of equations for the computation of the abutment thickness S_s (Corradi 1998)

Author	Equation
Semicircular arch	
Lesguillier	$S_s = (0.60 + 0.04 \cdot h) \cdot \sqrt{l}$
L'Éveillé	$S_s = (0.60 + 0.162 \cdot l) \cdot \sqrt{\dfrac{0.865 \cdot l \cdot (h + 0.25 \cdot l)}{H \cdot (0.25 \cdot l + e)}}$
German and Russian engineers	$S_s = 0.305 + \dfrac{5}{24} \cdot l + \dfrac{h}{6} + \dfrac{h_1}{12}$
Segmental arch bridges	
Lesguillier	$S_s = \left[0.60 + 1.10 \cdot \left(\dfrac{l}{f} - 2 \right) + 0.04 \cdot h \right] \cdot \sqrt{l}$
L'Éveillé	$S_s = (0.33 + 0.212 \cdot l) \cdot \sqrt{\dfrac{l \cdot h}{H \cdot (f + e)}}$
German engineers	$S_s = 0.305 + 0.125 \cdot l \cdot \left(\dfrac{3 \cdot l - f}{l + f} \right) + \dfrac{2 \cdot h + h_1}{12}$
Italian engineers	$S_s = 0.05 \cdot h + 0.20 \cdot l + \left(\dfrac{10 + 0.5 \cdot l}{100} \right) \cdot \left(\dfrac{l}{f} \right)$
Semi-elliptical arches	
Lesguillier	$S_s = \left[0.60 + 0.05 \cdot \left(\dfrac{l}{f} - 2 \right) + 0.04 \cdot h \right] \cdot \sqrt{l}$
L'Éveillé	$S_s = (0.43 + 0.154 \cdot l) \cdot \sqrt{\left(\dfrac{h + 0.54 \cdot f}{H} \right) \cdot \left(\dfrac{0.84 \cdot l}{0.65 \ f + e} \right)}$
German engineers	$S_s = 0.05 \cdot h + 0.20 \cdot l + \left(\dfrac{10 + 0.5 \cdot l}{100} \right) \cdot \left(\dfrac{l}{f} \right)$
Manuale del'Ingenere	$S_s = \sqrt{l} \cdot \left(0.42 + \dfrac{0.17 \cdot l}{2 \cdot f} + 0.44 \cdot h \right)$ bei $h_1 < 1.50$ m
	$S_s = \sqrt{l} \cdot \left(0.42 + \dfrac{0.17 \cdot l}{2 \cdot f} + 0.44 \cdot h \right) + 0.0185 \cdot H \cdot \sqrt{h_1}$
	bei $h_1 \geq 1.50$ m und mit $H = h + f + e$

Author	Equation
Hütte	$$S_s = \frac{l}{8} \cdot \left(\frac{3 \cdot l - f}{f + l} \right) + 1.00 + \frac{h}{6}$$ and for semi-circular arches $$S_s = \frac{5}{24} \cdot l + 1.00 + \frac{h}{6}$$
Croizette-Desnoyers	$$S_s = 0.33 + 0.212 \cdot l \cdot \left(\frac{l \cdot h}{H \cdot (f + e)} \right)$$
Italian railway engineers	$$S_s = 0.20 + 0.030 \cdot (\rho + 2 \cdot e) + 0.10 \cdot h$$

A discussion of some formulas is given in Corradi (1998).

Besides the abutment thickness, the pier width also had to be estimated. Here, some general rules such as the pier width have to have at least the thickness of the abutment, and the pier width has to have at least twice the thickness of the crown joint to be fulfilled. Furthermore, possible hydro dynamic water pressure had to be considered. Finally, some equations for the computation of the pier width are given in Table 3-8. The variables are h as the height from the foundation springing line to the haunch joints of rupture, H represents the distance between the foundation line and the bridge road line, e is the thickness of the crown joint, f is the distance between the crown joint extrados and the haunch joint of rupture, d is the horizontal distance between the haunch joints of rupture, P is the weight of the section of arch up to the joint of rupture, ω is the weight of the bridge construction material, and k is a safety factor (Corradi 1998).

Table 3-8. Computation of the pier width according to different authors (Corradi 1998)

Author	Formula and arch shape
L'Éveillé	Circular arch $$E = (0.33 + 0.212 \cdot D) \cdot \frac{\sqrt{\dfrac{h}{H}}}{\sqrt{\dfrac{f + e}{D}}}$$ Half-circular arch $$E = (0.60 + 0.162 \cdot D) \cdot \frac{\sqrt{\dfrac{h + 0.25 \cdot D}{H}}}{\sqrt{\dfrac{0.25 \cdot D + e}{0.865 \cdot D}}}$$ Basket arch

Author	Formula and arch shape
	$E = (0.42 + 0.154 \cdot D) \cdot \dfrac{\sqrt{\dfrac{h + 0.54 \cdot D}{H}}}{\sqrt{\dfrac{0.465 \cdot D + e}{0.84 \cdot D}}}$
Séjourne	$E > \dfrac{1}{5} \cdot l$
	$E = 0.4 + 0.15 \cdot l$
	$E = 0.8 + 0.1 \cdot l$
Perronet	$E = 2.25 \cdot e$
Castigliano	$E = \dfrac{S_s}{2}$
Colombo	$E = 0.20 \cdot h' + 0.6$
	$E = \dfrac{1}{6} \cdot l,\ E = \dfrac{1}{10} \cdot l$
	h' as pier height from foundation to springing
Further applied formula	$E = 0.292 + 2 \cdot e$
Further applied formula	$E = 2.50 \cdot e$ für $l \leq 10$ m
	$E = 3.50 \cdot e$ für $l > 10$ m
Further applied formula	$\dfrac{1}{10} \cdot l < E < \dfrac{1}{5} \cdot l$
Minimum value	$E > \sum e_1$, if several arches

3.2.1.5 Maximum arch length

Weber (1999) has estimated the maximum arch length of certain materials based on the breaking length of the material. The breaking length of the material is the maximum length of a bar of a certain material where the tensile or compression strength of the material and the stress caused by the self-weight of the bar are equal. In Table 3-9, the breaking length for different construction materials is summarized.

$$\lim l = \pi \cdot \frac{\sigma}{\gamma_w}. \tag{3-37}$$

Table 3-9. Theoretical maximum span of arch bridges for different materials (Weber 1999)

Material	$\lim l$ in m
St 37	6,467
GS-52	7,585

Material	lim l in m
Brick masonry	205–748
Natural stone masonry	145–785
Reinforced concrete	228–1870
Lightweight concrete	524–2,244
Coniferous wood	6,912–8,639

3.2.1.6 Coefficient of boldness

Very interesting parameters describing the capacity utilization of an arch bridge are so-called coefficients of boldness. Such boldness numbers were developed by different authors. First, the number originating from Résal is introduced. The number is computed as

$$A = \frac{l \cdot \rho}{2}.$$

$$(3\text{-}38)$$

The variables are explained in Table 3-10 and in the footnotes of Table 3-3. Furthermore, examples for different bridges are given there.

Table 3-10. Résal's coefficient of boldness taken from Corradi (1998)

Bridge	Span l in m	Rise f in m	Arch thickness e in cm	A
Fouchard Bridge	26.0	2.6	33.8	439
Nogent Bridge	50.0	25.0	25.0	625
Tournon Bridge	49.2	17.7	25.9	639
Antoinette Bridge	50.0	15.9	27.6	690
Lavaur Bridge	61.5	27.5	30.9	931
Moskwa Bridge	44.8	5.6	7.6	1066
Cloris Bridge	50.0	7.4	46.0	1150
Chester Bridge	61.0	12.8	42.9	1308
Cabin John Bridge	67.0	17.6	40.7	1363
Trezzo Bridge	72.3	20.7	42.0	1517
Soupees arch	37.9	2.1	88.0	1667

Another such performance parameter is Spangenberg's coefficient of boldness:

$$k = \frac{l^2}{f}.$$

$$(3\text{-}39)$$

Spangenberg's number is widely known. However, Weber (1999) has shown that for a parabolic arch second order, Spangenberg's number corresponds with eight times the curvature radius of the arch at the crown.

The number can also be adapted to other arch curvatures as shown in Table 3-11.

Table 3-11. Spangenbergs boldness number (Weber 1999)

Arch shape	Spangenberg's number represents
Parabola second order	$8 \cdot \rho$
Parabola fourth order	$8 \cdot \rho \cdot (1 - c)$
Catenary First type	$\approx 8 \cdot \rho \cdot \left(1 - \dfrac{l^2}{48 \cdot \rho^2} \right)$
Segmental circular arch	$8 \cdot r \cdot \left(1 - \dfrac{f}{2 \cdot r} \right)$
Sinus curvature (half wave)	$\pi^2 \cdot \rho$
Upper elliptic half	$4 \cdot \rho$

l arch span, f arch rise, ρ curvature radius

Weber (1999) criticizes both definition and content of Spangenberg's number. For example, the consideration of the arch thickness would be compelling. Based on the Hugi's self-capacity utilization

$$\omega = \frac{g}{g + p}.$$

(3-40)

Weber (1999) defines

$$\chi = \frac{g}{p} \text{ and } \omega = \frac{\chi}{1 + \chi}.$$

(3-41)

Weber (1999) has then introduced a boldness number that considers the

- Static system
- Cross-section geometry
- Load pattern
- Construction material
- Slenderness
- Structural size

The number is given as

$$\omega = \frac{\dfrac{3}{2} \cdot \lambda + \dfrac{1}{k_0}}{16 \cdot \theta + \dfrac{3}{2} \cdot \lambda - \dfrac{1}{k_0}}$$

(3-42)

with

$$k_0 = \frac{f}{l}, \quad \theta = \frac{\bar{l}}{l}, \quad \lambda = \frac{l}{d_4}, \quad \bar{l} = \frac{f_d}{\gamma_w} \tag{3-43}$$

and d_4 as arch thickness at the quarter point. Examples are given in Table 3-12.

For further analytical models of the ultimate load-bearing behaviour, see Audenaert et al. 2007.

Table 3-12. Weber's ω and Spangenberg's boldness number k for certain bridge structures including further parameters (Weber 1999)

Bridge[1]	Year	l in m	f/l	d_4	l/d_4	M.[2]	ω	k
Thur Bridge Kurmmenau (S)	1911	63.3	1/4.57	2.26	27.99	LM	0.65	289
Iller Bridge St. Buchlow-Kempten	1906	63.8	1/2.48	1.94	32.89	TC	0.51	158
Iller Bridge St. Buchlow-Lindau	1906	64.5	1/2.34	1.98	32.58	AM	0.51	151
Prut Bridge Karemča (U)	1892[3]	65.0	1/3.63	2.60	25.0	AM	0.48	236
Rummel Bridge Sidi-Rached (A)	1912	68.0	1/3.09	1.52	44.74	AM	0.65	210
Adda Bridge bei Morbegno (I)	1904	70.0	1/7.00	1.73	40.46	AM	0.61	490
Roizonne Bridge, La Mure (F)	1916	79.5	1/2.10	1.60	49.66	AM	0.70	169
Valserine Bridge, Montages (F)	1910	80.3	1/4.01	2.00	40.15	AM	0.69	322
Pétruss Bridge (L)	1903	84.6	1/2.71	1.80	47.03	AM	0.71	229
Soča Bridge, Solkan (SL)	1927	85.0	1/3.90	2.45	34.69	AM	0.59	332
Friedens Bridge in Plauen	1904	90.0	1/5.00	1.65	54.55	LM	0.69	495
Lot Bridge in Villeneuve (F)	1920	96.3	1/6.23	1.45	66.38	TC	0.82	600
Nanpan Bridge Changhong (C)	1961	112.5	1/5.43	2.21	50.90	AM		611
Chang Bridge Jiuxigou, Feng.(C)	1972	116.0	1/80	1.87	62.03	AM		928
Wuchao Bridge, Fong-Huan (C)	1990	120.0	1/50	1.50	80.00	AM	0.88	600
120 m Bridge Project (A-H)		120.0	1/3.6	5.10	23.53	AM	0.69	432

[1] S – Switzerland, SL – Slovenia, U – Ukraine, C – China, A–H – Austria–Hungry, L – Luxemburg, F – France, A – Algeria, names without country identification are in Germany located

[2] LM – Layered masonry, TC – tamped concrete, AM – Ashlar masonry, M. – Material

[3] 1916 destroyed by the Russian army, 1927 reconstructed

3.2.2 Modern Rules

3.2.2.1 MEXE method

The MEXE method (Military Engineering Experimental Establishment) is probably the best known simple method for the load-bearing assessment of historical arch bridges. The method is heavily applied not only in Great Britain (Huges and Blackler 1997), the country of origin, but also in many other countries since it was included in UIC-Codex (1995).

Early works for the development of this method were carried out by Pippard in the 1930s. The basic assumption was linear-elastic behaviour of the material. Especially during the World War II, Pippard's method found wide application for the assessment of historical arch bridges under military loads. However, since military loads changed later due to the new NATO, the method had to be revised. The revision based on the work by Pippard yielded to the first-generation MEXE method. The testing of several arch bridges in Great Britain in the 1950s provided newer results, which could be considered again for a revision of the method. Therefore, in the 1960s, also caused by changing live loads, the method was modified (Das 1995).

The application of the method is simple and fast. An example will show that. First, the conditions found have to comply with the boundary conditions of the method. The boundary conditions are given in Table 3-13.

Table 3-13. Boundary conditions of the MEXE method

Requirements	Example
Span smaller then 20 m	10 m
Rise greater then ¼ of the clear span	¼ ×10 = 2.5>1.8 m
Filling above crown is between 30 and 105 cm	1.2–0.5 = 0.7 m

The load-bearing behaviour is then computed with

$$Q_{adm} = Q_p \cdot f \tag{3-44}$$

with

$$q_{adm} = \frac{Q_{adm}}{1.5 \text{ m}} \tag{3-45}$$

and

$$f = f_S \cdot f_M \cdot f_J \cdot f_C \cdot f_N \cdot \frac{1}{f_\Phi} . \tag{3-46}$$

The factor f considers a variety of parameters. The first one is the arch shape factor:

$$\frac{r_q}{r_c} = \frac{1.4}{1.8} = 0.78 \rightarrow f_S = 2.3 \cdot \frac{(r_c - r_q)^{0.6}}{r_c} = 2.3 \cdot \frac{(1.8-1.4)^{0.6}}{1.4} = 0.95 . \tag{3-47}$$

Further factors, which have to be taken from diagrams and tables, are the following:

Material factor

$$f_M = 1.0 . \tag{3-48}$$

Joint factor

$$f_J = f_W \cdot f_{mo} . \tag{3-49}$$

Joint thickness factor

$$f_W = 0.8 \ \rightarrow \text{Joint thickness} > 12.5 \ \text{mm} . \tag{3-50}$$

Mortar factor

$$f_{mo} = 0.9 \ \rightarrow \text{weak, crumbly mortar} . \tag{3-51}$$

Factor for the arch condition

$$f_c = 0.85 \ \rightarrow \text{Longitudinal cracks in the middle third of the arch} \tag{3-52}$$

Factor for the numbers of arches

$$f_N = 0.8 \ \rightarrow \text{Arch supported by two piers} . \tag{3-53}$$

And, f can be computed as

$$f = 0.95 \cdot 1.0 \cdot 0.8 \cdot 0.9 \cdot 0.85 \cdot 0.8 \cdot \frac{1}{1.25} = 0.372 . \tag{3-54}$$

Using this value, one can read from a diagram (UIC-Codex 1995)

$$Q_p = 425 \ \text{kN} \tag{3-55}$$

$$Q_{adm} = 0.425 \ \text{MN} \cdot 0.372 = 0.158 \ \text{MN} \tag{3-56}$$

$$q_{adm} = \frac{0.158 \text{ MN}}{1.5 \text{ m}} = 0.105 \text{ MN/m}. \tag{3-57}$$

As shown, the method can be quickly applied. This explains the wide application. However, some authors such as Brencich et al. (2001) have criticized the method that produces unsafe results under some circumstances.

Therefore, it is not surprizing that beside the successful and wide application of the MEXE method in the last few years, major efforts have been undertaken to develop successors of the MEXE-method. A few such methods will be introduced in the following sections.

3.2.2.2 FILEV method

Martín-Caro and Martínez (2004) have developed the FILEV method based on nearly 800 bridge models with different parameters for which the ultimate load bearing was investigated with the FE-program Sofistic. Table 3-14 gives a summary of the investigated parameter combinations for single loads and Table 3-15 for uniform loads. The meaning of some of the parameters is shown in Fig. 3-6. The computation database was then used for derivation of approximation equations. However, of course, such equations include some simplifications. For example, the live load was considered either only as single load with changing position or as a uniform load over half entire arch span, respectively. The position of the single load was modified between 1/5 and 1/2 of the arch span. Furthermore, for the single load, only the development of a four-hinge mechanism or shear failure was considered as failure. For the uniform load, additional compression failure was allowed.

Table 3-14. Investigated parameter combinations for single loads

Spannweite l in m	c/l	f/l	h_o in m	h_p in m	b_p/l	μ	γ in kN/m^3	γ_{fill} in kN/m^3
5.00	0.10	1/2	0.25	2.0	1/4	0.60	20.0	18.0
7.50	0.09	1/4	0.50	5.0	1/6	0.80		
10.00	0.07	1/6	2.00	10.0	1/8			
12.50	0.06							
15.00	0.05							
17.50								
20.00								

Table 3-15. Investigated parameter combinations for uniform loads

Spannweite l in m	f/l	h_o in m	f_c in MPa	γ in kN/m³
5.00	1/2	0.40	4.0	20.0
10.00	1/6		6.0	
20.00			8.0	
			10.0	

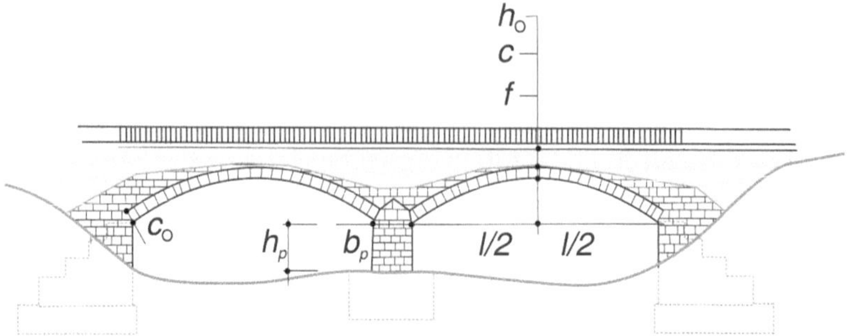

Fig. 3-6. Introduction of variables of the FILEV method

The arch width was chosen constant with 3.0 m, and the load spreading in the backfill was considered with 30°. Dead load was included with a partial safety factor of 1.0, and uniform live load was considered with a partial safety factor of either 1.0 or 1.35. Further assumptions are the following:

- The bridge is straight and neither skewed nor bended
- The bridge is only a one span arch bridge
- Adjacent arches have the same span
- Abutments are considered as infinite stiff
- The bridge is not damaged
- The backfill is not hollowed
- The span is between 2 and 20 m
- The minimum ratio of rise to span is $f/l > 1/6$
- The ratio of arch thickness at the crown should follow the values given in Table 3-15
- Backfill exists above extrados between the abutments. The height of the backfill is computed based on the ratio of rise to span for high arches ($f/l = 1/2$) with $h_{\text{backfill}} = 0.6 \times f$ and for flat arches ($f/l = 1/6$) with $h = 0.3 \times f$
- The filling above crown is assumed between 0.25 and 2.0 m
- The maximum pier height is considered with 10 m
- The maximum pier width is at least 1/6 of the arch span

Table 3-16. Ratio between arch thickness at crown c and arch span l

l in m	2.0–5.0	5.0–7.5	7.5–10.0	10.0–15.0	15.0–20.0
$c/l \geq$	0.10	0.09	0.07	0.06	0.05

3.2.2.3 Method by Harvey et al.

Currently, Harvey (2007) and Harvey et al. (2007a) have developed a substitute method for the MEXE method. Unfortunately, not much was known about the new method at the time this book was printed. However, new material will probably be published in the years to come.

In general, Harvey et al. (2007) promised that the method will be simpler to use and will behave robustly. The method can be applied by artisans and will probably include nomograms such as shown in Fig. 3-7.

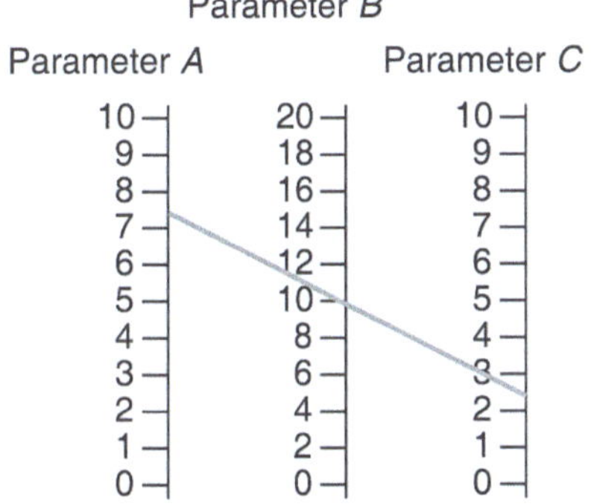

Fig. 3-7. General outline of the Harvey method (Harvey et al. 2007)

3.2.2.4 Method by Purtak et al.

Purtak et al. (2007) have developed curvatures for the ultimate load of historical arch bridges based on finite element simulations using the program ANSYS. The simulation considers not only material nonlinear behaviour of the stone material using the Mohr–Coulomb model, but also openings of joints. The load-bearing curvatures have been constructed for a great variety of different materials and geometrical parameters, such as stone width to height, joint thickness, stone and mortar compression strength, stone tensile strength, and eccentricity. Due to the variety of different parameters, the curvatures are able to cover a wide range of different conditions, such as compression failure or development of mechanisms. Figure 3-8 shows only a simple example of such load-bearing curvatures.

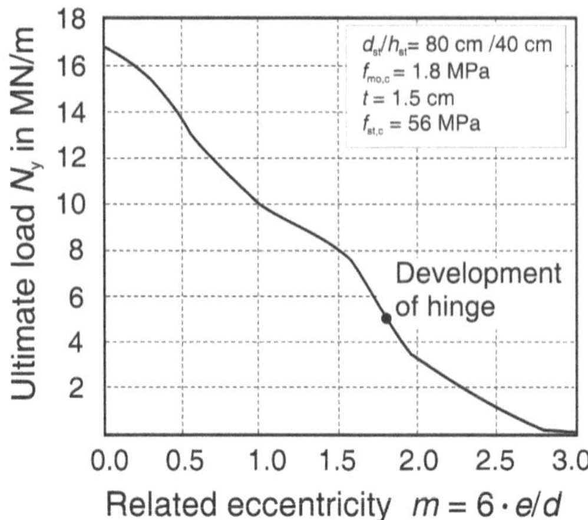

Fig. 3-8. Example of ultimate load versus eccentricity functions (Purtak et al. 2007)

3.2.2.5 Method by Martinez et al.

Martinez et al. (2001) have also developed ultimate load-bearing curvatures for the crown arch thickness depending on the stone material's compression strength, the ratio between rise and span, and the span. Again, this method permits a very quick estimation of the ultimate load-bearing behaviour of the arch bridge.

3.2.2.6 Method by Aita et al.

Aita et al. (2007) discuss the application of two different methods for the assessment of stress levels in arch systems based on geometrical and material properties. They apply the methods to semicircular and pointed arches and receive the so-called stability area curvatures. Such stability area curvatures can be extended towards diagrams including maximum arch-wall system height related to arch thickness or maximum height versus volume.

3.2.2.7 Method of Pauser

The consulting company of Pauser in Vienna has developed a method of estimating the load-bearing capacity for arch bridges in a fast way. However, it has been strongly suggested that a computer program should be provided for the computation since several correction parameters considering different conditions at the bridge have to be estimated.

The program is based on linear-elastic failure concept. It uses pre-computed tables comparable to the technique by Purtak. Finally, the load-bearing capacity of the arch bridge can be related to diagrams shown in Fig. 3-9 (Heinlein 2008, Pauser 2005).

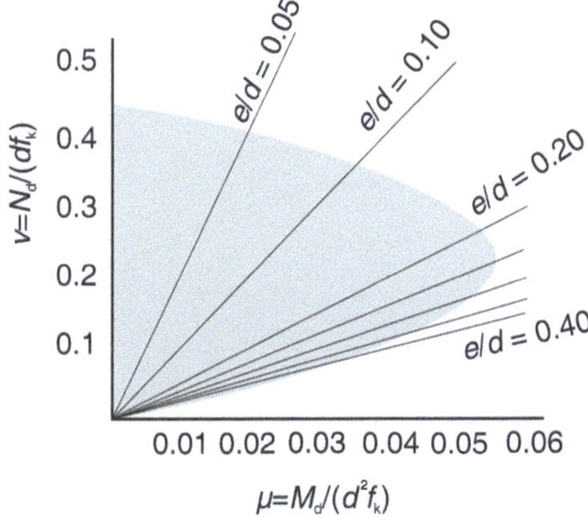

Fig. 3-9. Interaction diagram for an arch cross section (Heinlein 2008)

3.3 Beam Models

3.3.1 Single Beam Models

With the introduction of analytical models, the epoch of beam models was also started. Whereas the first beam models had to comply with the re-requirement of simple hand computation (static determined structures), this requirement does not hold nowadays. With the loss of this requirement, beam models were more and more extended to consider more and more effects on the arch bridge load behaviour yielding more precise results. This evolution is shown in Table 3-17. Whereas the first models simply chanced the number of predefined hinges, the backfill, elastic foundation, and roadway structures were also considered later. The last model from Gocht (1978) is mainly suited for arch bridges with a reinforced roadway or railway slab.

Table 3-17. Evolution of beam models for arch bridges

Type of model	Visualization
Fixed arch without backfill	
Two hinge arch without backfill	
Three hinge arch without backfill according to Ril 805 or UIC Codex	
Backfill considered	
Model according to Voigtländer (1971)	
Model according to Model (1977)	
Model for arch with roadway slab according to Gocht (1978)	

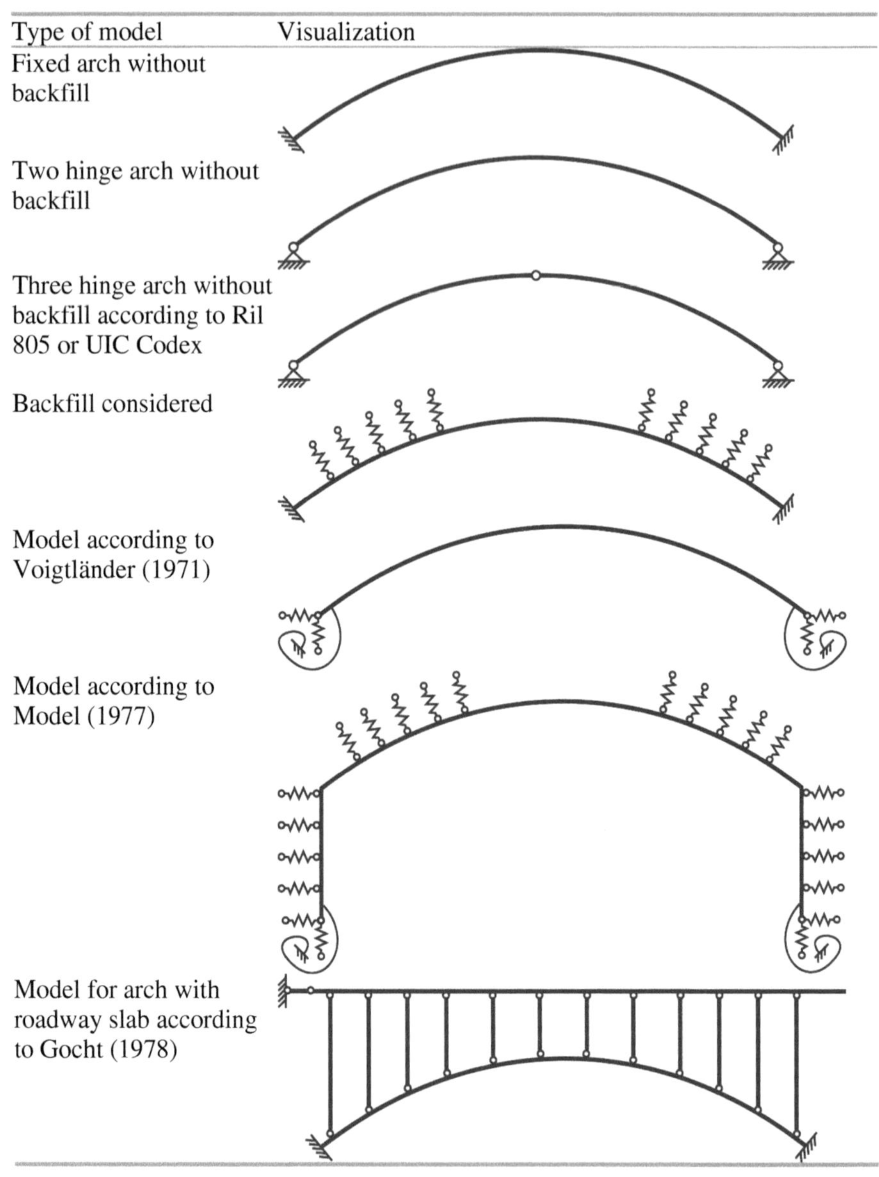

The consideration of further elements of arch bridges is required since the arch itself, although the biggest single contribution to the load-bearing behaviour of arch bridges, is only one element of several. Weber (1999) has quantified the contribution of the single elements of an arch super-structure based on a test. In general, he proposed a factor of 1.5 increased load-bearing capacity from the single-span pure arch to the single-span arch bridge. Based on Fig. 3-10, this factor is wide on the safe side. Another rough measure considering further elements in an arch superstructure was given by the former German national railway. Without any further proof the railway permitted an increase of 20% in the load-bearing capacity of historical arch bridges by building reinforced concrete slabs on the bridges (DR 1985). This corresponds with the model from Gocht (1978).

The load-increasing effects of roadway and pavement concrete are also known from other bridge types. For example, Gutermann (2002) has investigated the influence of the three elements and found experimentally an increase of the load-bearing capacity by the road asphalt layer of about 3%, by the pavement concrete of about 12%, and the protection concrete up to 10%. However, the investigation of Gutermann (2002) was carried out on serviceability levels and in terms of stresses and deformations.

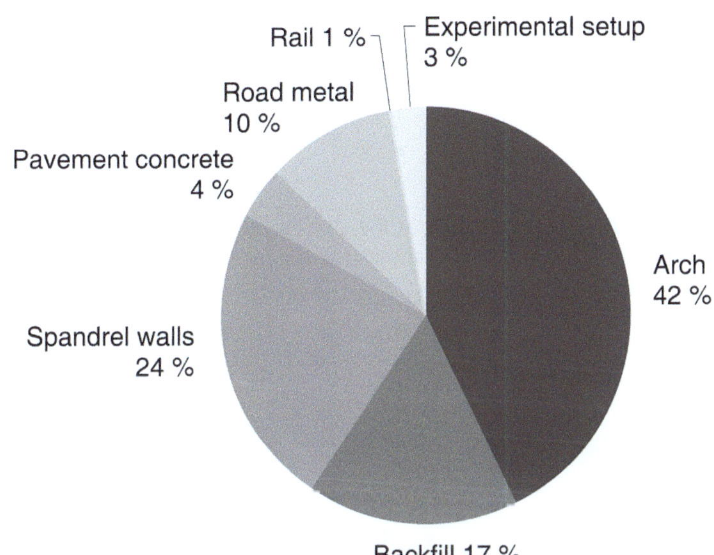

Fig. 3-10. Contribution of different structural elements to the load bearing based on measurements on the railway arch bridge Schauenstein (Weber 1999)

The contribution of further structural elements in arch bridges has been known for a long time. For example, Craemer (1943) already stated a change of the line of trust caused by the backfill in arch bridges. Fischer (1940–1942) has investigated the shear stress between the arch and the backfill. For simplification reasons, he only considered three-hinge arches. Jäger (1938) proved an increased load-bearing capacity of arch bridges when shear forces were transferred to the backfill. Herzog (1962) also showed an higher load-bearing capacity. He found a lower eccentricity when backfill was acting. However, he only considered the same Young modulus for the arch and the backfill. Bienert (1959–1960) and Bienert et al. (1960–1962) also discussed the effects on the load-bearing behaviour of arch bridges caused by backfill. A very intensive list of references can be found in Gocht (1978).

Furthermore, Gocht (1978) has developed some models that should be shown here. First, he developed a model that if no shear forces are transferred between the arch and the backfill, a roadway slab still exists. Later, he discussed the effects of an active height of the backfill. Figures 3-8 and 3-9 show the height considering a solid joint or a sliding joint between the arch and the backfill. Also, Fig. 3-11 considers whether a joint at the springing exists. Figure 3-12 shows that a solid joint between the arch and the backfill can also have negative effects. Then, a flat arch can develop inside the backfill. These effects also change the span of the arch.

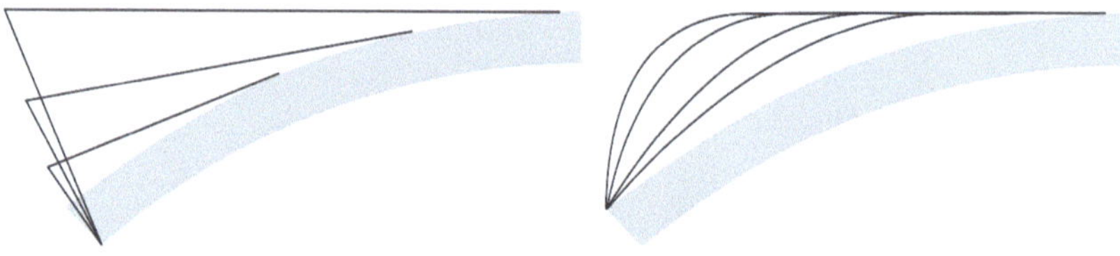

Fig. 3-11. Model of the effective height of the arch by consideration of the backfill according to Gocht (1978). *Left figure* shows the height without a springing joint and *right figure* with springing joint

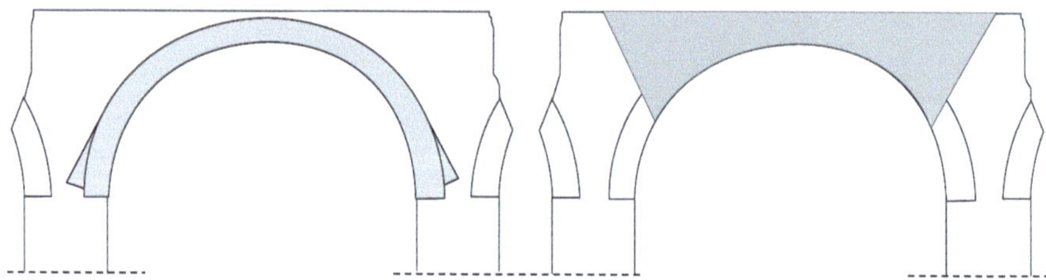

Fig. 3-12. Effective height for a half-cycle arch based on Gocht (1978)

Indeed, the collapse of the Italian Traversa railway bridge between Turin and Genoa was related to such a change of system (Brencich and Colla 2002). Figure 3-13 shows the plan and front view of the collapsed bridge. Please compare this figure with Fig. 3-12 (right).

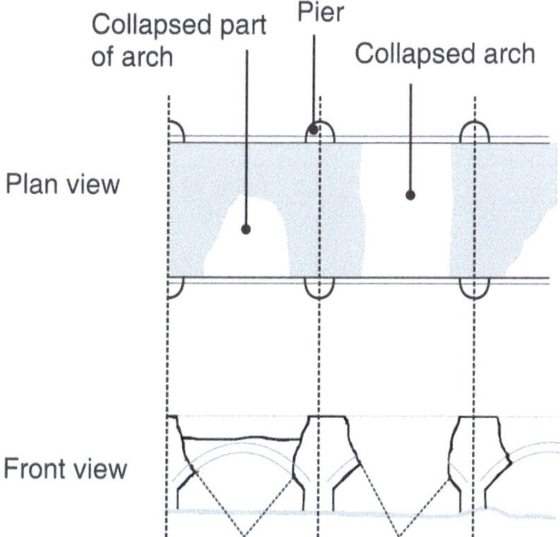

Fig. 3-13. Traversa Bridge in Italy after partial collapse according to Brencich and Colla (2002)

Smith et al. (2004) have given an estimation of the increase of the load-bearing behaviour by backfill and coverage, respectively, as shown in Fig. 3-14.

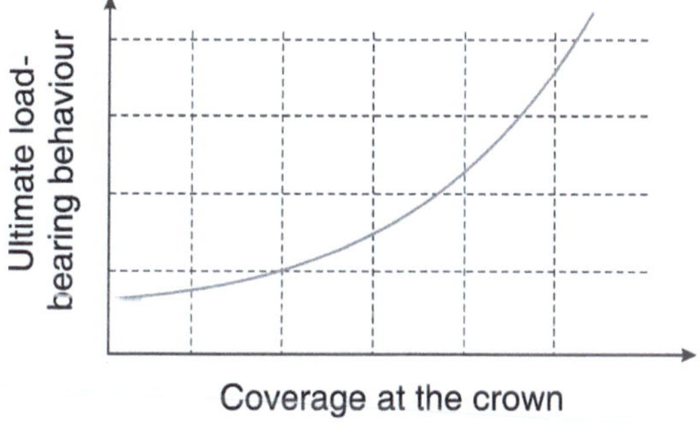

Fig. 3-14. Increase of the load-bearing behaviour if coverage is considered (Smith et al. 2004)

Molins and Roca (1998) not only have shown the influence of the backfill, but also of other structural elements. For two arches, they have computed the failure load to 15 and 16 kN, respectively, if only the arch itself is considered. If the model is extended with spandrel walls, the failure load increases to 46 and 48 kN, respectively. However, only if tensile bars (tie) are considered, the experimental failure loads in the range of 90 kN can be achieved by numerical investigation. That means that the arch itself contributes only by 25%, which fits very well with the results from Weber (1999). The results from Molins and Roca (1998) are summarized in Table 3-18.

Table 3-18. Results from Molins and Roca (1998)

	Arch ①	Arch ②
Ultimate load measured	60 kN	95 kN
Ultimate load computed with		
arch and backfill at arch 2	15 kN	16.1 kN
Arch and spandrel walls	46 kN	48 kN
Arch, spandrel wall, and tie bars considered	60 kN	91 kN

Comparable results were published by Cavicchi and Gambarotta (2004). They have investigated the influence of the backfill at the real size tests in the United Kingdom at the Prestwood Bridge. The bridge had a span of 6.55 m, a rise of 1.428 m, and the thickness of the arch at the key of 0.22 m. The coverage at the key was 0.165 m. The load was introduced at ¼ point over the entire width of 3.0 m. The bridge collapsed by a four-hinge mechanism at a load of 228 kN. The masonry compression strength was assumed to be 4 MPa. Cavicchi and Gambarotta (2004) computed the load capacity of the pure arch without backfill in the range of 46 kN. This would correspond to 20% of the overall load-bearing capacity. Another example by Cavicchi and Gambarotta (2004) also gave a load contribution by the pure arch between 33 and 50% of the overall load-bearing capacity. See also Cavicchi and Gambarotta (2005).

Royles and Hendry (1991) have also investigated the influence of the backfill, masonry backup, and spandrel walls on the overall load-bearing capacity of arch bridges. They reach an improvement compared to the pure arch of factor 2–12, which would correspond to 8–50%. Becke (2005) has given an increase of the load by backfill compared to the pure arch of factor 3 (33%). However, Becke distinguishes between effective and noneffective backfills. This distinction depends on the stiffness of the backfill: If the stiffness of the backfill is less than 1/100 of the arch, the backfill is not effective and cannot be considered.

If the backfill is not continuous over the arch, it still can contribute to the load bearing. Braune (1980) mentions that even transverse and vaults support the arch.

Latest works about the influence of the backfill and the spandrel walls are Cavicchi and Gambarotta (2007) and Harvey et al. (2007b).

Besides the backfill, the spandrel walls can also significantly contribute to the load bearing. For example, Voigtländer (1971) has given a decrease of stresses of the arch, if the spandrel wall is functioning, of about 20%. However, this decrease is not found in all areas of the arch over the arch length: usually experiments only found a decrease in stresses at the springing, not at the crown. At the crown, the spandrel walls usually do not reach good bond conditions. Triebecker has developed a method for the consideration of the spandrel wall in beam models: he simply considers the arch as fixed beam and supports this beam with an additional beam. This second beam should consider the positive effects of the spandrel wall. Schreyer (1960) also has mentioned the active functioning of the spandrel walls.

The description of the load-bearing effects of further structural elements besides the arch itself is already an indicator for the difficulty of the development of a beam model for arch bridges. In many cases, the models are heavily disputed. Some codes give general indications such as the Ri 805, which simply states that every model that yields to an equilibrium can be applied. On the one hand, sometimes models are claimed such as shown in Fig. 3-15. Here, the limitations of developed hinges during collapse are considered in that way that under all conditions a moment has to be applied.

On the other hand, if hinges develop, then the stresses decrease. Model (1977) suggests a 30% drop of the stresses in the arch after cracks were initiated. As a consequence, complete hinges could not develop. However, Model (1977) also considers rather large cracked areas that would correspond with smeared crack models in FE models.

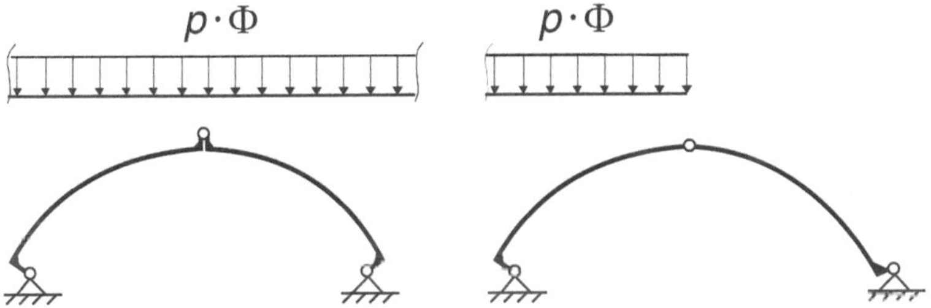

Fig. 3-15. Minimum moments in springing hinges according to Ril 805

In contrast, historical and current research has shown that historical masonry arch bridges under ultimate loads do not show large cracked areas, but do show great single cracks in regions of maximum loading. In this region, usually much higher rotations occur than linear-elastic models show. Figure 3-16 shows test results from Jagfeld and Barthel (2004). Brencich and Gambarotta (2005) also found such hinges on a bridge over the Scrivia in Alessandria/Italy.

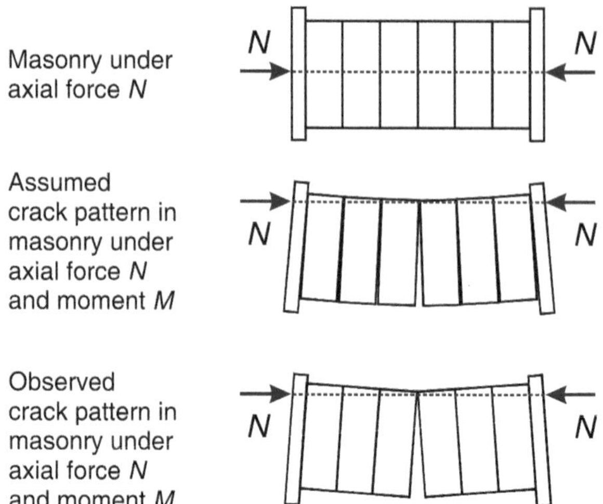

Fig. 3-16. Development of hinges in masonry according to Jagfeld and Barthel (2004)

This observation is the reason for the introduction of hinges in arch beam models even if no hinges have been designed and built. Figure 3-17 visualizes this concept. The plasticity theory application for masonry arches was originally introduced by Kooharian (1952) and Heyman (1966). An excellent overview of the development of arch models can be found in Gilbert (2007).

Both methods, the search of an equilibrium inside the arch as static modelling and the consideration of the kinematic chains as kinematic criteria, are used for the load evaluation process of arch bridges. They can be seen as upper and lower bound solutions (Fig. 3-18).

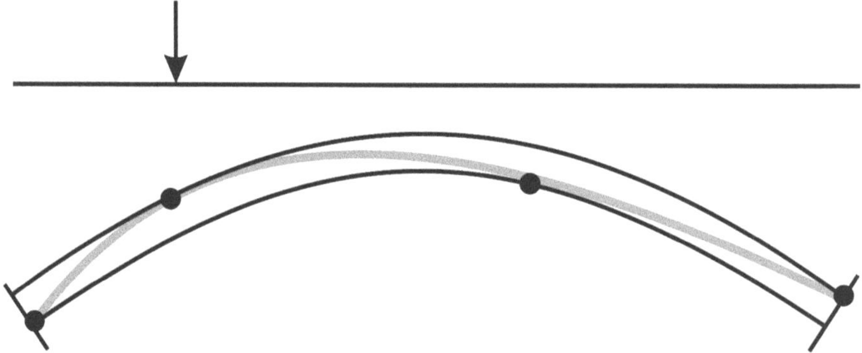

Fig. 3-17. Application of plasticity theory for masonry arch bridges (Heyman 1966, Kooharian 1952)

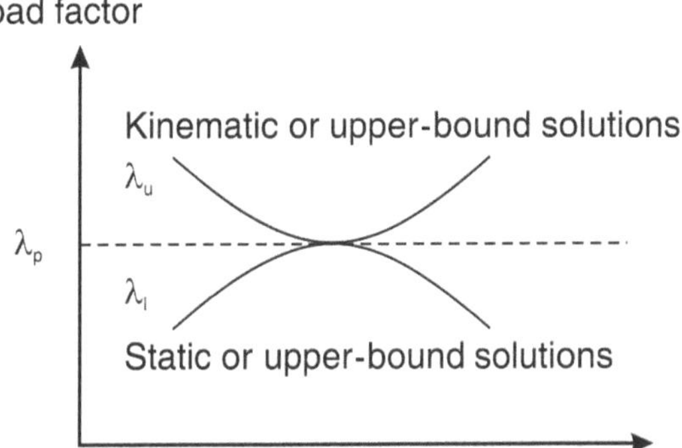

Fig. 3-18. Relationship between upper- and lower-bound solutions (Gilbert 2007)

The MAFEA code in the UIC codes is using nonlinear beam elements, permitting the development of hinges. Such elements have also been used in the program Sofistik by Bothe et al. (2004), Herrbruch et al. (2005), and Mildner (1996). The program STATRA has been used by Möller et al. (2002). Highly sophisticated programs for the computation of stone arch bridges are RING (LimitState Ltd 2008) and Archie-M (Harvey 2008).

3.3.2 Compound Beam Models

An introduction into the theory and computation of compound beam cross sections can be found in Hannawald (2006). In general, for the consideration of compound cross sections, a distinction between solid joints and sliding joints has to be undertaken. If relative sliding inside joints is prohibited, this is called a solid joint. The properties of this compound cross section are based on geometrical and stiffness properties of single cross sections. For example, this model can be used for arch bridges by simply increasing the arch cross section by a certain factor for the partial load transfer into the backfill. The drawback of this simple model is the fact that the thickness of the backfill may vary in the length of the arch and therefore the factor has to be altered over the arch length.

In the following paragraphs, the computation of such a factor will be shown for solid joints in terms of an effective cross section. Such formulae have found wide application in structural wood design.

The position a of the line of thrust in such a compound cross section with solid functions can be computed as

$$a = \frac{1}{2} \cdot \frac{(EA)_2 \cdot (h_1 + h_2)}{\sum (EA)_i} , \qquad (3\text{-}58)$$

where E_1 and E_2 are Young's modulus for the arch and the backfill, A is the cross-sectional area, h are the corresponding heights, and the variable a is shown in Fig. 3-19. If the arch and the backfill have the same height, then the formula becomes

$$a = h_1 \cdot \frac{E_2}{E_1 + E_2} . \qquad (3\text{-}59)$$

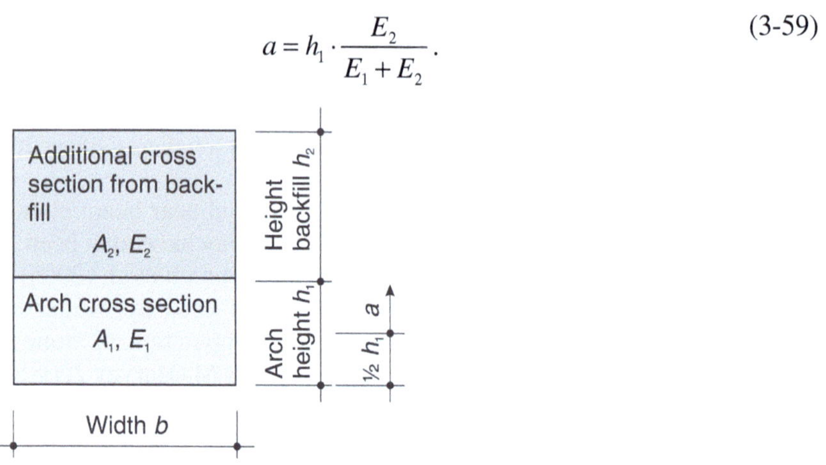

Fig. 3-19. Variables at the compound cross section

The effective bending stiffness of the compound cross section becomes

$$(EI)_{eff} = \sum \left((EI)_i + (EA)_i \cdot a^2 \right) \qquad (3\text{-}60)$$

$$(EI)_{eff} = (E_1 + E_2) \cdot \frac{b \cdot h_1^3}{12} + b \cdot h_1 \cdot (E_1 \cdot h_1^2 + E_2 \cdot a^2) , \qquad (3\text{-}61)$$

where I is the bending stiffness of the single cross section and b is the width according to Fig. 3-19.

However, if sliding occurs in the joint, the contribution of the backfill to the load bearing will decrease compared to the solid joint. Such a sliding is more probable than a solid joint and depends on many factors, such as the shape of the extrados and material properties of the backfill. The sliding can be covered in a simplified way by the consideration of a sliding factor γ. This factor can be computed according to

$$\gamma = \frac{1}{1+k_1}$$

(3-62)

$$k_1 = \frac{\pi \cdot (EA)_2 \cdot s_1}{K_1 \cdot l^2}$$

(3-63)

with $(EA)_2$ as the product of Young's modulus and the cross section size for cross section 2, which is connected to cross section 1

- K_1/s_1 as stiffness of the joint between cross sections 1 and 2
- s_1 as distance of dowels between cross sections 1 and 2
- l as distance of the moment zero point (l is the span for single-span beams, $0{,}8 \times l_i$ for continuous beams, $2 \times l_k$ for cantilevers)

The position of the line of thrust by the same height for the arch and the backfill and the consideration of the sliding factor becomes

$$a = h_1 \cdot \frac{\gamma \cdot E_2}{E_1 + \gamma \cdot E_2} \cdot$$

(3-64)

The effective bending stiffness also depends on the sliding factor:

$$(EI)_{eff} = (E_1 + E_2) \cdot \frac{b \cdot h_1^3}{12} + b \cdot h_1 \cdot (E_1 \cdot a^2 + \gamma \cdot E_2 \cdot (h_2 - a)^2).$$

(3-65)

The effective normal force stiffness is computed as

$$(EA)_{eff} = E_1 \cdot A_1 + \gamma \cdot E_2 \cdot A_2.$$

(3-66)

The line of thrust is changing over the length of the arch due to the consideration of the backfill (Fig. 3-20). It can even come out of the arch. These results fit very well with results by Gocht (1978).

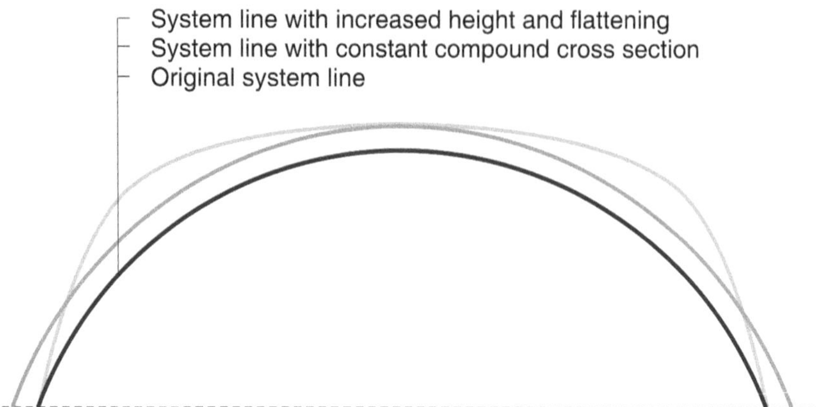

Fig. 3-20. Resulting line of thrust by consideration of the backfill according to Becke (2005)

The resulting forces inside the arch can be computed as

$$N_1 = N \cdot \frac{(EA)_1}{\sum (EA)_i} + \gamma \cdot M \cdot \frac{(EA)_1 \cdot a}{(EI)_{eff}} = N \cdot \frac{E_1}{E_1 + E_2} + \gamma \cdot M \cdot \frac{(EA)_1 \cdot a}{(EI)_{eff}} \quad (3\text{-}67)$$

$$M_1 = M \cdot \frac{(EI)_1}{(EI)_{eff}} . \quad (3\text{-}68)$$

The effects of the backfill on the load-bearing behaviour of arches by using beam models can be done either by applying springs and additional beams for the backfill or by the introduction of compound cross sections as shown here. A further model is the application of the finite element method. This technique not only permits the consideration of the backfill and possible sliding inside the joints but also allows a general consideration of material and geometry nonlinear behaviour. Therefore, in contrast to linear-elastic beam models, wherein hinges chosen in advance represent areas of nonlinear behaviour, this will be done automatically in finite element models.

3.4 Finite Element Method (FEM)

Besides the application of simple beam elements for the modelling of arch bridges, finite element models of masonry and concrete arch bridges have become more and more popular. The first finite element analysis of arch bridges was probably carried out by Towler (1985) and Crisfield (1985).

Nowadays, an increasing number of professional finite element programs include modules for realistic material description of masonry and are used for the simulation of arch bridges. Choo et al. (1991), Dialer (1991), Loo (1995), Mojsilović (1995), Lourenço (1996), Baker (1997), Seim and Schweizerhof (1997), Seim (1998), Parikh and Patwardhan (1999), Huster (2000), Schlegel and Rautenstrauch (2000), Schlegel et al. (2003), Schlegel (2004), Schlegel and Will (2007), and Schlegel (2008) have supported this development by developing such realistic material models for FE programs. Others have simply used already included material description modules (Drucker-Prager, Willam Warnke 1974). Such modules in ANSYS has been used, for example, by Weigert (1996), Wedler (1997), Jagfeld (1998), Melbourne and Tao (1998), Trautz (1998), Frunzio and Monaco (1998), Fanning and Boothby (2001), Fanning et al. (2001), Hänel and Reintjes (2001), Mildner and Mildner (2001), Reintjes (2002), Droese and Bodendiek (2002), Proske (2003), Brencich et al. (2002), Purtak (2001, 2004), Witzany and Jäger (2005), Purtak et al. (2007), and Bién et al. (2008). The program ATENA has been used by Schueremans (2001), Cervenka (2004), and Slowik et al. (2005). Healey and Counsell (1998), Brencich and Colla (2002), Brencich and de Francesco (2004), and LUSAS (2005) have used the LUSAS program. Stavrouli and Stavroulakis (2003) and Mura et al. (2003) used the program MARC for the investigation of historical arch bridges. Rots and van Zijl (2005) have used DIANA. The program SAP 2000 was used by Toker and Ünay (2004) and Prader et al. (2008) and the program Strand7 FE analysis was employed by Ford et al. (2003).

Further examples of finite element models of arch bridges can be found in Aoki et al. (2004), Aoki and Sabia (2003), Aoki and Sato (2003), Orlando et al. (2003), Cavicchi and Gambarotta (2006), Drosopoulos et al. (2007), and Knoblauch et al. (2008). An extension of finite element models is mentioned by Fuhlrott (2004). The requirement of extensive modelling of the foundation and the ground is required by Rombach (2007). And finally, some authors see general limitations in the application of finite element models for masonry, such as Chiostrini et al. (1989) and Dialer (2002).

In general, masonry requires material nonlinear modelling and therefore finite element models have to adapt to this (Figs. 3-21 to 3-24).

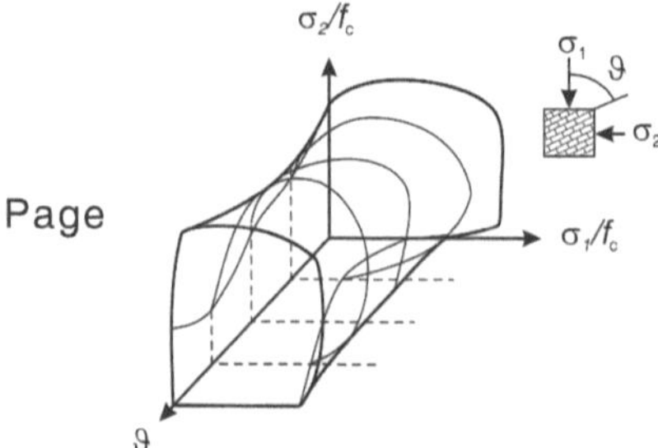

Fig. 3-21. Failure surface of masonry based on tests according to Page (1983)

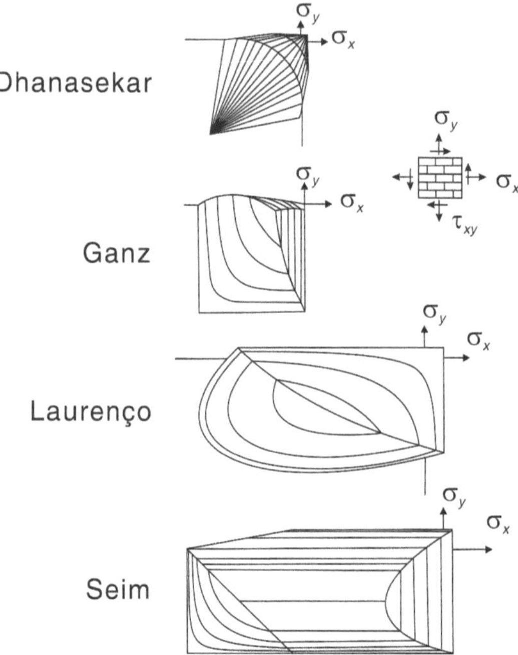

Fig. 3-22. Failure surfaces of masonry according to Seim (1998) and Lourenço (1996)

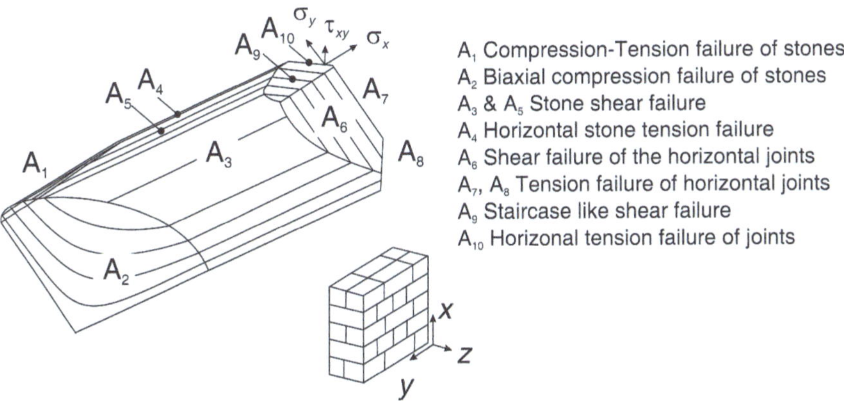

Fig. 3-23. Failure surface of masonry according to Schlegel (2004)

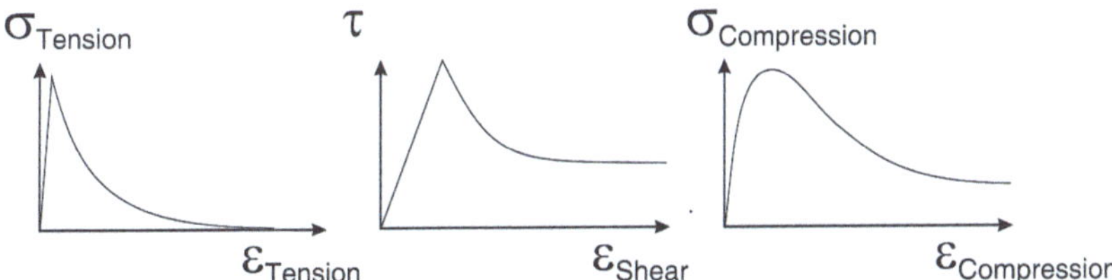

Fig. 3-24. Stress-strain behaviour of masonry for tension, shear and compression according to Schlegel (2004)

A summary of different computation strategies is given by Lourenço (2002), including DEM. However, and that is quite important, the more complicated the simulation techniques become, the higher a numerical safety factor gets, as Table 3-19 shows. This increased safety factor reflects the increasing uncertainty of an increasing number of input variables.

Table 3-19. Safety factors for different computation strategies (Lourenço 2002)

Approach/analysis type	Safety factor
Allowable stress (f_{ta} = 0.2 MPa)	0.31
Kinematic limit analysis	1.8
Geometric safety factor	1.2
Physical nonlinear and no tensile strength	1.8
Physical and geometrical nonlinear and no tensile strength	1.7
Physical nonlinear and tensile strength of 0.2 MPa	2.5
Physical and geometrical nonlinear and tensile strength of 0.2 MPa	2.5

3.5 Discrete Element Method (DEM)

The finite element method can show convergence problems especially under great cracks. Then, the advantage of finite element method by assuming homogenous material properties over certain space regions cannot hold anymore. An alternative is the application of the discrete element method (DEM)—sometimes called distinct element method.

In general, the DEM is a procedure for the simulation of movements of a limited number of bodies with any shape subject to certain interactions. Single bodies can freely move in space, however, contacts between bodies are considered. Both static and kinematic boundary conditions are fulfilled. The static boundaries consider the following in detail:

- Material behaviour functions
- Contact behaviour functions
- External field conditions (Neuberg 2002)

The material behaviour functions describe the behaviour of the single bodies under external forces. The contract behaviour functions describe the behaviour of single bodies during interaction and external fields describe forces such as gravitation that act on all single bodies.

Cinematic conditions are the movement equations of Newton's second law and the extension to rotation. Furthermore, for quasistatic solutions, numerical damping is required.

The overall computation includes the alternate computation of the static and cinematic relationships over a discrete time. The discrete time also allows the change of the external field forces if required (Neuberg 2002).

Early applications of DEM were done by Cundall (1971) and Cundall and Strack (1979). Since then, DEM has found wide application in the computation of masonry elements and masonry arch bridges. For example, Chiostrini et al. (1989), Mamaghani et al. (1999), Bićanić et al. (2002), Dialer (2002), and Keip and Konietzky (2005) and Lemos (2006) have used DEM for masonry structures in general.

Maunder (1993), Lemos (1995), Owen et al. (1998), Roberti and Calvetti (1998), Thavalingam et al. (2001), Brookes and Collings (2003), Bićanić et al. (2003), Jackson (2004), Schlegel (2004) and Rouxinol et al. (2007) have used DEM for historical masonry arch bridge computation.

Although DEM is a very general and robust method, the problem for practical application is still an extensive computation time and a great multitude of different material parameters that are often unknown or difficult to measure on the structure.

3.6 Comparison of Testing and Modelling

3.6.1 Load Tests on Arches

The quality of numerical models requires the comparison with observed data—in this case, the comparison with the investigated ultimate load-bearing behaviour of real arches. Therefore, for more than a century, tests on arch bridges were carried out.

First test reports were published by the Austrian engineer and architecture association in 1895. The testing of reinforced and non-reinforced concrete arches with a maximum span of 23 m has been described (Huerta 2001). In Germany, the first tests were carried out in 1908 on a three-hinge road vault with tamped concrete. The bridge was originally constructed in 1902 for an industry exhibition in Düsseldorf. The bridge was designed for a streamroller with 230 kN, and for a crown with 4 kN/m². The vault reached a span of 28 m. The rise was 2 m. The thickness of the arch reached 0.75 m at the springing, 0.85 m in the quarter points, and 0.65 m at the crown. The backfill of the arch was removed. Unfortunately, the maximum load of the test equipment of 4,230 kN was reached without failure. Therefore, the maximum load was applied for two days continuously; however, the bridge did not fail. The ratio between observed and computed ultimate load was more than 18.

In 1935, tests on arch bridges were carried out in the United States (Jäger 1935). In 1936, Pippard and colleagues carried out tests on arch bridge models. They discovered that arches develop hinges during failure and therefore simple linear elastic computations are insufficient for ultimate load-bearing assessment since they do not consider the development of such hinges. Pippard and Chitty therefore developed a technique called "instability analysis." This method showed great affinity with the plastic methods introduced by Heyman (Das 1995).

At the end of the 1980s and the beginning of the 1990s, about ten arch bridges were tested in the United Kingdom (eight stone arch bridges and two models) (Das 1995, BE 16/97). Some of the tests were carried out in the Bolton Laboratory. These tests, and additional tests ordered by the Transport and Road Research Laboratory (TRRL, later TRL), resulted in considerable amount of material for recomputation of the ultimate load of the bridges. This is especially true for the tests on real arch bridges.

Most of the bridges collapsed by the development of four-hinge mechanisms. Only a minority failed by buckling or exceeding the masonry compression strength. A summary of some parameters of the tested bridges is shown in Table 3-20.

In connection with the tests, the investigation by Royles and Hendry (1991) should be mentioned. They have classified the tested arches as follows:

- Compression arch alone
- Compression arch with backfill
- Compression arch with backfill and backup masonry
- Compression arch with backfill, backup masonry and spandrel walls

The tests clearly showed some positive effects of additional structural elements. For example, load distribution caused by the backfill increased the load bearing and also backup masonry, stiffer backfill, and stiffer spandrel walls increased the load capacity of the arches between factor 2 and factor 12 in comparison to the arch only (Royles and Hendry 1991).

Table 3-20. Parameters of the tested bridges (BE 16 1997, RING 2005)

Bridge	Span in m	Rise in m	Arch thickness in mm	Width in m	Skewness in degrees
Bridgemil	18.30	2.85	711	8.3	0
Bargower	10.36	5.18	588	8.68	16
Preston	5.18	1.64	360	8.7	17
Prestwood	6.55	1.43	220	3.6	0
Torksey	4.90	1.15	343	7.8	0
Shinafoot	6.16	1.19	542	7.03	0
Strathmashie	9.42	2.99	600	5.81	0
Barlae	9.86	1.69	450	9.8	29
Bundee	4.00	2.00	250	6.0	0
Bolton	6.00	1.00	220	6.0	0

In 1989, the German railway in Schauenstein (Germany) tested the load capacity of a single track natural stone arch bridge with tamped concrete. The tests are intensively presented in Weber (1999) and are also mentioned in Hoch and Schmitt (1999). The bridge was constructed in 1919 with a span of 14.0 m and a pier height of 4.20 m. The width of the vault reached 3.65 m. The spandrel walls reached a height of 0.83 m. The arch shape followed a basket arch with either three or five circles. The railway line was closed in 1976 (Weber 1999).

A single load was applied on the railway slab at a quarter point using a reinforced concrete traverse. The bridge reached an ultimate load of 3,840 kN. At that load, the test had to be stopped since the ultimate load of the ground anchors used for the anchoring of the concrete traverse had reached the limit. However, the results showed that the arch overleaped the development of a three-hinge stadium and went directly into a four-hinge mechanism. The arch showed a horizontal deformation of 4.7 mm and a vertical deformation of 9.0 mm at the load introduction point. At the unloaded quarter point, the arch moved upwards about 0.1 mm and horizontally by 3.0 mm. It should be mentioned that at up to 30% of the ultimate load the spandrel walls continued to function together with the arch and only afterwards partly separated. The rail track also contributed to the load resistance with a normal force of about 140 kN (Weber 1999).

After the first test, the bridge was partly disassembled: the rail track and the backfill were cleared. Until a load of 3,450 kN, the development of a four-hinge mechanism was completed. However, then the load was reduced again and the arch returned to the original shape. All failure joints closed. A further reloading of up to 2,398 kN caused the collapse of the bridge (Weber 1999).

Since then, many tests have been carried out and the following example should not be understood as a complete list. Roca and Molins (2004) presented tests on two bridges with a span of 3.2 m and a rise of 0.65 and 1.6 m, respectively. The bridges were loaded in the quarter points. Details about the bridges are given in Table 3-21. The bridges failed under 60 and 95 kN, respectively, due to the development of a four-hinge mechanism. Two hinges already developed at early loading states—one directly under the load and the other one at the close springing. At 60–75% of the ultimate load, the spandrel wall separated from the arch. The third hinge developed at 80–90% of the ultimate load, and finally the development of the fourth hinge at the crown yielded to the collapse. The arch with the lower rise showed a maximum vertical deformation at the crown of 4.5 cm and the second arch reached about 2.0 cm.

Vermeltfoort (2001) reports about tests on a masonry arch bridge without backfill in the Netherlands. In the last years Purtak et al. (2007) have carried out tests at the University of Technology in Dresden, Germany.

Table 3-21. Properties of the bridges tested by Roca and Molins (2004)

Property	Bridge ①	Bridge ②
Span	3.20 m	3.20 m
Rise	0.65 m	1.60 m
Overall length	5.20 m	5.20 m
Overall height	0.95 m	2.15 m
Width	1.00 m	1.00 m
Arch thickness	0.14 m	0.14 m
Thickness spandrel wall	0.14 m	0.14 m
Height of backfill at crown	0.10 m	0.10 m
Maximum height of backfill	0.78 m	1.17 m

Property	Bridge ①	Bridge ②
Number of tendons for horizontal restrains	6	8
Loading at	¼ Point	¼ Point
Compression strength of stones (longitudinal)	56.8 MPa	56.8 MPa
Young's modulus of the stone (longitudinal)	12.750 MPa	12.750 MPa
Compression strength of stones (transversal)	51.0 MPa	51.0 MPa
Young's modulus of the stone (transversal)	10.450 MPa	10.450 MPa
Mortar compression strength ($40 \times 40 \times 80$ mm)	8.34 MPa	8.34 MPa
Mortar flexural bending strength ($40 \times 40 \times 160$ mm)	2.68 MPa	2.68 MPa
Young's modulus of the mortar	780 MPa	780 Mpa
Friction coefficient of joint arch backfill	0.33–0.36	0.33–0.36
Compression strength of masonry	21.0	14.0
Density backfill	18 kN/m^3	18 kN/m^3

3.6.2 Comparison Results

As already mentioned, such tests enable us to assess the performance of computer models of arch bridges. Such comparisons are visualized in Fig. 3-25 and Table 3-22. However, under normal conditions, the engineer lacks data of real tests. Therefore, usually other programs are used to estimate the performance of our programs or models. For example, the FILEV method introduced earlier was tested with results from the program RING. The results were quite impressive, as shown in Table 3-23 and Fig. 3-26.

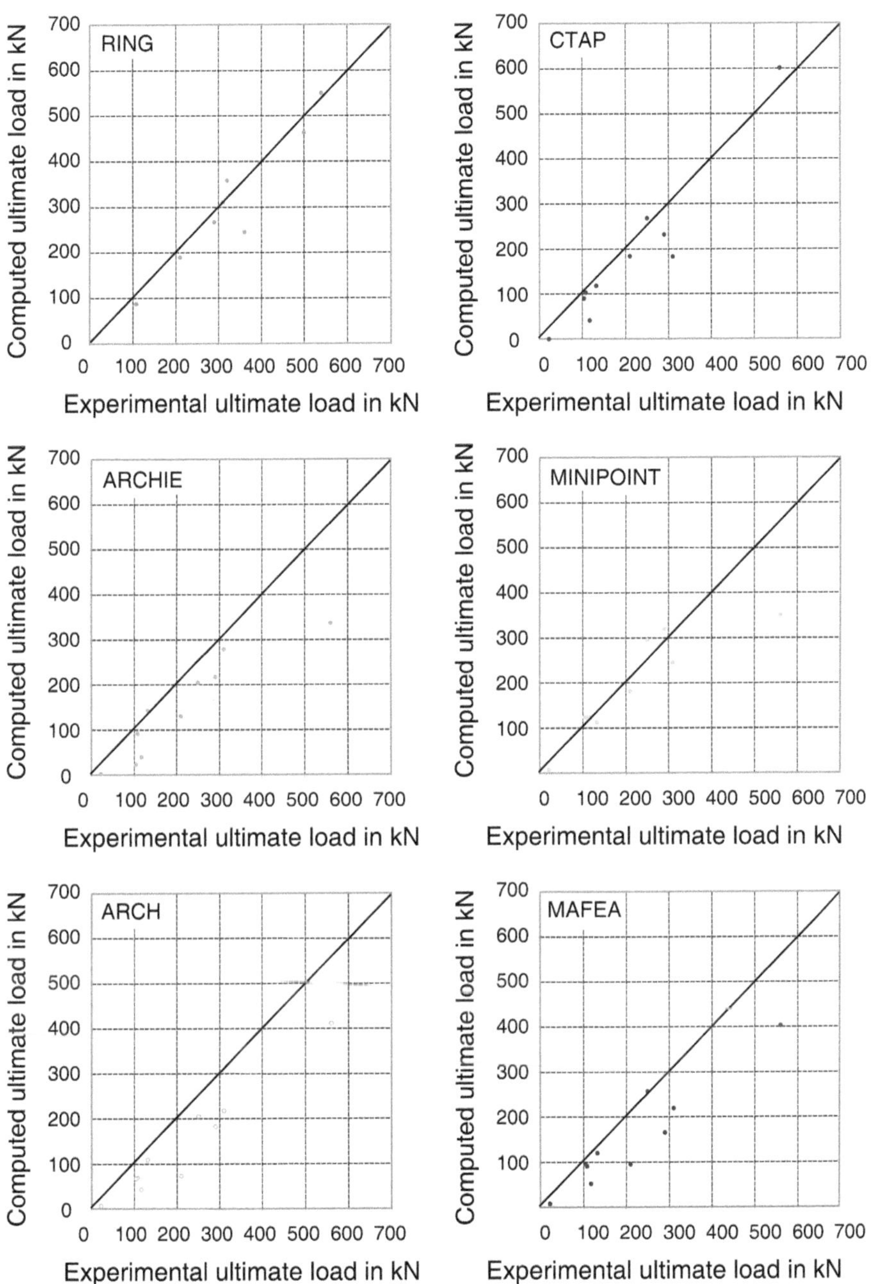

Fig. 3-25. Comparison of experimental and numerical ultimate load-bearing investigations (B 16 1997, RING 2005)

Table 3-22. Comparison of experimental and numerical ultimate load-bearing investigations (B 16 1997, RING 2005)

	Experimental ultimate load	Computed ultimate load using the program				
		CTAP	ARCHIE	MINIPOINT	ARCH	MAFEA
Bridgemill	310	183	278	245	217	219
Bargower	560	601	336	350	411	403
Preston	210	184	130	181	73	95
Prestwood	22	0	2	7	6	8
Torksey	108	103	91	124	69	91
Shinafoot	250	268	204	295	205	257
Dundee	104	90	23	123	67	96
Bolton	117	41	39	124	43	52
Strathmashie	132	118	142	112	109	120
Barlae	290	232	216	320	182	165

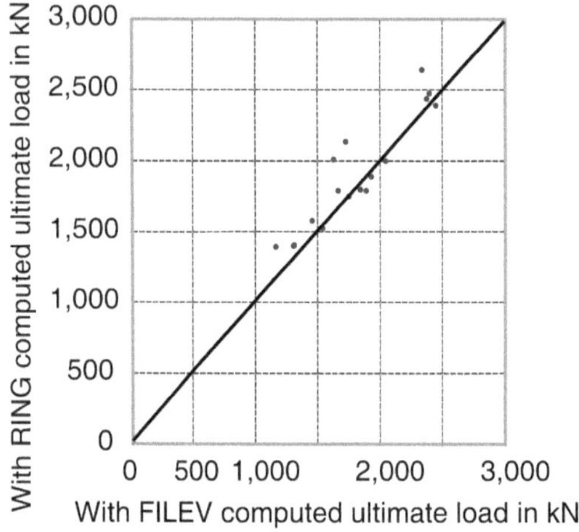

Fig. 3-26. Comparison of the numerical ultimate load-bearing investigation using FILEV and RING software (Martín-Caro and Martínez 2004)

Table 3-23. Comparison of the numerical ultimate load-bearing investigation using FILEV and RING software (Martín-Caro and Martínez 2004)

Nr.	Span in m	Rise in m	Arch thickness in m	Coverage at crown in m	Ultimate load according to FILEV in kN	Ultimate load according to RING in kN	Ratio
1	5.0	2.50	0.500	0.5	1,541	1,525	1.0105
2	5.0	1.25	0.500	0.5	1,726	2,135	0.8084
3	7.5	3.75	0.750	0.5	2,376	2,438	0.9746
4	7.5	1.88	0.750	0.5	2,662	3,175	0.8384
5	7.5	3.75	0.675	0.5	1,456	1,580	0.9215

Nr.	Span in m	Rise in m	Arch thickness in m	Coverage at crown in m	Ultimate load according to FILEV in kN	Ultimate load according to RING in kN	Ratio
6	7.5	1.88	0.675	0.5	1,630	2,010	0.8109
7	10.0	5.00	0.700	0.5	1,165	1,393	0.8363
8	12.5	2.50	0.700	0.5	1,305	1,400	0.9321
9	12.5	6.25	0.750	0.5	1,312	1,405	0.9338
10	12.5	2.50	0.750	0.5	1,666	1,790	0.9307
11	12.5	6.25	0.875	0.5	1,841	1,798	1.0239
12	12.5	2.50	0.875	0.5	2,338	2,640	0.8856
13	15.0	7.50	0.900	0.5	1,928	1,890	1.0201
14	15.0	3.00	0.900	0.5	2,449	2,390	1.0247
15	18.0	9.00	0.900	0.5	1,750	1,750	1.0000
16	18.0	3.60	0.900	0.5	2,046	2,000	1.0230
17	20.0	10.00	1.000	0.5	1,887	1,790	1.0542
18	20.0	4.00	1.000	0.5	2,397	2,475	0.9685

The program ARCHIE was originally developed in 1983 at the University of Dundee. It heavily relates to works by Heyman. However, additionally earth pressure can be included in the computation. The program first investigates the location of the hinges in the arch under live and dead load. Then the line of thrust is computed. ARCHIE then gives a lower limit of the load-bearing capacity. The program did not give deformations, multi-layered arches could not be considered, the failure criterion was always the development of a four-hinge mechanism, multi-span arches could not be investigated, and concrete backfill and vaults also could not be modelled (Brookes and Collings 2003).

In 1999, the program ARCHIE-M was presented by Obvis Ltd in the United Kingdom. The program first showed only slight differences from the ARCHIE, such as the consideration of ground pressure and the distribution of single loads in longitudinal direction (Brookes and Collings 2003). However, since then, the program has been permanently extended and updated. A demo version can be downloaded at Harvey (2008) or at Obvis Ltd (2006).

The program RING was originally developed during the 1990s at the Bolton Institute and the University of Sheffield. In 2007, the second edition was released. It considers the development of a mechanism, but shear failure can also be considered. Furthermore, some special conditions can be considered, such as multi-ring arches. The original version could not consider concrete backfill and vaults. Furthermore, no deformation was given and buckling was not considered (Brookes and Collings 2003). A trial version of RING 1.5 can be downloaded at the University of Sheffield (2008), and a demo version of RING 2.0 at LimitState Ltd (2008).

The Italian program SAV (Stabilita`di Archie Volte in Matura) has been introduced by AEDES (2008). Another Italian program is called ARCO

(Analysis of Masonry Arch and Vaults) and it can be downloaded under Gelfi (2006). Smars (2006) has developed the program Calipous for the computation of gothic vaults and structures of masonry arches in Belgium.

The German railway often uses the program GETRA for the load estimation of historical railway arch bridges (Ludwig 2000), and the Saxon road departments often use Excel sheets developed by Dibeh et al. (1997) based on linear-elastic arch computation using the theory of Strassner. Also, the consultant company Pauser has developed a program (Heinlein 2008, Pauser 2007). Teaching programs for students regarding the mechanisms of arches were developed by Greenwold et al. (2008) during the Active Statics project and by Block (2006).

3.7 Transverse Direction (Effective Width)

The presented beam models have not considered the transverse action of loads inside the arch bridges: they simply considered only longitudinal effects. However, the consideration of transversal effects may decrease the load on the arch and offer an opportunity to permit higher traffic loads on arch bridges. Therefore, the load spreading in transverse direction is usually considered in terms of effective width, as shown in Fig. 3-27.

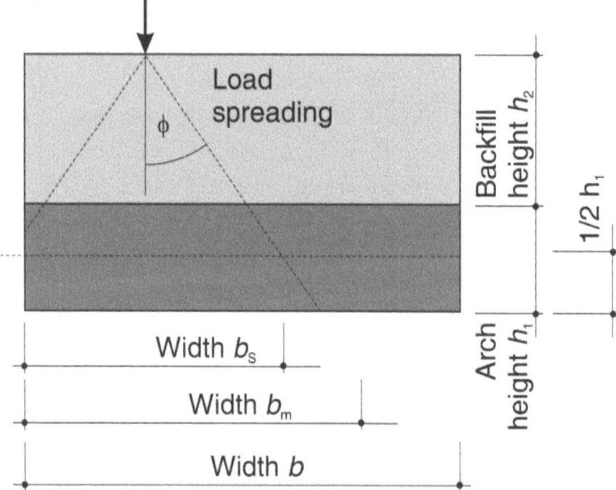

Fig. 3-27. Effective width based on the height of backfill and height of the arch

The spreading angle of the load can be taken from several codes of practice or recommendations. For example,

- For gravel or sand, an angle of 30° is used according to the DIN 1055-1
- For concrete or masonry, 45° is assumed by Haser and Kaschner (1994)
- For masonry, about 30° is assumed according to DIN 1052-1 (1996)

However, these differences are negligible for usual covering heights. Whereas the current German codes (DIN 1075 or DIN-report 101) for road bridges lack information about the effective width, former codes included some recommendations, such as

- TGL 12999 (3/1977) "Re-computation of existing bridges"
- SBA-regulation 169/89 (9/1989) "Bridges for traffic – recomputation of road traffic bridge built of concrete and masonry"

Based on these former recommendations Haser and Kaschner (1994) give, the following rules:

One-lane road	Two-lane road
$b_m = 4$ m	$b_m = 7$ m
$b_m = 0.25 \times l$	
$b_m \leq b$	

However, the width must not be wider than the bridge.

For railway bridges, the Ril 804 gives a spreading function. The general assumption for such a spreading is the coverage height of bridges. Bridges are considered as covered, if the coverage is greater than 60 cm. Figure 3-28 shows an example of a railway bridge with and without concrete plate. As shown in Fig. 3-28, the effective width alters with the length of the arch. For railway bridges, therefore, the Ril 804/805 permits the computation of a middle effective width b_m

$$b_m = \frac{b_c + b_s}{2} \tag{3-69}$$

with b_c as the effective width at crown and b_s as the effective width at springing.

However, if the force spreading exceeds the bridge width, the force creates a horizontal loading on the spandrel walls as shown in Fig. 3-29.

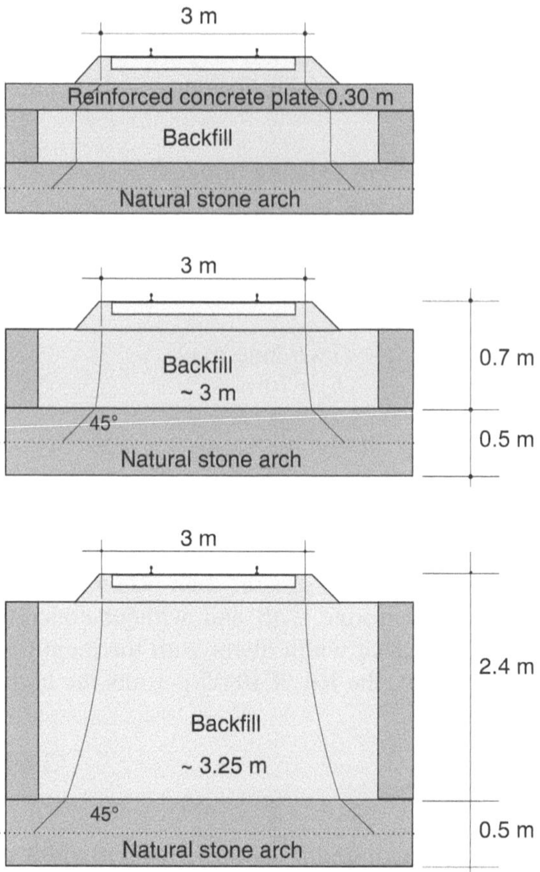

Fig. 3-28. Example of load spreading in an arch bridge. In the *top* subfigure, the bridge is shown with a railway plate, and in the subfigure in the *middle* and at the *bottom*, the bridge is in the original state at the crown and at the springing

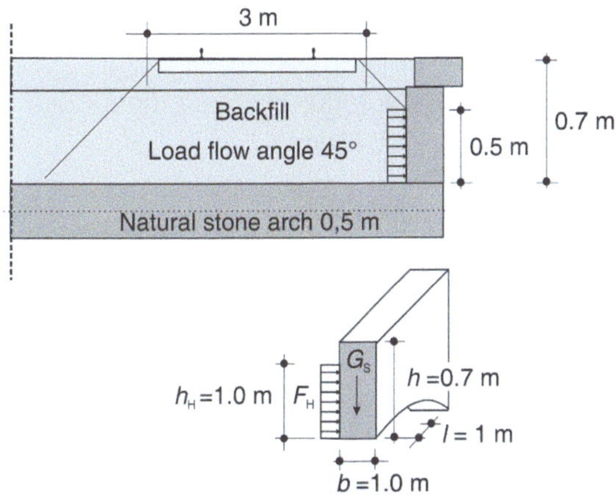

Fig. 3-29. If the load spreading reaches the spandrel wall, the load can cause horizontal pressure on the spandrel wall as shown in this example

For more detailed investigations of the load spreading, further advanced models can be used such as the models by Harvey (2006) or by the use of FEM as shown by Nautiyal (1993) and Frenzel (2004). Figure 3-30 shows the different load-spreading models by Harvey, whereas Fig. 3-31 shows the results from a FEM computation in terms of vertical stresses inside an arch bridge caused by traffic load.

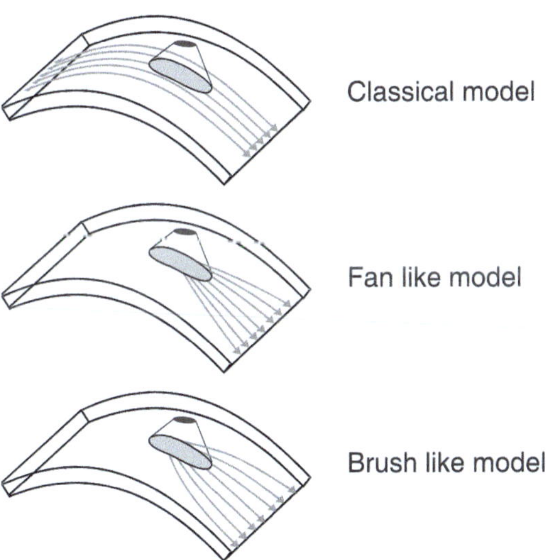

Fig. 3-30. Force flow models in longitudinal and transversal direction according to Harvey (2006)

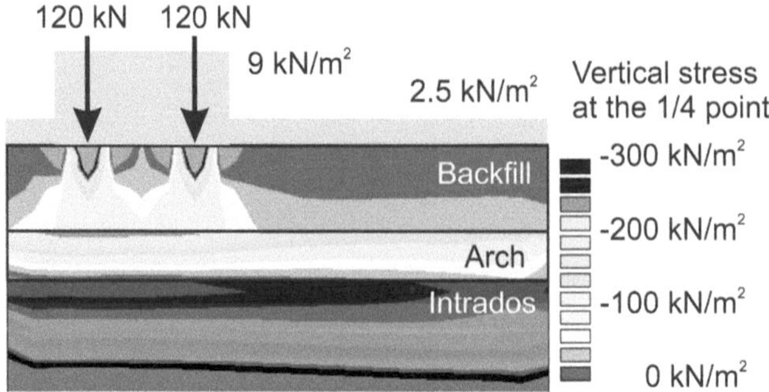

Fig. 3-31. Vertical stress at the ¼ point of the arch according to Frenzel (2004)

References

AEDES (2008) SAV (Stabilita`di Archie Volte in Matura) http://www.aedes.it/

Aita D, Foce F, Barsotti R & Bennati S (2007) Collapse of masonry arches in Romanesque and Gothic constructions. Proceedings of the Arch`07 – 5th International Conference on Arch Bridges, PB Lourenco, DV Oliveira & A Portela (Eds), 12–14 September 2007, Madeira, pp 625–632

Aoki T & Sabia D (2003) Collapse Analysis of Masonry Arch Bridges, Proceedings of the 9th International Conference on Civil and Structural Engineering Computing, Egmond aan Zee, The Netherlands, 2–4, September 2003, BHV Topping (Ed), Civil Comp Press, Stirling, United Kingdom, 2003, pp 1–14 (CD-ROM)

Aoki T & Sato T (2003) Application of Bott-Duffin Inverse to Static and Dynamic Analysis of Masonry Structures, Structural Studies, Repairs and Maintenance of Heritage Architecture VIII, pp 277–286

Aoki T, Komiyama D, Sabia D & Rivella R (2004) Theoretical and Experimental Dynamic Analysis of Rakanji Stone Arch Bridge, Honyabakei, Oita, Japan, Proceedings of 7th International Conference on Motion and Vibration Control MOvIC'04, St. Louis, 8.–11. August, 2004, pp 1–9 (CD-ROM)

Audenaert A, Peremans H & Reniers G (2007) Journal of Engineering Mathematics, 59, pp 323–336

B 16 (1997) Comparison of Masonry Arch Bridge Assessment Methods. Annex E, Volume 3, Section 4, Part 4, http://www.archive2.official-documents.co.uk/document/deps/ha/dmrb/vol3/sect4/ba1697h.pdf, 2005

Baker M (1997) FE analysis of Masonry and Application to masonry Arch Bridges. PhD Thesis, Cardiff University

BE 16/97 (2005) Comparison of masonry arch bridge assessment methods. Annex E, Volume 3, Section 4, Part 4, BE 16/97, www.archive2.official-documents.co.uk/document/deps/ha/dmrb/vol3/sect4/ba1697h.pdf

Becke A (2005) Entwicklung eines ingenieurmäßigen Modells zur Berücksichtigung der Mitwirkung der Hinterfüllung bei historischen Natursteinbogenbrücken. Diplomarbeit Technische Universität Dresden

Bićanić N, Stirling C & Pearce CJ (2002) Discontinuous Modelling of Structural Masonry. Fifth World Congress on Computational Mechanics, HA Mang, FG Rammerstorfer & J Eberhardsteiner (Eds), July 7–12 2002, Vienna, Austria

Bićanić N, Stirling C & Pearce CJ (2003) Discontinuous modelling of masonry bridges. Computational Mechanics, 31 (1–2), May 2003, pp 60–68

Bién J, Kamiński T & Trela C (2008) Numerical analysis of old masonry bridges supported by field tests. Bridge Maintenance, Safety, Management, Health Monitoring and Informatics, H Koh & DM Frangopol (Eds), Taylor & Francis Group, London, pp 2384–2391

Bienert G (1959–60) Aufgaben der Modellstatik bei der Konstruktion und Unterhaltung der Verkehrsbauwerke. Wissenschaftliche Zeitschrift der Hochschule für Verkehrswesen „Friedrich List" Dresden, 1959/60, Heft 1

Bienert G, Sauer & Schmidt (1960–62) Beitrag zur Ermittlung von Tragreserven in massiven Brückengewölben. Wissenschaftliche Zeitschrift der Hochschule für Verkehrswesen "Friedrich List" Dresden, 1960/61, Heft 2 und 9, 1961/62, Heft 1

Block P (2006) InteractiveTHRUST. http://web.mit.edu/masonry/interactiveThrust/examples.html

Bothe E, Henning J, Curbach M, Bösche T & Proske D (2004) Nichtlineare Berechnung alter Bogenbrücken auf der Grundlage neuer Vorschriften. Beton- und Stahlbetonbau 99, Heft 4, pp 289–294

Braune W (1980) Über die Tragwirkung von Gewölbebrücken mit längsgegliederten massiven Aufbauten. Die Bautechnik, 57 (2), pp 37–45 and (3), pp 93–100

Brencich A & Colla C (2002) The influence of construction technology on the mechanics of masonry railway bridges, Railway Engineering 2002, 5th International Conference, 3–4 July 2002, London

Brencich A & de Francesco U (2004) Assessment of Multispan Masonry Arch Bridges: Simplified Approach. ASCE Journal of Bridge Engineering, November/December 2004, pp 582–590

Brencich A & Gambarotta L (2005) Mechanical response of solid clay brickwork under eccentric loading. Part I. Unreinforced masonry. Materials and Structures, 38(2), pp 257–266

Brencich A, Corradi C, Gambarotta G, Mantegazza G & Strerp E (2002) Compressive Strength of Solid Clay Brick Masonry under Eccentric Loading. Department of Structural and Geotechnical Engineering, University of Genoa, Italy

Brencich A, De Francesco U, Gambarotta L (2001) Elastic no tensile resistant – plastic analysis of masonry arch bridges as an extension of castigliano's method. 9th Canadian masonry symposium

Brookes C & Collings M (2003) ARCHTEC – Verification of Structural Analysis. Gifford and Partners Document No: B1660A/V10/R03

Brühwiler E (2008) Rational approach for the management of medium size bridge stock. Bridge Maintenance, Safety, Management, Health Monitoring and Informatics, H Koh & DM Frangopol (Eds), Taylor & Francis Group, London, pp 2642–2649

Busch P & Zumpe G (1995) Tragfähigkeit, Tragsicherheit und Tragreserven von Bogenbrücken. 5. Dresdner Brückenbausymposium, Fakultät Bauingenieurwesen, Technische Universität Dresden, pp 153–169

Cavicchi A & Gambarotta L (2004) Upper bound limit analysis of multispan masonry bridges including arch-fill interaction. Arch Bridges IV – Advances in Assessment, Structural Design and Construction, P Roca & C Molins (Eds), CIMNE, Barcelona, pp 302–311

Cavicchi A & Gambarotta L (2005) Collapse analysis of masonry bridges taking into account arch-fill interaction. Engineering Structures, 27, pp 605–615

Cavicchi A & Gambarotta L (2006) Two-dimensional finite element upper bound limit analysis of masonry bridges. Computers & Structures, 84, pp 2316–2328

Cavicchi A & Gambarotta L (2007) Load carrying capacity of masonry bridges: numerical evaluation of the influence of fill and spandrels. Proceedings of the Arch`07 – 5th International Conference on Arch Bridges, PB Lourenco, DV Oliveira & A Portela (Eds), 12–14 September 2007, Madeira, pp 609–616

Cervenka V (2004) Cervenka-Consulting: http://www.cervenka.cz

Chiostrini S, Foraboschi P & Sorace S (1989) Problems connected with the arrangement of a non-linear finite element method to the analysis of masonry structures. Structural Repair and Maintenance of Historical Buildings. Computational Mechanics Publications Birkhäuser

Choo BS, Coutie MG & Gong NG (1991) Finite-element analysis of masonry arch bridges using tapered elements. Proceedings of the Institution of Civil Engineers, Part 2, December 1991, pp 755–770

Corradi M (1998) Empirical methods for the construction of masonry arch bridges in the 19th century. Arch Bridges: history, analysis, assessment, maintenance and repair. Proceedings of the second international arch bridge conference, Edr A. Sinopoli, Venice 6–9 October 1998, Balkema Rotterdam, pp 25–36

Corradi M & Filemio V (2004) A brief comparison between mechanical aspects and construction of arch bridges during the XVIIIth and XIXth centuries. Arch Bridges IV – Advances in Assessment, Structural Design and Construction, P Roca & C Molins (Eds), CIMNE, Barcelona, pp 79–86

COST-345 (2004) European Commission Directorate General Transport and Energy: COST 345 – Procedures Required for the Assessment of Highway Structures: Numerical Techniques for Safety and Serviceability Assessment – Report of Working Groups 4 and 5

Craemer H (1943) Die "erzwungene Drucklinie" als Ausdruck der versteifenden Wirkung der Übermauerung von Gewölben. Beton- und Stahlbetonbau, Heft 9/10

Crisfield M A (1985) Finite Element and Mechanism Methods for the Analysis of Masonry and Brickwork Arches. Transport and Road Research Laboratory, London

Cundall PA & Strack ODL (1979) A discrete numerical model for granular assemblies. Géotechnique, 29, pp 47–65

Cundall PA (1971) A computer model for simulating progressive large scale movements in blocky rock systems. In Proceedings of the Symposium of the International Society of Rock Mechanics. Vol. 1, Paper II-8, Nancy, France

Das PC (1995) The Assessment of Masonry Arch Bridges. The Aesthetics of Load bearing Masonry Arch Bridges. Arch Bridges, C. Melbourne (Ed), Proceedings of the 1st International Conference on Arch Bridges, Bolton, Thomas Telford, London, pp 21–27

Dialer C (1991) Bruch- und Verformungsverhalten von schubbeanspruchten Mauerwerksscheiben. Mauerwerkskalender 1992, Ernst & Sohn, Berlin

Dialer C (2002) Ausgewählte Schäden und Schäden an Instandsetzungen von Massivbauwerken. Massivbau 2002, Forschung, Entwicklungen und Anwendungen. Hrsg. Konrad Zilch, Technische Universität München, pp 218–237

Diamantidis D (2001) Assessment of Existing Structures, Joint Committee of Structural Safety JCSS, Rilem Publications S.A.R.L.: The publishing Company of Rilem, Volume 159

Dibeh T, Pfeifer T & Hein S (1997) Statische Berechnung von Bogenbrücken & Statische Berechnung von Natursteinbogenbrücken. Saxon institute for road construction, Rochlitz

Dischinger F (1949) Massivbau. Taschenbuch Bauingenieure. F Schleicher (Ed), Reprint 1949, Springer Verlag Berlin, 1943

DR (1985) Deutsche Reichsbahn: Richtlinie Einbau von Fahrbahnwannen. year uncertain

Droese S & Bodendiek P (2002) Die Unterfangung des Bahrmühlenviaduktes, Rechnerischer Nachweis des Viaduktes für Lasten und die Zwangseinwirkung aus dem Bauvorgang. Die Bautechnik 79, pp 455–463

Drosopoulos GA, Stavroulakis GE & Massalas CV (2007) FRP reinforcement of stone arch bridges: Unilateral contact models and limit analysis. Composites: Part B 38, pp 144–151

Enevoldsen I (2008) Practical implementation of probability based assessment methods for bridges. Bridge Maintenance, Safety, Management, Health Monitoring and Informatics, H Koh & DM Frangopol (Eds), Taylor & Francis Group, London, pp 83–104

Fain JS (1953) Beispiele zur Berechnung von Stein- und Bogenbrücken. VEB Verlag Technik, Berlin

Falter H (1998) Untersuchung historischer Wölbkonstruktionen – Auswirkungen der Herstellung und des Baumaterials. Dissertation an der Universität Stuttgart

Fanning PJ & Boothby TE (2001) Three-dimensional modelling and full-scale testing of stone arch bridges. Computers and Structures, 79, pp 2645–2662

Fanning PJ, Boothby TE & Roberts BJ (2001) Longitudinal and transverse effects in masonry arch assessment. Construction and Building Materials, 15, pp 51–60

Fischer U (1940–41) Ausnutzung des Zusammenwirkens von Bogen und Aufbau. Internationaler Verein für Brücken- und Hochbau, Abhandlung Band 61

Fischer U (1942) Die Mitwirkung des Aufbaus massiver Bogenbrücken. Beton und Eisen, 37. Jahrgang, Heft 19, pp 310–315

Ford TE, Augarde CE & Tuxford SS (2003) Modelling masonry arch bridges using commercial finite element software, 9th International Conference on Civil and Structural Engineering Computing, Egmond aan Zee, The Netherlands, 2–4 September 2003

Frenzel M (2004) Untersuchung zur Quertragfähigkeit von Bogenbrücken aus Natursteinmauerwerk. Diplomarbeit Technische Universität Dresden

Frunzio G & Monaco M (1998) An approach to the structural model for masonry arch bridges: Pont Saint Martin as a case study. Arch Bridges: History, Analysis, Assessment, Maintenance and Repair. Proceedings of the Second International Arch Bridge Conference, A Sinopoli (Ed), Venice 6.–9. October 1998, Balkema Rotterdam, pp 231–240

Fuhlrott A (2004) Grundlagen der Extended Finite Elemente Methode (X-FEM). Studienarbeit am Institut für Strukturmechanik, Fakultät Bauingenieurwesen, Bauhaus-Universität Weimar

Gelfi P (2006) Arco.exe http://dicata.ing.unibs.it/gelfi/arco.htm

Gilbert M (2007) Limit analysis applied to masonry arch bridges: state-of-the-art and recent developments. Proceedings of the Arch`07 – 5th International Conference on Arch Bridges. PB Lourenco, DV Oliveira & A Portela (Eds), 12–14 September 2007, Madeira, pp 13–28

Gilbrin G (1913) Störungen des normalen Zustandes in Brückengewölben. Dissertation an der damaligen Kgl. TH zu München, Verlag Ernst und Sohn, Berlin 1913

Gocht R (1978) Untersuchungen zum Tragverhalten rekonstruierter Eisenbahngewölbebrücken, Dissertation Hochschule für Verkehrswesen "Friedrich List" Dresden

Greenwold S, Allen E & Zalewski W (2008) Active Statics. http://acg.media.mit.edu/people/simong/statics/Start.html

Gutermann M (2002) Ein Beitrag zur experimentell gestützten Tragsicherheitsbewertung von Massivbrücken, Dissertation, Fakultät Bauingenieurwesen

Haase H (1899) Theorie des parabolischen Brückengewölbes oder das Grundgesetz des Horizontalschubs in seiner Anwendung auf das Brückengewölbe unter der ausschließlichen Wirkung vertikaler Aussenkräfte, entwickelt an dem Beispiel einer gewölbten Bahnüberbrückung. Kommissionsverlag der Nationalen Verlagsanstalt, Regensburg

Hänel D & Reintjes KH (2001) Die Unterfangung des Bahrmühlenviaduktes, Planung, Entwurf und Vergabe. Die Bautechnik, 78, pp 543–555

Hannawald F (2006) Zur physikalisch nichtlinearen Analyse von Verbund-Stabtragwerken unter quasi-statischer Langzeitbeanspruchung. Dissertation, Technische Universität Dresden, Institut für Stahl- und Holzbau

Harvey B (2006) Some problems with arch bridge assessment and potential solutions. Structural Engineer, 84 (3), 7th February 2006, pp 45–50

Harvey B (2008) Archie-M. http://billharvey.typepad.com/archi_m/

Harvey B, Ross K & Orbán Z (2007a) A simple, first filter assessment for arch bridges. Proceedings of the Arch`07 – 5th International Conference on Arch Bridges, PB Lourenco, DV Oliveira & A Portela (Eds), 12–14 September 2007, Madeira, pp 275–280

Harvey WJ (2007) Rule of thumb method for the assessment of arches. UIC Report (draft)

Harvey WJ, Maunder EAW & Ramsay ACA (2007b) The influence of the spandrel wall construction on arch bridge behaviour. Proceedings of the Arch`07 – 5th International Conference on Arch Bridges, PB Lourenco, DV Oliveira & A Portela (Eds), 12–14 September 2007, Madeira, pp 601–608

Haser H & Kaschner R (1994) Spezielle Probleme bei Brückenbauwerken in den neuen Bundesländern. Berichte der Bundesanstalt für Straßenwesen, Verlag für neue Wissenschaft, Bergisch Gladbach

Healey TN & Counsell JHW (1998) Widening and strengthening of Londons Kingston Bridge. Arch Bridges: history, analysis, assessment, maintenance and repair. Proceedings of the second international arch bridge conference, Edr. A. Sinopoli, Venice 6.–9. October 1998, Balkema Rotterdam, pp 389–398

Heinlein M (2008) Pauser Ingenieurbüro. Berechnung von gemauerten Gewölbebrücken. One-on-one interview on the 16.12.2008

Heinrich B (1983) Brücken – Vom Balken zum Bogen. Rowohlt Taschenbuch Verlag GmbH, Hamburg

Herrbruch J, Groß JP & Wapenhans W (2005) Wirklichkeitsnähere Tragwerksanalyse von Gewölbebrücken. Mauerwerk 9, Heft 3, pp 89–92

Herzog R (1962) Wechselbeziehungen zwischen Bogen und Bogenüberbauten von Brücken. Technisch-Wissenschaftlicher Verlag des Ministeriums für Automobiltransport und Chaussee-Straßen der RSFSR, Moskau

Heyman J (1966) The stone skeleton. International Journal of Solids and Structures, 2, pp 249–279

Heyman J (1998) The assessment of strength of masonry arches. Arch Bridges: History, Analysis, Assessment, Maintenance and Repair. Proceedings of the Second International Arch Bridge Conference, Edr A Sinopoli, Venice 6.–9. October 1998, Balkema Rotterdam, pp 95–98

Hoch G & Schmitt B (1999) Nebenbahnen in der Oberpfalz, Verlag Michael Resch, Coburg

Huerta S (2001) Mechanics of masonry vaults: The equilibrium approach. Historical Constructions, PB Lourenço & P Roca (Eds), Guimarães, pp 47–69

Huerta S (2004) Arcos, bóvedas y cúpulas. Instituto Juan de Herrera. Escuela Técnica Superior de Arquitetura Madrid.

Huges TG & Blackler MJ (1997) A review of the UK masonry arch assessment methods. Proceedings of Institution Civil Engineering Structures and Buildings, 122, August, pp 305–315

Huster U (2000) Tragverhalten von einschaligem Natursteinmauerwerk unter zentrischer Druckbeanspruchung: Entwicklung und Anwendung eines Finite-Elemente-Programmes. Kassel University Press: Kassel

Jackson P (2004) Highways Agency BD on New Masonary Arch Bridges. 27th February 2004. http://www.smart.salford.ac.uk/February_2004_Meeting.php

Jäger A (1935) Amerikanische Versuche an den Modellen eines eingespannten Bogens und einer Bogenreihe. Beton- und Eisen, 34. Jahrgang, Heft 21, pp 338–344

Jäger A (1938) Untersuchung eines eingespannten Gewölbes und einer Gewölbereihe unter Berücksichtigung der Hintermauerung als mittragende Masse. Verlag Ernst & Sohn, Berlin

Jagfeld M & Barthel R (2004) Zur Gelenkbildung in historischen Tragsystemen aus Mauerwerk. Bautechnik (81), Heft 2, pp–96102

Jagfeld M (1998) Tragverhalten gemauerter Wölbkonstruktionen bei großen Auflagerverschiebungen – Vergleich verschiedener FE Modelle. 16. CAD-FEM User's Meeting, 7.–9. Oktober 1998 in Bad Neuenahr

Jensen JS, Plos M, Casas JR, Cremona C, Karoumi R & Melbourne C (2008) Guideline for load and resistance assessment of existing European railway bridges. Bridge Maintenance, Safety, Management, Health Monitoring and Informatics, H Koh & DM Frangopol (Eds), Taylor & Francis Group, London, pp 3658–3665

Keip MA & Konietzky H (2005) Mauerwerk unter dynamischen Belastungen – Untersuchungen mittels der DEM, Mauerwerkstag Essen, 3. Mai 2005

Knoblauch F-J, Laubach A, Sauer W, Schmidt M & Sprinke P (2008) Erneuerung des Burtscheider Viadukts in Aachen. Beton- und Stahlbetonbau 103, Heft 3, pp 195–203

Kooharian A (1952) Limit analysis of voussoir (segmental) concrete arches. Journal American Concerte Institute, 24, pp 317–328

Kurrer K-E (2002) Geschichte der Baustatik. Ernst und Sohn Verlag für Architektur und technische Wissenschaften GmbH, Berlin 2002

Lachmann H (1990) Über die Standsicherheit gemauerter Gewölbebrücken, Bautechnik 67, Heft 2, pp 61–63

Leliavsky S (1982) Arches and short span bridges. Design Textbook in Civil Engineering: Volume VII, Chapman and Hall, London

Lemos JV (1995) Assessment of the ultimate load of a masonry arch using discrete elements. Computer Methods in Structural Masonry – 3, GN Pande & J Middleton (Eds), Books Journals International 1995, pp 294–302

Lemos JV (2006) Modelling of historical masonry with discrete elements. Computational Mechanics – Solids, Structures and Coupled Problems, Springer, pp 375–391

LimitState Ltd (2008) http://www.masonryarch.com

Loo Y-C (1995) Collapse load analysis of masonry arch bridges. Arch Bridges. Proceedings of the 1st International Conference on Arch Bridges, C. Melbourne (Ed), Bolton, Thomas Telford, London, pp 167–174

Lourenço PB (1996) Computational Strategies for Masonry Structures. Dissertation, Technische Universität Delft, Delft University Press

Lourenço PB (2002) Computations on historic masonry structures, Structure Engineering Materials, 4, pp 301–319

Ludwig D (2000) Programm zur Ermittlung der Tragfähigkeit und der Beförderungsbedingungen sowie zur Einstufungsberechnung an Gewölbebrücken der Bewertungsstufe 2, DB Netz AG, Magdeburg

LUSAS (2005): Medieval masonry arch passes assessment with LUSAS www.lusas.com/case/bridge/masonry_arch.html

Mamaghani IHP, Aydan Ö & Kajikawa Y (1999) Analysis of masonry structures under static and dynamic loading by discrete finite element method. Journal of

Structural Mechanics and Earthquake Engineering, Japan Society of Civil Engineers (JSCE), Japan, No. 626/I-48, 16, pp 75–86

Martín-Caro JA & Martínez LJ (2004) A first level structural analysis tool for the Spanisch Railways Masonry Arch Bridges. Arch Bridges IV – Advances in Assessment, Structural Design and Construction, P Roca & C Molings (Eds), CIMNE, Barcelona, pp 192–201

Martinez JA, Revilla J & Aragon Torre A (2001) Critical thickness criteria on stone arch bridges with low rise/span ratio and current traffic loads. Historical Constructions, PB Lourenço & P Roca (Eds), Guimarães, pp 609–616

Maunder EAW (1993) Limit analysis of masonry structures based on discrete elements. Structural Repair and Maintenance of Historical Building III (STREMAH), CA Brebbia & RJB Frewer (Ed), Computational Mechanics Publication, Glasgow, pp 367–374

Melbourne C & Tao H (1998) The behaviour of open spandrel brickwork arch bridges. Arch Bridges: History, Analysis, Assessment, Maintenance and Repair. Proceedings of the Second International Arch Bridge Conference, A Sinopoli (Ed), Venice 6.–9. October 1998, Balkema Rotterdam, pp 263–269

Mildner K & Mildner A (2001) Bauwerksmessungen vor und nach der Instandsetzung einer Eisenbahn-Gewölbebrücke und statische Modellbildung. Ehrenkolloquium Prof. Dr.-Ing. Möller zum 60. Geburtstag, Lehrstuhl für Statik und Dynamik der Tragwerke, Technische Universität Dresden, Dresden

Mildner K (1996) Tragfähigkeitsermittlung von Gewölbebrücken auf der Grundlage von Bauwerksmessungen. 6. Dresdner Brückenbausymposium, Fakultät Bauingenieurwesen, Technische Universität Dresden, pp 149–162

Model P (1977) Beitrag zur Berücksichtigung der Gewölbehintermauerung, der Rissbildung und der elastischen Stützung durch den Baugrund und die Dammschüttung bei der statischen Berechnung einfeldriger Gewölbebrücken, Dissertation Hochschule für Verkehrswesen "Friedrich List" Dresden

Modena C, Valluzzi MR, da Porto F, Casarin F & Bettio C (2004) Structural upgrading of a brick masonry arch bridge at the Lido (Venice). Arch Bridges IV – Advances in Assessment, Structural Design and Construction, P Roca & C Molins (Eds), CIMNE, Barcelona, pp 435–443

Mojsilović N (1995) Zum Tragverhalten von kombiniert beanspruchten Mauerwerk. Dissertation, IBK Bericht Nr. 216, ETH Zürich, Birkhäuser Verlag, Basel

Molins C & Roca P (1998) Capacity of masonry arches and spartial structures. Journal of Structural Engineering ASCE, 124 (6), pp 653–663

Möller B, Graf W, Beer M & Bartzsch M (2002) Sicherheitsbeurteilung von Naturstein-Bogenbrücken. 6. Dresdner Baustatik-Seminar. Rekonstruktion und Revitalisierung aus statisch-konstruktiver Sicht. 18. Oktober 2002. Lehrstuhl für Statik + Landesvereinigung der Prüfingenieure für Bautechnik Sachsen + Ingenieurkammer Sachsen. Technische Universität Dresden, pp 69–91

Mura I, Odoni Z & Perra M (2003) Static analysis of the Turris Lybisonis Roman multispan stone arch bridge. Structural Studies, Repairs and Maintenance of Heritage Architecture VIII, CA Brebbia (Ed), Southampton, WIT-Press 2003, pp 667–674

Nautiyal SD (1993) Parallel Computing Techniques for Investigating Three Dimensional Collapse of Masonry Arch. PhD. Cambridge University

Neuberg C (2002) Ein Verfahren zur Berechnung des räumlichen passiven Erddrucks vor parallel verschobenen Trägern. Institut für Geotechnik, Technische Universität Dresden, Mitteilungen Heft 11, Dresden

O'Connor A, Pedersen C, Enevoldsen I, Gustavsson L & Hammarbäck J (2008) Probability Based Assessment of a Large Riveted Truss Railway Bridge. Bridge Maintenance, Safety, Management, Health Monitoring and Informatics, H Koh & DM Frangopol (Eds), Taylor & Francis Group, London, pp 1812–1819

Obvis Ltd (2006) http://www.obvis.com/

Orlando M, Spinelli P & Vignoli A (2003) Structural analysis for the reconstruction design of the old bridge of Mostar. Structural Studies, Repairs and Maintenance of Heritage Architecture VIII, Edr Brebbia, WIT-Press, pp 617–626

Owen DR, Peric D, Petric N, Brookes CL & James PJ (1998) Finite/discrete element models for assessment and repair of masonry structures. Arch Bridges: History, Analysis, Assessment, Maintenance and Repair. Proceedings of the Second International Arch Bridge Conference, A Sinopoli (Ed), Venice 6.–9. October 1998, Balkema Rotterdam, pp 173–180

Page AW (1983) The strength of brick masonry under biaxial tension-compression. International Journal of Masonry Construction, 3(1), pp 26–31

Parikh SK & Patwardhan YG (1999) Widening of an existing old masonry arch bridge through RCC box girder deck. Current and Future Trends in Bridge Design, Construction and Maintenance in Bridge Design Construction and Maintenance, PC Das, DM Frangopol & AS Nowak (Eds), Thomas Telford, London, pp 218–226

Pauser A (2005) Ingenieurbüro: Guide to the high-level assessment of masonry arch bridges. UIC Report (draft)

Prader J, Weidner JW, Hassanain H, Moon F, Aktan E, Jalinoos F, Buchanan B & Ghasemi H (2008) Load Testing and Analysis of Bridges Missing Critical Documentation. Bridge Maintenance, Safety, Management, Health Monitoring and Informatics, H Koh & DM Frangopol (Eds), Taylor & Francis Group, London, pp 1877–1885

Proske D (2003) Ein Beitrag zur Risikobeurteilung alter Brücken unter Schiffsanprall. Dissertation, Technische Universität Dresden

Purtak F (2001) Tragfähigkeit von schlankem Mauerwerk. Dissertation. Technische Universität Dresden, Fakultät Architektur

Purtak F (2004) Zur nichtlineare Berechnung von Bogenbrücken aus Natursteinmauerwerk. 22. CAD-FEM Users Meeting 2004 – International Congress on FEM Technology with ANSYS CFX & ICEM CFD Conference, 10.–12. November 2004, Dresden

Purtak F, Geißler K & Lieberwirth P (2007) Bewertung bestehender Natursteinbogenbrücken. Bautechnik 8, Heft 8, pp 525–543

Reintjes K-H (2002) Die Unterfangung des Bahrmühlenviaduktes bei Chemnitz – Stand des Bauvorhabens. 12. Dresdner Brückenbausymposium – Planung, Bauausführung und Ertüchtigung von Massivbrücken. 14. März 2002.

Lehrstuhl für Massivbau – Technische Universität Dresden, Dresden, pp 145–163

RING (2005) Technical details. http://www.shef.ac.uk/ring/technical.html

Roberti MG & Calvetti F (1998) Distinct element analysis of stone arches. Arch Bridges: History, Analysis, Assessment, Maintenance and Repair. Proceedings of the Second International Arch Bridge Conference, A Sinopoli (Ed), Venice 6.–9. October 1998, Balkema Rotterdam, pp 181–186

Roca P & Molins C (2004) Experiments on Arch Bridges. Arch Bridges IV - Advances in Assessment, Structural Design and Construction, P Roca & C Molins (Eds), Barcelona, pp 365–374

Rombach G (2007) Probleme bei der Berechnung von Stahlbetnkonstruktionen mittels dreidimensionaler Gesamtmodelle. Beton- und Stahlbetonbau 102, Heft 4, pp 207–214

Rots JG & van Zijl G (2005) Analysis of Concrete and Masonry Structures with DIANA 9. Weiterbildungsseminar 31.5.-3.6.2005, München

Rouxinol GAF, Providencia P & Lemos JV (2007) Bridgemill bridge bearing capacity assessment by a discrete element method. Proceedings of the Arch`07 – 5th International Conference on Arch Bridges. PB Lourenco, DV Oliveira & A Portela (Eds). 12–14 September 2007, Madeira, pp 669–676

Royles R & Hendry AW (1991) Model tests on masonry arches, Proceedings of the Institution of Civil Engineers, 91 (2), pp 299–321

Rücker W, Hille F & Rohrmann R (2006) Guideline for the Assessment of Existing Structures. SAMCO Final Report 2006

Ruiz ME, Castelli EA, Prato TA (2008) A new bridge management system for the National Department of Transportation of Argentina. Bridge Maintenance, Safety, Management, Health Monitoring and Informatics, H Koh & DM Frangopol (Eds), Taylor & Francis Group, London, pp 598–605

Schlegel R & Rautenstrauch K (2000) Ein elastoplastisches Berechnungsmodell zur räumlichen Untersuchung von Mauerwerkstrukturen. Bautechnik 77, Heft 6, pp 426–436

Schlegel R & Will J (2007) Sensitivitätsanalyse und Parameteridentifikation von bestehenden Mauerwerkstrukturen. Mauerwerk 11, Heft 6, pp 349–355

Schlegel R (2004) Numerische Berechnung von Mauerwerksstrukuren in homogenen und diskreten Modellierungsstrategien. Dissertation. Bauhaus-Universität Weimar

Schlegel R (2008) Möglichkeiten der numerischen Simulation von Mauerwerk heute anhand praktischer Beispiele. Mauerwerkskalender 2009, Ernst und Sohn

Schlegel R, Will J & Popp J (2003) Neue Wege zur realistischen Standsicherheitsbewertung gemauerter Brückenviadukte und Gewölbe unter Berücksichtigung der Normung mit ANSYS. 21. CAD-FEM User's Meeting 2003, 12.–14. November 2003, Potsdam

Schreyer C (1960) Praktische Baustatik – Teil 3., BG Teubner Verlagsgesellschaft Leipzig

Schueremans L (2001) Probabilistic evaluation of structural unreinforced masonry, December 2001, Dissertation, Katholieke Universiteit Leuven,

Faculteit Toegepaste Wetenschappen, Departement Burgerlijke Bouwkunde, Heverlee, Belgien,

Schueremans L, Van Rickstall F, Ignoul S, Brosens K, Van Balen K & Van Gemert D (2003) Continuous Assessment of Historic Structures – A State of the Art of applied Research and Practice in Belgium. KULeuven, Department of Civil Engineering

Seim W & Schweizerhof K (1997) Nichtlineare FE-Analyse eben beanspruchter Mauerwerksscheiben mit einfachen Werkstoffgesetzen. Beton- und Stahlbetonbau 92, Heft 8, pp 201–207 and Heft 9, pp 239–244

Seim W (1998) Numerische Modellierung des anisotropische Versagens zweiachsig beanspruchter Mauerwerksscheiben. Aus Forschung und Lehre, Institut für Tragkonstruktionen, Universität Karlsruhe, Heft 27

SIA 269 (2008) Grundlagen der Erhaltung von Tragwerken.

Slowik V, Fiedler LD & Kapphahn G (2005) Zur kombinierten Anwendung experimenteller und rechnerischer Methoden bei der Tragsicherheitsbewertung bestehender Massivbauwerke, 3. Symposium Experimentelle Untersuchungen von Betonkonstruktionen, Technische Universität Dresden, Institut für Massivbau, 23.Juni 2005, Tagungsband, pp 157–168

Smars (2006) Calipous: http://pierre.smars.dyndns.org/calipous/

Smith CC, Gilbert M & Callaway PA (2004) Geotechnical issues in the analysis of masonry arch bridges. Arch Bridges IV – Advances in Assessment, Structural Design and Construction, P Roca & C Molins (Eds), pp 344–352

Stavrouli ME & Stavroulakis GE (2003) Unilateral contact analysis and failure prediction in stone bridges. Structural Studies, Repairs and Maintenance of Heritage Architecture VIII, CA Brebbia (Ed), WIT-Press, pp 308–315

Straub H (1992) Die Geschichte der Bauingenieurkunst. Ein Überblick von der Antike bis in die Neuzeit. Birkhäuser Verlag Basel

Thavalingam A, Bicanic N, Robinson JI & Ponniah DA (2001) Computational framework for discontinuous modelling of masonry arch bridges. Computers and Structures, 79, pp 1821–1830

Toker S & Ünay A I (2004) Mathematical modelling and finite element analysis of masonry arch bridges. G.U. Journal of Science, 17 (2): pp 129–139

Towler KDS (1985) Applications of non-linear finite element codes to masonry arches, Proceedings of the 2nd International Conference on Civil and Structural Engineering Computing

Trautz M (1998) Zur Entwicklung von Form und Struktur historischer Gewölbe aus Sicht der Statik. Dissertation, Institut für Baustatik, Universität Stuttgart, Stuttgart

UIC-Codex (1995) Empfehlungen für die Bewertung des Tragvermögens bestehender Gewölbebrücken aus Mauerwerk und Beton. Internationaler Eisenbahnverband. 1. Ausgabe 1.7.1995

University of Sheffield (2008) http://www.ring.shef.ac.uk/

Vermeltfoort AT (2001) Analysis and experiments of masonry arches. Historical Constructions, PB Lourenço & P Roca (Eds), Guimarães, pp 489–498

Voigtländer J (1971) Beitrag zur Ermittlung der Schnittkraftumlagerung in Gewölbebrücken infolge Rissbildung, Dissertation Hochschule für Verkehrswesen "Friedrich List" Dresden

von Leibbrand K (1897) Gewölbte Brücken. Verlag von Wilhelm Engelmann, Leipzig

Weber WK (1999) Die gewölbte Eisenbahnbrücke mit einer Öffnung. Begriffserklärungen, analytische Fassung der Umrisslinien und ein erweitertes Hybridverfahren zur Berechnung der oberen Schranke ihrer Grenztragfähigkeit, validiert durch einen Großversuch. Dissertation, Lehrstuhl für Massivbau der Technischen Universität München

Wedler B (1997) Einfluß der Füllungsschichtung auf das Tragverhalten des mehrschaligen Mauerwerks in historischen Bauwerken. Diplomarbeit. Technische Universität Dresden. Lehrstuhl für Massivbau

Weigert A (1996) Untersuchung zum Einfluss der Steinhöhe auf das Tragverhalten des einschaligen Mauerwerks aus Elbsandstein. Diplomarbeit. Lehrstuhl für Massivbau. Technische Universität Dresden

Willam KJ & Warnke EP (1974) Constitutive Model for the Triaxial Behaviour of Concrete, IABSE Report Vol. 19, 1974, Colloquium on „Concrete Structures Subjected to Triaxial Stresses". ISMES Bergamot

Witzany J & Jäger W (2005) Die Karlsbrücke in Prag. Mauerwerk 9, Heft 3, pp 108–119

Wolf F (1989) Ein Beitrag zur Beurteilung des Tragverhaltens von massiven Eisenbahn-Gewölbebrücken. Dissertation, Hochschule für Verkehrswesen. Dresden

4 Masonry Strength

"Numerical models are best suited to problems within a known solution space."

M. F. M. Yossef, Delft 2005

4.1 Introduction

In the previous chapter, the numerical modelling of arch bridges was discussed intensively. During the discussion, it should have become clear that in most cases, historical arch bridges fail due to the development of a mechanism chain with hinges, by sliding, or by a combination of this. For these models, the compression strength is not of utmost importance. However, in two other cases one has to consider the masonry compression strength: first looking at the developing hinges where maximum compression forces are reached, and second considering arch bridges under maximum equal load. Maximum equal load can be reached by widening the road lane and therefore increasing the dead load of the bridge.

The estimation of the maximum strength of masonry is not a simple task. Masonry is a multi component building material. Due to the enormous variety of physical and chemical properties of the components (Table 4-1), many different numerical models have been developed over time to describe the strength of masonry. In general, the different numerical models can be classified into different types of models. The model classifications are shown in Fig. 4-1.

D. Proske, P. van Gelder, *Safety of Historical Stone Arch Bridges*, DOI 10.1007/978-3-540-77618-5_4,
© Springer-Verlag Berlin Heidelberg 2009

Table 4-1. Some major factors influencing the masonry strength according to Wenzel (1997)

Stone and brick	Mortar	Masonry	Structural element
Compression strength	Compression strength	Joint thickness	Dimensions
Flexural bending strength	Adhesion bond with stone	Joint filling	Slenderness
Stress–strain curve		Cavity ratio	Support conditions
Throatiness		Layer thickness	Stiffening
Surface working		Stone order	Connection to other structural elements
		longitudinal and transverse	Direction of loading
			Eccentricity

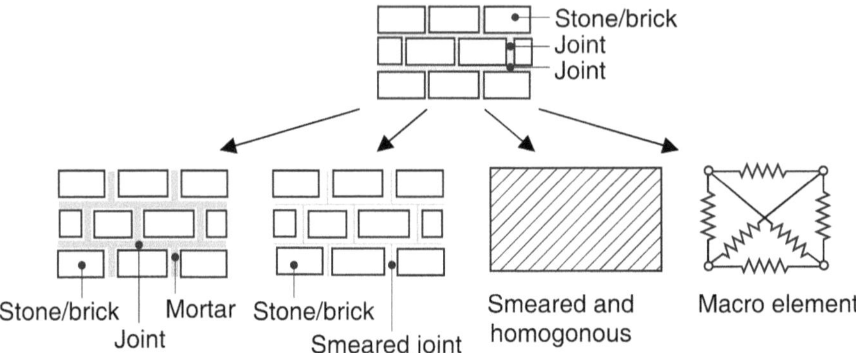

Fig. 4-1. Classifications of numerical masonry models according to Meskouris et al. (2004)

4.2 Masonry Elements

4.2.1 Masonry Stones

4.2.1.1 Types of natural stones and their properties

Stones are static homogenous, natural mineral conglomerates, which make up a compound caused by some geo dynamical processes (Herrbach 1996). Stones are classified according to their origin into genetic systems. This system distinguishes three major groups (Table 4-2). Stones evolve from mineral melt, magma and lava igneous. Stones originating from some diagenetic processes of deposited material are called sediment stones.

Stones that show new crystallization under high pressure and temperature are called metamorphic stones.

Table 4-2. Classification of stones

Major group	Sub-group	Examples
Igneous	Plutonic	Granite
	Volcanic	Basanite
	Matrix	Gabbro
Sedimentary	Clastic sediments	Sandstone
	Chemical sediment	Lime stone
	Biogenic sediment	Chert
	Residual stones	
Metamorphic		Mica schist

Indeed a complete classification by this system is not possible (Börner and Hill 1999). Further properties such as colour, structure, or technical properties have been used to identify the so-called varieties. Peschel (1984) clearly shows the differences of these technical properties. Technical properties and possible application of natural stones are given in Peschel (1984), Dienemann and Burre (1929), Gäbert et al. (1915), and Schubert (2004) for German natural stones. A summary is listed in Table 4-3.

Table 4-3. Technical properties of some natural stones according to Stein (1993)

	Weights in kN/m^3	Compression strength in MPa	Flexural bending strength in MPa	Thermal expansion coefficient in mm/mK
Granite, Syenite	28	160–240	10–20	0.80
Diorite, Gabbro	30	170–300	10–22	0.88
siliceous porphyry	28	180–300	15–20	
Basalt lava	24	80–150	8–12	
Diabase	29	180–250	15–25	0.75
Quartzite, Greywacke	27	150–300	12–20	
Siliceous sandstone	27	120–200	3–15	
Dense lime and dolomite stones	28	80–180	6–15	0.75
Other lime stones	28	20–90	5–8	
Travertine	26	20–60	4–10	0.68
Volcanic Tuff	20		2–6	
Gneiss	30	160–280	10–15	

The determination of the strength of natural stones depends on many factors that have to be considered and stated. Such factors are, for example, the size of the test specimen, the humidity of the stone material, or the layering of the stone material. Figure 4-2 illustrates the effect of the specimen size on the flexural bending strength of granite stone in terms of specimen height. The left side of the diagram shows data that have been taken from many different references, whereas on the right side predominantly its own data have been used. Furthermore, on the far right, data from uniaxial tensile tests have been included. A complete list of data references is given in Curbach and Proske (2003).

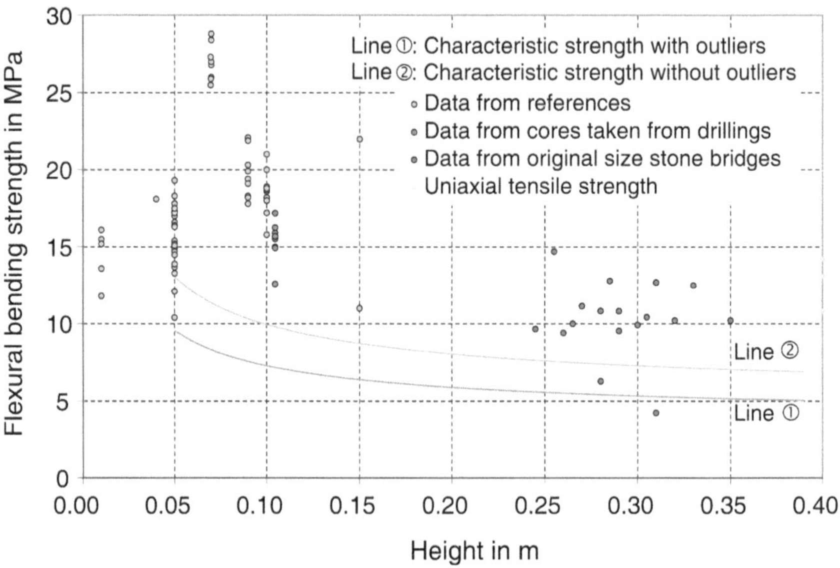

Fig. 4-2. Flexural bending strength of granite based on the test specimen height (Curbach et al. 2004)

Not only do different test conditions heavily influence the outcome of the stone properties, but natural stone material also shows, in most cases, strong deviation from the mean value. This is exemplarily shown in Figs. 4-3 and 4-4. Here, compression and splitting tensile test results are shown as frequencies for 500 samples of Saxon sandstone (from Lohmen). Studies by Curbach and Proske (1998) indicate a possible application of the normal distribution (Gauss distribution) or the beta distribution for the compression strength of natural stones. Furthermore, Fig. 4-5 shows a correlation analysis for the data. From this figure, it becomes clear that a correlation between the aforementioned strength values is almost negligible. Further, such investigations are shown in Proske (2003). For German natural stones, Peschel (1984) gives a nearly complete list of technical properties including not only mean values but also measurements for variance and deviation.

Sometimes, not only the maximum strength values of the natural stones are required but also the complete stress-strain relationship is needed. Although this topic will not be discussed here in detail, an example of a stress-strain relationship is shown in Fig. 4-6 for Silesia sandstone. More detailed information can be found in Alfes (1992) for sandstone.

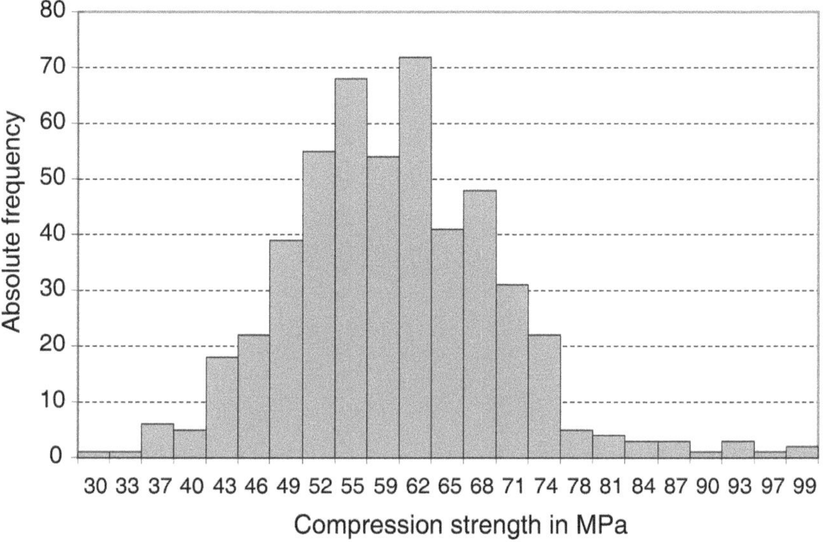

Fig. 4-3. Histogram for the compression strength of Saxon sandstone according to Curbach and Proske (1998)

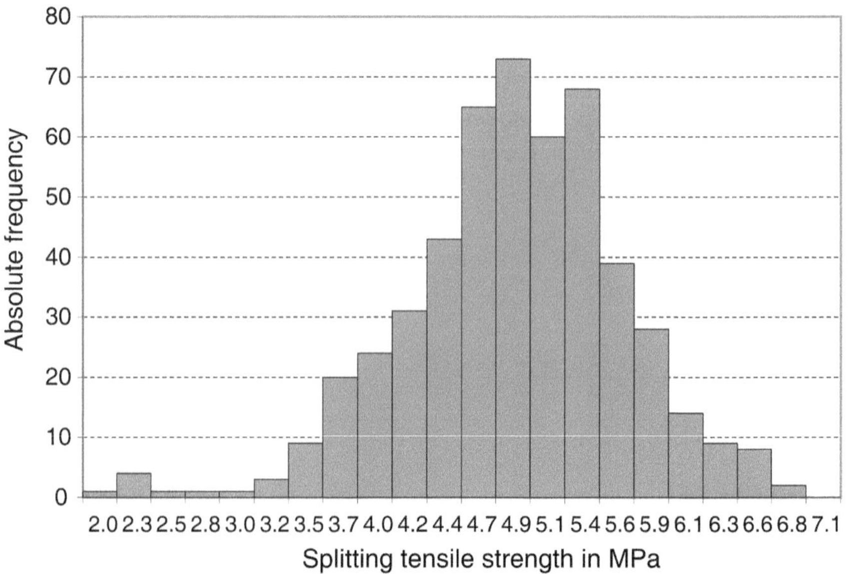

Fig. 4-4. Histogram for the splitting tensile strength of Saxon sandstone according to Curbach and Proske (1998)

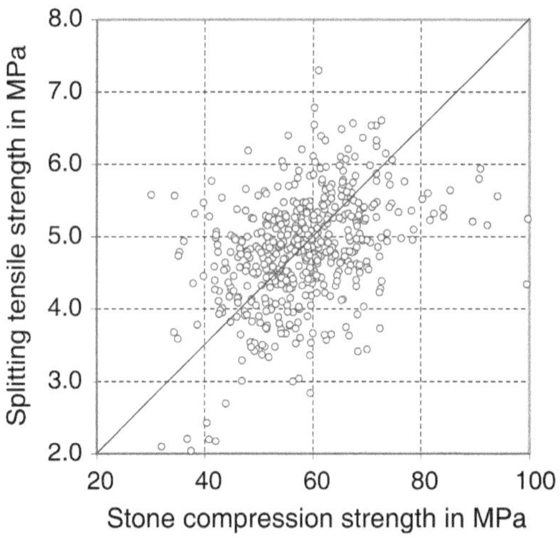

Fig. 4-5. Correlation analysis between sandstone compression strength and splitting tensile strength according to Curbach and Proske (1998)

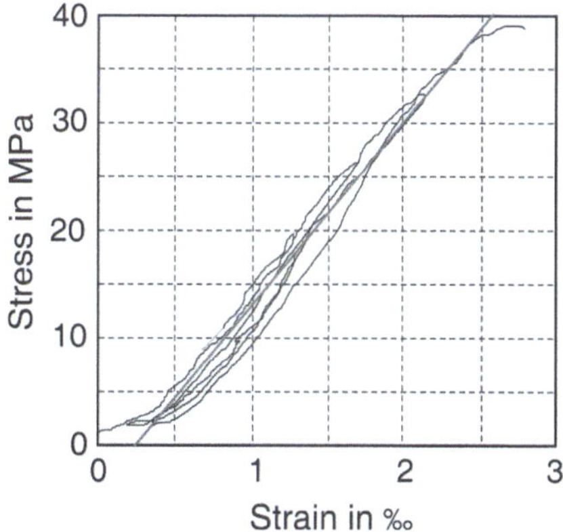

Fig. 4-6. Stress-strain relationship for Silesia sandstone (Frenzel 2004)

4.2.1.2 Working types of natural stones

The same types of stones are frequently used within a region. This effect is often visible if one compares geological maps, such as the one for the Free State Saxony (1992), with the geographical distribution of certain natural building materials.

This is especially true for sediment stones, which were heavily used in former centuries. This type of stone is easy to work with. Therefore, for example in Saxony, many buildings received a covering of sandstone.

Igneous stones were also used in early times for structures. Granite, for example, was already used as a building material in Germany by the Romans in the 2nd century (Müller 1977). In contrast to the sediment stones, working with igneous stones was much more difficult. Therefore, different working levels exist for this type of stone.

Quarry stones and cobblestones are usually rough and irregular in geometry. There are great deviations in the size of these stones: the size depends on their strength and workability. The only joints observed by geological processes acting on the stones themselves are parallel horizontal joints, and such joints can be bedding.

Cut stones show abrasive worked joints where the stones are mainly produced by cleaving. Horizontal and vertical joints are already in a rectangular angle. The chosen geometry still depends very much on the stone strength. Freestones or cut stones feature sight surfaces that are manufactured based on geometrical and artistic requirements.

The sight surfaces can furthermore be distinguished as shown in Table 4-4 and Fig. 4-7.

Table 4-4. Working types of stone sight surfaces according to Warnecke (1995)

Name	Pointed	Flattened	Pilled	Charring	Wide charring
Muster	1st action		2nd action		
Tool					
Time	Until middle of the 11th century	Until beginning of the 12th century	End 12th until End 13th century	Middle 15th to end 17th century	From middle 17th century

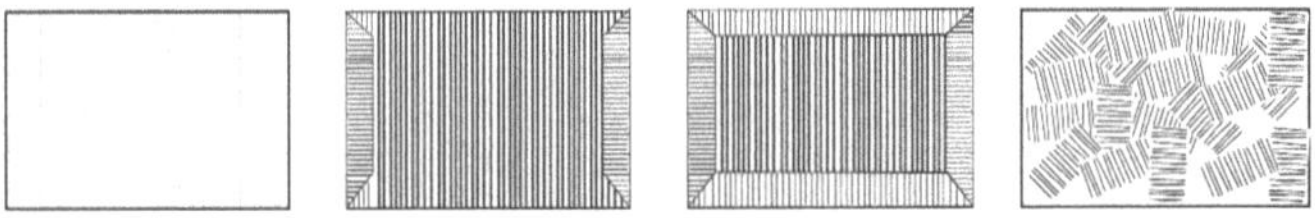

Fig. 4-7. Further stone sight surface working types, such as the Belgium and the Dutch type, according to Van der Vlist et al. (1998)

4.2.2 Mortar

Historically, masonry arch bridges can be found with and without mortar. Masonry can tolerate some loads even if it is fabricated with sand instead of mortar or without any joint material (this will be mentioned again later). However, historical bridges constructed without mortar usually had iron clamps to provide connection to the next stone (Straub 1992). An example of a historical bridge without masonry is the Pont du Gard in France (Garbrecht 1995). This bridge clearly shows that such structures can survive long lifetimes. Armaly et al. (2004) assume that the iron clamps can increase the load-bearing capacity up to 30%.

Besides those bridges without mortar, most historical arch bridges made of masonry have used, and still consist of, mortar. The properties of historical mortar are discussed in many publications such as in Baronio and Binda (1991), Huesmann and Knöfel (1991), Knöfel and Schubert (1991), Knöfel and Middendorf (1991), Wisser and Knöfel (1988), Warnecke (1995), Franken and Müller (2001), Freyburg (1994), Franken (1995), Gucci and Barsotti (1995), van Hees et al. (2004), Domède et al. (2008), and Bökea et al. (2006). Figure 4-8 gives an overview about the many different factors influencing the final strength of the mortar.

In general, historical mortar is weaker and much softer compared to modern mortar. Tables 4-5 and 4-6 show some values of the compression strength of historical mortar. Figure 4-9 shows the stress-strain relationships of different mortar types, and Fig. 4-10 gives a comparison of historical and modern mortars.

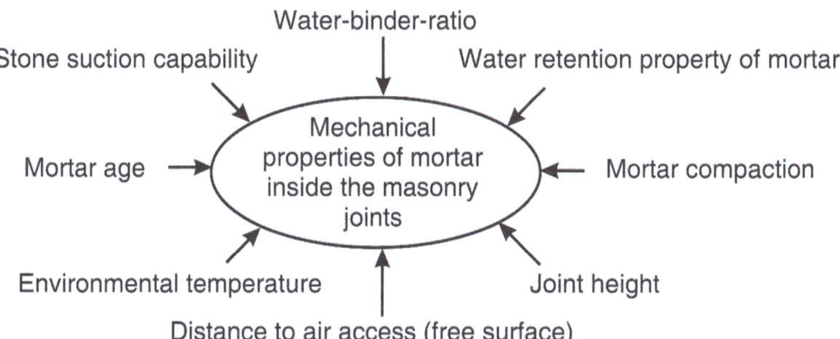

Fig. 4-8. Influences on the strength of mortar inside masonry according to Huster (2000)

Table 4-5. Compression strength of historical mortar according to Papayianni and Stefanidou (2003)

Structure	Period	Compression strength in MPa
Roman Forum	2nd century	2.5–4.0
Galerius Palace	3rd century	3.0–4.5
Acheropiitos	5th century	2.3–3.0
Hagia Sophia	7th century	2.0–6.0
Hagios Panteleimonas	14th century	1.0–1.4
Hagia Aikaterini	13th century	1.6–2.0
Bezesteni	16th century	2.5–3.5
Old house Mouson	19th century	1.5–2.0

Table 4-6. Compression strength of historical mortars according to the COST 345 (2006)

Mortar class	Mixture ratio cement : lime : sand (volume)	Compression strength in MPa
I	1:0–0.25:3	11–16
II	1:0–5:4.5	4.5–6.5
III	1:1:5–6	2.5–3.6
IV	1:2:8–9	1.0–1.5
V	1:3:10–12	0.5–1.0
VI	0:1:2–3 (hydraulic lime)	0.5–1.0
VII	0:1:2–3 (pure lime)	0.5–1.0

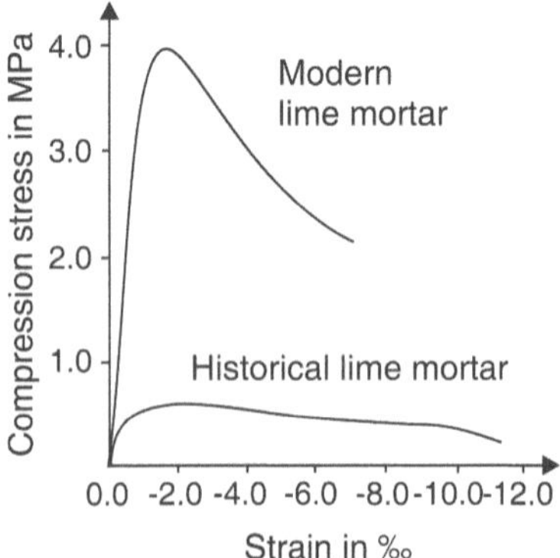

Fig. 4-9. Comparison of the stress-strain relationship for historical and modern mortar according to Warnecke (1995)

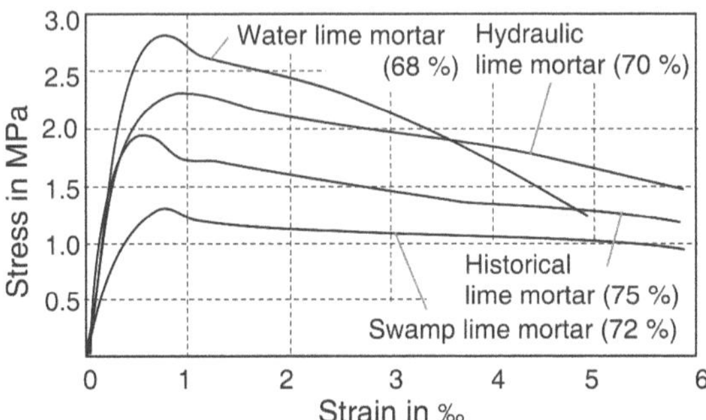

Fig. 4-10. Stress-strain relationships for different lime mortars according to Frenzel (2004)

Large deformations in combination with creeping of the masonry may cause failure of historical structures. An example of the failure of a structure where mortar could potentially have been the cause was the city tower in Pavia. However, mortar did not cause the failure of this medieval tower, resulting in four fatalities, as studies showed later. Other examples of the failure of historical structures in relation to the creep of masonry, with a significant contribution by mortar, are given in Verstrynge et al. (2008).

Furthermore, the strength of historical mortar shows great deviations as illustrated in Figs. 4-11 and 4-12. Figure 4-11 shows the distribution of the mortar strength for the city tower in Pavia. The data from Fig. 4-12 originate from a bridge built in 1875. In the face of statistical analysis, it should be mentioned here that very often the mortar data from historical structures are heavily censored. This means that very often mortar cores or specimens do not survive the exploitation process and only the strongest can be finally tested. This should be considered and kept in mind while in-interpreting mortar data from such structures. The deviation of the mortar strength is related not only to the exploitation process but also to the loss of binder and to the grading curves as shown in Fig. 4-13.

For the testing of mortar, many different codes and recommendations exist, such as DIN 18555 1-9 (1982).

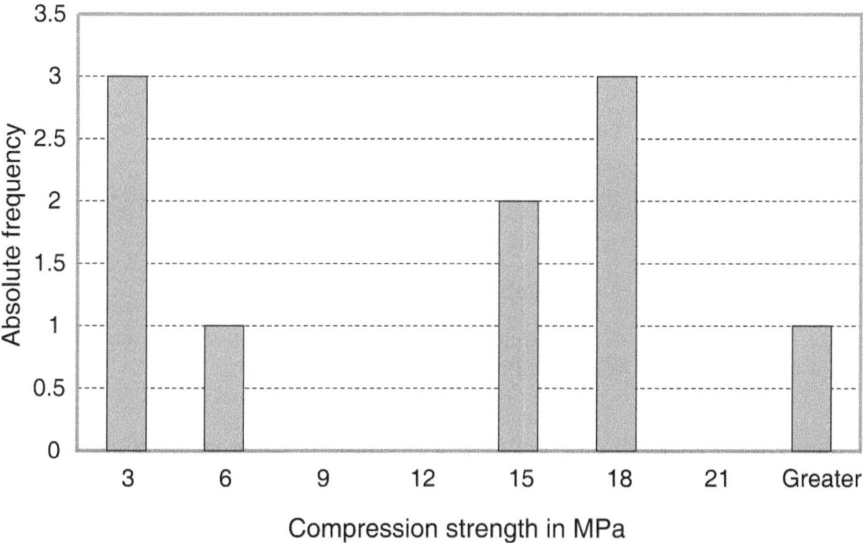

Fig. 4-11. Example of the strength distribution of historical mortar based on Baronio and Binda (1991)

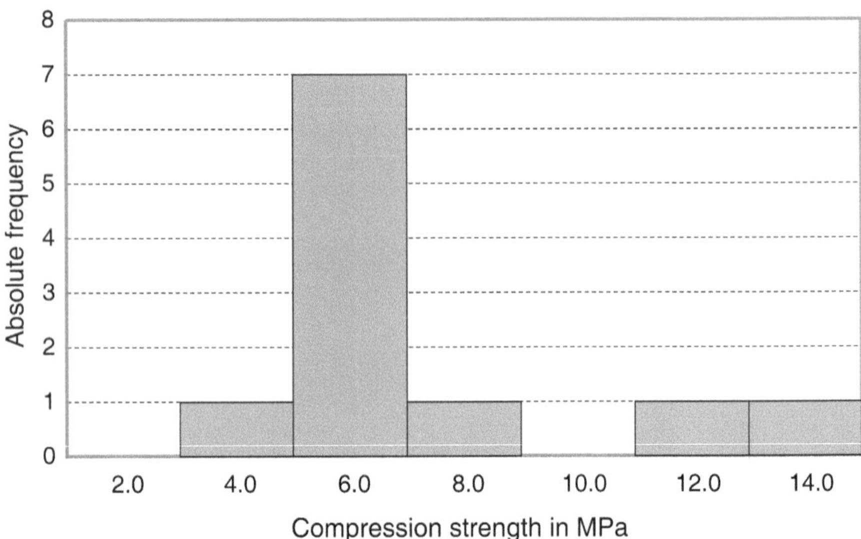

Fig. 4-12. Example of the strength distribution of historical mortar based on measurements of one bridge by Proske (2003)

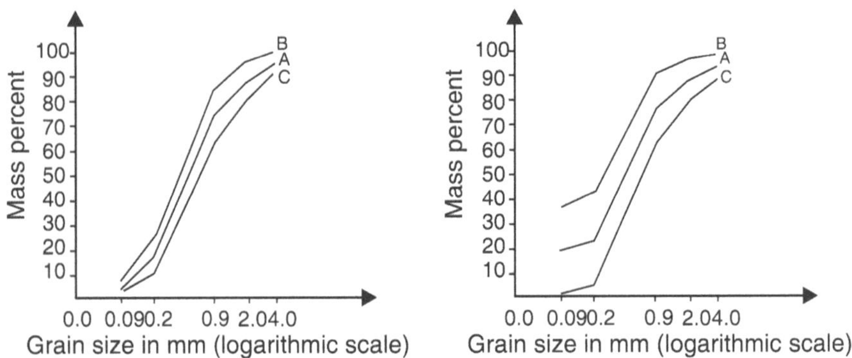

Fig. 4-13. Comparison of the grading curves of some historical (*left*) and some Roman mortar (*right*) according to Wisser and Knöfel (1988)

4.3 Maximum Centric Masonry Compression Strength

Based on the known properties of the single elements of masonry, the estimation of the properties of the masonry should be theoretically possible. Although this trial has been carried out frequently, an entire theory about the estimation of the strength properties based on the mechanical properties of the elements is still missing. Most models are based on simple empirical investigations and are usually strongly related to the special conditions of the investigated masonry type. A chronological list of models partially taken from Purtak (2001), Schulenberg (1982), and Mann (1983) for the maximum masonry compression strength is given by the following:

Krüger (1916), Graf (1926), Drögsler (1933), Voellmy (1937), Drögsler (1938), Hansson (1939), Hermann (1942), Kreüger (1943), Nylander (1944), Svenson (1944), Haller (1947), Ekblad (1949), Oniszczyk (1951), Bröcker (1961), Hilsdorf (1965/69), Monk (1967), Francis et al. (1970), Khoo and Hendry (1972), Brenner (1973), Schnackers (1973), Kirtschig (1975), Probst (1981), Schulenberg (1982), Rustmeier (1982), Mann (1982/83), Atkinson et al. (1985), Ohler (1986), Berndt (1992/96), Sabha and Pöschel (1993), Babylon (1994), and Ebner (1996).

However, the simple naming of these models is not sufficient because most models are only based on the analysis of some tests, whereas other models include some theoretical considerations. The models based on limited tests are strongly related to special conditions or properties of the tests, such as stone type. Therefore, in the following, some of the models will be discussed in more detail.

4.3.1 Model According to DIN 1053-100

The new German code of practice DIN 1053-100 (2004) is the follower of the DIN 1053-1 (1996). The new code gives some rough measures for the natural stone masonry compression strength based on the stone strength and the mortar type. The application is simple since only two tables have to be used (Tables 4-7 and 4-8). First, the masonry has to be classified, and second, according to the classification, the stone strength, and mortar class, the masonry strength can be estimated.

Table 4-7. Classification of natural stone masonry according to DIN 1053-100

Quality category	General classification	Joint height to stone length	Angle of joint in tan α	Transfer factor η
N1	Quarry stone masonry	≤ 0.25	≤ 0.30	≥ 0.50
N2	Hammered coursed rubble masonry	≤ 0.20	≤ 0.15	≥ 0.65
N3	Coursed rubble masonry	≤ 0.13	≤ 0.10	≥ 0.75
N4	Ashlar masonry	≤ 0.07	≤ 0.05	≥ 0.85

Table 4-8. Characteristic mortar compression strength based on stone strength and mortar class according to DIN 1053-100 (2004)

Quality category	Stone compression strength f_{bk}	Mortar compression strength f_k in MPa subject to the mortar group			
		I	II	IIa	III
N 1	≥ 20 Mpa	0.6	1.5	2.4	3.6
	≥ 50 Mpa	0.9	1.8	2.7	4.2
N 2	≥ 20 MPa	1.2	2.7	4.2	5.4
	≥ 50 MPa	1.8	3.3	4.8	6.0
N 3	≥ 20 MPa	1.5	4.5	6.0	7.5
	≥ 50 MPa	2.1	6.0	7.5	10.5
	≥ 100 MPa	3.0	7.5	9.0	12.0
N 4	≥ 5 MPa	1.2	2.0	2.5	3.0
	≥ 10 MPa	1.8	3.0	3.6	4.5
	≥ 20 MPa	3.6	6.0	7.5	9.0
	≥ 50 MPa	6.0	10.5	12.0	15.0
	≥ 100 MPa	9.0	13.5	16.5	21.0

4.3.2 Model According to DIN 1053

The former German code of practice DIN 1053 (1996) also gave a table for the estimation for the natural stone masonry compression strength (Table 4-9). It is likely that the table was based on the model from Mann, because the stone compression strength has only a minor influence on the masonry compression strength, and the stone tensile strength has not been considered.

Table 4-9. Characteristic mortar compression strength based on stone strength and mortar class according to DIN 1053 (1996)

Quality category	Stone compression strength f_{bk}	Mortar compression strength σ_0 in MPa subject to the mortar group			
		I	II	IIa	III
N 1	≥ 20 MPa	0.2	0.5	0.8	1.2
	≥ 50 MPa	0.3	0.6	0.9	1.4
N 2	≥ 20 MPa	0.4	0.9	1.4	1.8
	≥ 50 MPa	0.6	1.1	1.6	2.0
N 3	≥ 20 MPa	0.5	1.5	2.0	2.5
	≥ 50 MPa	0.7	2.0	2.5	3.5
	≥ 100 MPa	1.0	2.5	3.0	4.0
N 4	≥ 20 MPa	1.2	2.0	2.5	3.0
	≥ 50 MPa	2.0	3.5	4.0	5.0
	≥ 100 MPa	3.0	4.5	5.5	7.0

4.3.3 Empirical Exponential Models

The tables mentioned in the German codes are a very simple way to assess the compression strength of natural stone masonry. Pure empirical models of the masonry compression strength $f_{mas,c}$ have also found wide application due to their simple application. Furthermore, they are easy to develop by simple regression analysis. A common type are exponential equations using the stone compression strength $f_{st,c}$ and the mortar compression strength $f_{mo,c}$ such as developed by Schubert and Krämer:

$$f_{mas,c} = a \cdot f_{st,c}^{b} \cdot f_{mo,c}^{c}. \tag{4-1}$$

This type is, for example, used in the Eurocode 6. The 5% quantile of the masonry compression strength $f_{mas,c,k}$ is then computed using average compression strength values of the stone $f_{st,c,m}$ and the mortar $f_{mo,c,m}$:

$$f_{mas,c,k} = 0.40 \cdot f_{st,c,m}^{0.75} \cdot f_{mo,c,m}^{0.25}. \tag{4-2}$$

Mann (1983) gives the following parameters:

$$f_{mas,c,m} = 0.83 \cdot f_{st,c,m}^{0.66} \cdot f_{mo,c,m}^{0.18}. \tag{4-3}$$

However, Mann's parameters may not fit very well for natural stone masonry. Therefore, the Ril 805 (1999) suggests the following exponents:

$$f_{mas,c,m} = 0.80 \cdot f_{st,c,m}^{0.70} \cdot f_{mo,c,m}^{0.20}. \tag{4-4}$$

4.3.4 Model According to Hilsdorf

In contrast to the former simple regression model, Hilsdorf (1969) developed a model for the estimation of the compression strength of masonry based on the multi axial stress conditions in the stone and the mortar. The model therefore includes some theoretical considerations. In general, the model assumes that the low Young's modulus of the mortar restrains the deformation of the stone posited in the masonry. These restraints cause transverse tensile forces inside the stone and transverse compression forces inside the mortar (Fig. 4-14). First, Hilsdorf simply assumed that there was a perfect bond between the mortar and the stone, but later dismissed this assumption. The model was primarily developed for brick masonry, yet was later adapted to natural stone masonry by a so-called asymmetry factor. The major advantage of this model is the theoretical consideration, but a drawback is the estimation of the asymmetry factor (Weigert 1996, Wedler 1997, and Warnecke et al. 1995).

The masonry compression strength is given as

$$f_{mas,c} = \frac{\dfrac{f_{st,c}}{u}}{f_{st,sp} + a \cdot f_{st,c}} \cdot (f_{st,sp} + a \cdot f_{mo,c}) \tag{4-5}$$

with $f_{mas,c}$ as masonry compression strength, $f_{st,c}$ as stone compression strength, $f_{st,sp}$ as stone splitting tensile strength, $f_{mo,c}$ as mortar compression strength, u as asymmetry factor, and a as

$$a = \frac{\dfrac{t}{h}}{4.1} \tag{4-6}$$

with t as joint height and h as stone height.

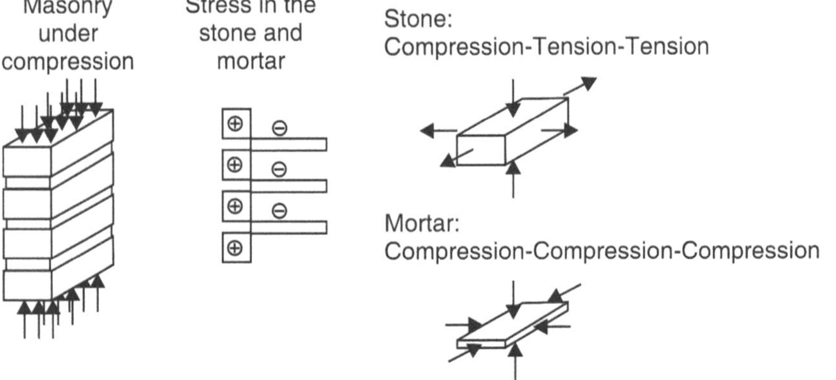

Fig. 4-14. Stress stages inside the masonry according to Hilsdorf (Warnecke et al. 1995)

4.3.5 Model According to Mann

Mann (1983) observed that the behaviour of masonry made of artificial stones (bricks) differs significantly from that of natural stones. The asymmetry and roughness of the stones and the joints yield to a qualitatively different load-bearing mechanism. Furthermore, natural stones mainly show a higher tensile strength compared to bricks. Therefore,

Mann assumes that the failure of natural stone masonry will be dominated by the failure of the mortar inside the masonry. This assumption, however, is in complete contradiction of test results with masonry built with sand in-instead of mortar. Usually, the uniaxial compression strength of the sand is virtually negligible. Then, the masonry is still able to take considerable loads whereas the formula from Mann would give the masonry compression strength of zero. The masonry failed by tearing of the stones (Warnecke et al. 1995).

$$f_{mas,c} = f_{mo,c} \cdot f \cdot \ddot{u} \tag{4-7}$$

$$f = \frac{8}{9} \cdot \frac{1}{1 - \left[1 - \frac{2}{3} \cdot \frac{t}{b}\right]^2 \cdot \cos^4 \alpha} \tag{4-8}$$

$$\ddot{u} = \frac{A_S}{A_{MW}} \tag{4-9}$$

In this, b is the width of the stone, $\ddot{u}$ is the effective cross section and α is the angle of the joints. Mann's formula gives good results for rubble masonry with weak mortar. It can be related to the works by Rustmeier (1982).

4.3.6 Model According to Berndt

Berndt (Berndt 1996, Berndt and Schöne 1991, and Wenzel 1997) has developed a concept for coursed rubble masonry from the so-called Elbe-sandstone. Berndt assumes a splitting tensile failure of the stone. As an extension to the work from Hilsdorf, Berndt not only considers tensile forces inside the stone due to constrained deformation of the mortar caused by the stone, but also tensile forces caused by force direction changes due to the unequal cross-sectional areas of the mortar and of the stone. The estimation of the maximum masonry compression strength is given as

$$f_{ma,c} = \frac{f_{st,c}}{\left[\frac{t}{h} \cdot \frac{v}{1-v} + k \cdot \frac{b}{h} \cdot \frac{d'}{b}\right] \cdot \frac{f_{st,c}}{f_{st,sp}} + 0.7} \tag{4-10}$$

with

$$k = 0.3...0.5 \tag{4-11}$$

$$d' \approx t + \frac{t}{\tan\left(45 + \dfrac{\rho}{2}\right)} \tag{4-12}$$

and

$$h' = \min\left\{\begin{matrix} h \\ 10 \ \text{cm} \end{matrix}\right\}. \tag{4-13}$$

Besides the shear formula, Berndt and Schöne (1991) have furthermore introduced a safety concept for the application of this formula. This is very useful, since for many formulas it is unclear what characteristic value of the masonry compression strength is computed. Table 4-10 shows the single elements of this concept.

Table 4-10. Safety elements in the safety concept for the evaluation of the compression masonry strength using the model by Berndt and Schöne (1991)

Factor	Description
m_1	Considers the change from mean masonry compression strength to characteristic masonry strength, usually the 5% fractile value
m_2	Considers the slenderness oft the test specimen
m_3	Considers the impossibility of load flow changes in piers
m_4	Considers the change from mean stone compression strength to characteristic stone compression strength, usually the 5% fractile value: $$m_{4,1} = 1 - 1.645 \cdot \frac{s}{f_{st,c}}$$ Considers the change from mean stone splitting tensile strength to characteristic stone splitting tensile strength, usually the 5% fractile value: $$m_{4,1} = 1 - 1.645 \cdot \frac{s}{f_{st,sp}}$$
m_5	Considers the joint thickness influence on the load-bearing behaviour: $m_5 = 0.85$
m_6	Considers the Sprödbruch behaviour of the masonry: $m_6 = 0.85$
m_7	Considers the influence of the stone layering on the masonry compression strength: $m_7 = 0.90$
m_8	Considers the long-term loading strength of the masonry: $m_8 = 0.90$

For a realistic case, some values are given in the following:

$$m_1 = m_{4,1} \cdot m_{4,2} \cdot m_5 \cdot m_6 \cdot m_7 \cdot m_8 \tag{4-14}$$
$$m_1 = 0.426 \cdot 0.604 \cdot 0.85 \cdot 0.85 \cdot 0.90 \cdot 0.90 = 0.151 \cdot$$

The characteristic masonry compression strength then reaches

$$f_{ma,c,k} = f_{ma,c,m} \cdot \prod m = 30.9 \cdot 0.151 = 4.66 \text{ MPa} \tag{4-15}$$

In terms of a global safety factor, a value between 4 and 5 is reached.

4.3.7 Model According to Sabha

The models of Berndt and Sabha have both been developed in Dresden, Germany, and are both strongly connected to the Elb sandstone found in this region. Furthermore, both authors consider mechanisms that cause a splitting tensile failure of the stones when masonry fails under maximum compression forces. However, as an extension to Berndt, Sabha (Sabha and Schöne 1994, Sabha and Weigert 1996, and Wenzel 1997) considers the locations of regions with maximum splitting tensile forces for the two mechanisms causing such forces. Whereas the maximum tensile force due to force change direction is approximately in the middle of the stone height, the maximum tensile force due to strain restraints of the mortar is reached in the stone heights close to the mortar. Therefore, Sabha does not add both the tensile forces that should reach higher masonry compression forces in comparison to Berndt:

$$f_{ma,c} = \frac{2 \cdot k \cdot f_{st,c} + f_{st,sp}}{k + \dfrac{f_{st,sp}}{f_{st,c}}} \tag{4-16}$$

with

$$k = 1.6 \frac{t}{b}\left(1.45 \frac{f_{st,sp}}{f_{st,c}} + 1\right) \cdot \tag{4-17}$$

Boye (1998) gives an extension of the Sabha model for flat stones.

4.3.8 Model According to Ohler

The UIC-Codex (1995) for the recomputation of the load bearing of historical railway bridges uses a model that is based on works by Ohler (1986). The formula of Ohler (1986) is given as

$$f_{ma,c} = 0.5 \cdot f_{mo,c,m} + \frac{a \cdot 0.5 \cdot f_{st,c,m} - 0.5 \cdot f_{mo,c,m}}{1 + \frac{b \cdot h_F \cdot 0.5 \cdot f_{st,c,m}}{2 \cdot h_S \cdot 0.05 \cdot f_{st,c,m}}} \qquad (4\text{-}18)$$

with h_F as mortar joint thickness and h_s as stone height. The splitting tensile strength of the stones inside the formula is considered as 5% fractile value.

4.3.9 Model According to Stiglat

Based on some experiments on historical stone masonry, Stiglat (1984) has developed a simple model that only considers the density of the stones γ and the mortar quality in terms of mortar groups (MG). The model is dominated by the stone failure according to Huster (2000).

$$f_{ma} = \begin{cases} 0.007 \cdot (18.7 \cdot \gamma - 355.2 \text{ MPa}) & \text{for MG I} \\ 0.017 \cdot (18.7 \cdot \gamma - 355.2 \text{ MPa}) & \text{for MG II} \\ 0.024 \cdot (18.7 \cdot \gamma - 355.2 \text{ MPa}) & \text{for MG III} \end{cases} \qquad (4\text{-}19)$$

4.3.10 Model According to Francis, Horman and Jerrems

The model of Francis et al. (1970) is based on works by Hilsdorf (Purtak 2001 and Simon 2002). The masonry compression strength is computed as

$$f_{ma,c} = f_{st,c} \cdot \frac{1}{1 + \dfrac{\dfrac{f_{st,c}}{f_{st,sp}} \cdot \left(\dfrac{E_{st}}{E_{mo}} \cdot \mu_{mo} - \mu_{st} \right)}{\dfrac{h_S}{t} \cdot \dfrac{E_{st}}{E_{mo}} \cdot (1 - \mu_{mo})}} \cdot \qquad (4\text{-}20)$$

4.3.11 Model According to Khoo and Hendry

The model of Khoo and Hendry (1972) uses a cubic equation for the estimation of the failure curves of stones and masonry. The masonry compression strength is then given by

$$(0.997 \cdot f_{st,sp} + 0.162 \cdot \frac{h_s}{t} \cdot f_{mo,c}) +$$

$$(0.203 \cdot \frac{f_{st,sp}}{f_{st,c}} + 0.113 \cdot \frac{h_s}{t}) \cdot f_{mas,c} +$$

$$(1.278 \cdot \frac{f_{st,sp}}{f_{st,c}^2} - 0.053 \cdot \frac{h_s}{t \cdot f_{mo,c}}) \cdot f_{mas,c}^2 +$$

$$(0.249 \cdot \frac{f_{st,sp}}{f_{st,c}^3} - 0.002 \cdot \frac{h_s}{t \cdot f_{mo,c}^2}) \cdot f_{mas,c}^3 = 0. \tag{4-21}$$

The works by Khoo and Hendry were extended by Probst (Simon 2002).

4.3.12 Model According to Schnackers

Furthermore, the model by Schnackers (1973) is only roughly mentioned:

$$f_{mas,c} = \frac{1}{\mu_{mas}} \cdot \frac{\frac{h_s}{2} \cdot f_{st,sp} + t \cdot f_{mo,sp}}{h_s + t}. \tag{4-22}$$

4.3.13 Model According to Ebner

Finally, the model by Ebner (1996) is given here as

$$f_{mas,c} = \left[1 - 1.2 \cdot \frac{t}{d} \cdot (1 - 2 \cdot \tan \varphi) \cdot \left(\frac{\sigma_y}{c} \right)^{0.4} \right] \cdot \left(1 - \frac{n \cdot b_s}{l} \right). \tag{4-23}$$

4.3.14 Further Masonry Compression Models

Many further models for the computation of masonry compression strength are known as mentioned in the beginning of this chapter. Such models, not discussed here, are (for example) models by Atkinson et al. (1985), Rustmeier (1982), or Pöschel. The comparison of all models based on different items such as model deviation, robustness, convergence, possible measurement of the input data, and minimum of required input data would exceed the capacity of this book. For the interested reader, the works by Huster (2000), Purtak (2001) or Warnecke, Rostasy and Budel-Budelmann (1995) can be recommended.

4.4 Stress-strain Relationship

The maximum compression strength of masonry is only one part in the estimation of masonry structural elements. Usually, structural elements not only are exposed to axial forces but also have to bear moments and shear forces. For such investigations, usually the stress-strain relationship for masonry under axial forces is required.

Several models of such relationships can be found in literature. An overview of current models has been given by Glock (2004), Lissai (1986), Becker and Bernard (1991), and Walthelm (1990). Glock lists the following models:

- Angervo (mineralic no-tensile materials)
- Becker and Bernard (masonry)
- Lewicki (concrete)
- Sargin (concrete)
- Jäger (masonry)
- DIN 1045 (concrete)
- Eurocode 6 (masonry), see Fig. 4-15.

A second look at the models reveals that only a minority is related to masonry and most stress-strain relationships originate from concrete.

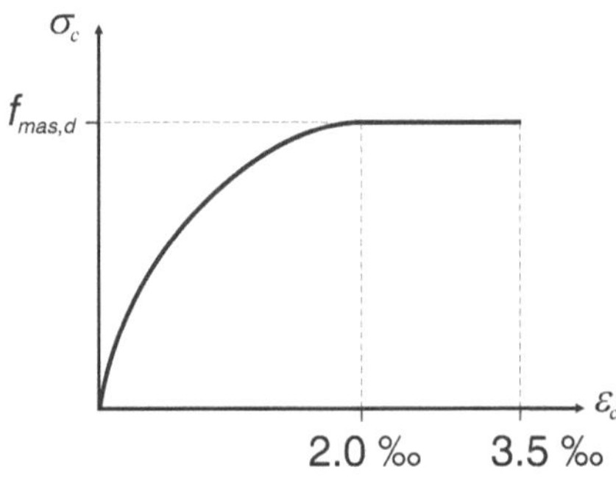

Fig. 4-15. Stress-strain relationship for masonry according to the Eurocode 6

The application of nonlinear stress-strain relationships for the computations of the ultimate load-bearing behaviour of masonry structural elements offers an increase in the numerical load by up to 25%, according to Becker and Bernard (1991). Figure 4-16 shows the development of different stress distributions in a cross section during the computation of the axial force with eccentricity.

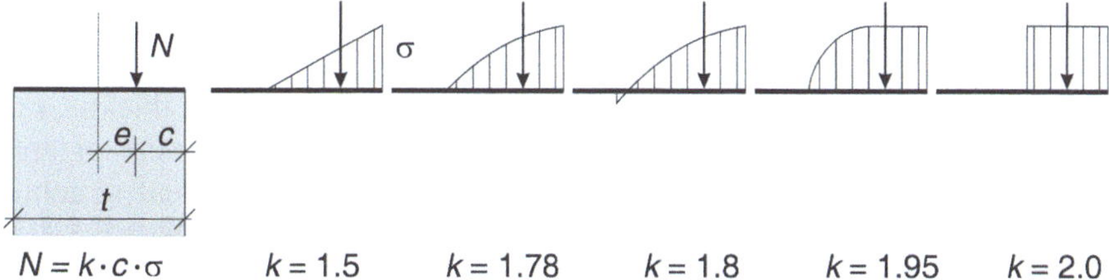

Fig. 4-16. Different stress distribution in a cross section according to Mann (1991)

4.5 Moment-Axial Force Diagrams

Besides the application of maximum stress measures for masonry or the application of stress-strain relationships, moment-axial force diagrams can also be used. The advantage of this diagram is the consideration of the nonlinear behaviour of the stress-strain relationship of masonry in a simple way, while the disadvantage is numerous computations to prepare such diagrams. However, if such diagrams are available, they can be easily used by practitioners. Such diagrams have been developed and published by Purtak (2001) – Fig. 4-17 for masonry walls and for arch bridges (Purtak et al. 2007). Furthermore, Lissai (1986) and Pauser (2005) have also prepared such moment-axial force diagrams.

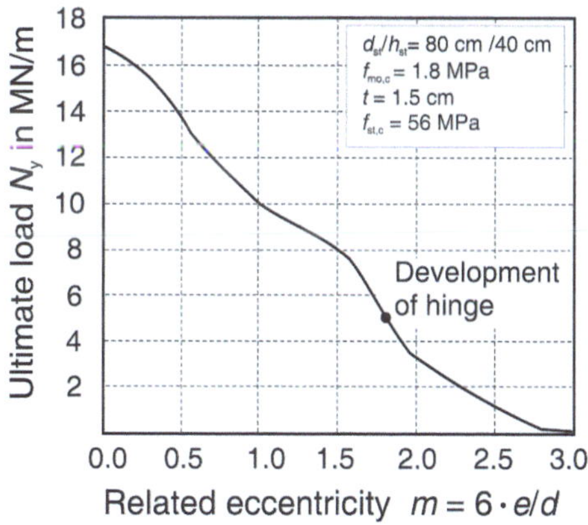

Fig. 4-17. Example of a moment-axial force diagram from Purtak et al. (2007)

4.6 Additional-leaf Masonry

4.6.1 Introduction

Besides the single-leaf masonry discussed so far, historical masonry consists in most cases of additional leafs due to the significant thickness of the masonry structural elements. This is also true for the piers of arch bridges or other elements of the arch bridges. Such a multi-leaf structure can often be proven by horizontal drillings into the piers.

Again, there exist many different models for such multi-leaf masonry. An introduction this field is given in Warnecke et al. (1995). In this chapter, only the models by Warnecke and Egermann are introduced.

4.6.2 Model According to Warnecke

Warnecke (1997) has introduced diagrams for the computation of maximum forces for multi-leaf masonry elements. He assumes that a correct estimation of the strength of masonry elements alone from drillings is not possible. Furthermore, a cohesive inner layer is considered. The following formulas show further assumptions, such as

$$v_{Mo} + v_{St} + v_{Hohlraum} = 1 \tag{4-24}$$

$$\frac{1}{E_i} = \frac{(1 - v_{St})^2}{E_{mo} \cdot v_{mo}} + \frac{v_{St}}{E_{St}}. \tag{4-25}$$

The strength of the inner masonry layer can be computed as

$$f_{mas,c,i} = f_{mo,c} \cdot \frac{v_{mo}}{1 - v_{St}}. \tag{4-26}$$

The strength of the outer masonry layers can be done equal to single-leaf masonry.

4.6.3 Model According to Egermann

The model according to Eggermann (1995) uses the following assumptions:

- External leaf with brick masonry and stretcher bond, the slenderness is less than 13.3

- Existence of cohesive internal lead
- There exists a plain surface between the external and the internal leaf
- Bernoullis hypothesis is valid (even strain distribution of cross section)
- Symmetrical support conditions at base and crown (usually fixed)
- Rigid foundation for the entire cross section

The basis value can be evaluated according to a singe-leaf masonry. However, this value has to be adapted according to

$$f_{DA} = \alpha_\lambda \cdot \alpha_\varphi \cdot f_{mas,c} \tag{4-27}$$

with

f_{DA} Masonry compression strength of the external leaf

$f_{mas,c}$ Masonry compression strength of the external leaf computed as single-leaf masonry

α_φ Consideration of the direction of pre-stressing

       $\alpha_\varphi = 1$ pre-stressing direction parallel to the loading direction

       $\alpha_\varphi = 2$ pre-stressing direction rectangular to the loading direction

A_A Cross section of the external leaf

I Moment of inertia for the non-cracked cross section

0.7 Decrease factor for cracking

s_k Effective length

$$\alpha_\lambda = 1 \text{ for } N_{cr} \geq \frac{1}{2} N_{W,0} \tag{4-28}$$

$$\alpha_\lambda = 2 \cdot \frac{N_{cr}}{N_{W,0}} \text{ for } N_{cr} < \frac{1}{2} N_{W,0} \tag{4-29}$$

$$N_{W,0} = A_A \cdot \sigma_{D,MW} \tag{4-30}$$

$$N_{cr} = 0.7 \cdot \pi^2 \cdot \frac{E \cdot I}{s_k^2} \tag{4-31}$$

$$E \approx 1,000 \cdot \sigma_{D,MW} \tag{4-32}$$

Finally the masonry compression strength for the overall cross section can be computed as

$$f_{mas} = 0.75 \cdot f_{mas,c,1} \frac{A_{A1}}{A} + 0.75 \cdot f_{mas,c,2} \frac{A_{A2}}{A} + 1.3 \cdot f_{mas,c,3} \frac{A_I}{A} \cdot \tag{4-33}$$

As the last equation clearly shows for the external leafs, the compression strength is decreased compared to a single-leaf masonry and for the internal leaf, it is increased due to multi axial compression state. However, practice has shown that the computation under common conditions does not yield to a significant change in the compression strength compared to a single-leaf cross-section assumption.

4.7 Shear Strength

As we have seen, it was already mentioned in the introduction that arch bridges may not only fail due to the development of hinges and chains, but sliding can also occur in the arch itself. To evaluate the permitted shear stresses inside the masonry, the failure surfaces discussed in Chapter 3 can be used. However, it is often desired to apply a more simple proof comparable to the computation of the maximum compression strength of the masonry.

Although Mann and Müller (Baier 1999) have developed an excellent theory for the shear failure of natural stone masonry, the approach here by Berndt (1996) will be recommended since this approach permits a continuous technique in combination with the model for the computation of the maximum compression force.

Identically to Mann and Müller (Baier 1999), Berndt (1996) has introduced three regions of failure. These three regions can also be compared to the shear failure of concrete beams.

The first region is simply the Coulombs friction:

$$\tau = f_{HS} + \mu \cdot \sigma_x \tag{4-34}$$

The second region is characterized by a tensile failure of the stones. The comparable situation in reinforced concrete is the tensile tie failure under shear force with an insufficient amount of reinforcement. The shear force forms a plateau and can be computed with

$$\max \tau = \frac{f_{st,c}}{1.4} \cdot \frac{\frac{1+k_\sigma}{2}}{\sqrt{\left(\frac{f_{st,c}}{f_{st,sp}} + 0.7 \cdot k_\sigma\right) \cdot \left(\frac{f_{st,c}}{f_{st,sp}} \cdot k_\sigma + 0.7\right)}} \cdot \tag{4-35}$$

Finally, the third region describes the failure of the stones by compression. This can be compared to the compression strut failure in concrete beams under high shear forces and high shear force reinforcement:

$$f_{mas,c} \approx \frac{f_{st,c}}{\dfrac{f_{st,c}}{f_{st,sp}} \cdot k_\sigma + 0.7} - \frac{1}{2} \cdot \left(\frac{1.4 \cdot \tau}{f_{st,c}}\right)^2 \cdot \frac{\left(\dfrac{f_{st,c}}{f_{st,sp}} + 0.7 \cdot k_\sigma\right) \cdot f_{st,c}}{\dfrac{1+k_\sigma}{2}} \cdot \tag{4-36}$$

If the three equations are used to construct a failure curve, the following figure can be drawn (Fig. 4-18).

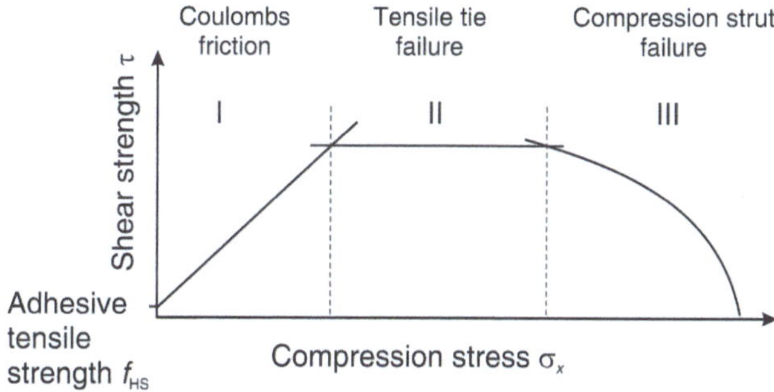

Fig. 4-18. Failure curve for sandstone masonry under shear and axial forces

Very often, either the minimum or the maximum shear forces are under discussion. According to the German code DIN 1053-1, only a maximum shear stress of 0.3 MPa can be applied. However, other works have shown that even the 5% fractile values of the maximum shear strength can reach values up to 2 or 3 MPa (Baier 1999).

4.8 Proof Equations

The computed stresses can be used for static proofs in the limit state of the ultimate load, and in the limit state of serviceability. However, the proof concepts differ significantly according to the different generations of codes of practice (Table 4-11). This is mainly based on different safety concepts as later discussed in Chapter 7.

Table 4-11. Different proof concepts for structural elements under axial forces

Code of practice	Loading $\leq$ resistance
EC 6	$N_d \leq R_d$
DIN 1053-2	$\gamma \cdot \sigma_R \leq \beta_R$
DIN 1053-1 (Feb. 1990)	$\sigma \leq$ zul σ_D
DIN 1053-100	$N_d \leq R_d$

Currently, no special requirements exist in Germany for historical arch bridges. Besides that, the German railway does not recommend the application of nonlinear computations of the arch under serviceability. International codes, such as a former version of the Eurocode, with some special remarks concerning proofs of historical arch bridges even for the limit state of serviceability, can be found:

- The stress in the extreme fibre should not exceed 65% of the maximum compression strength
- The computed deformations of the arch under the traffic load at the vertex (crown) should not exceed 1/1000 of the arch span

The British BABTIE draft (Jackson 2004) recommends for the serviceability proof:

- Crack depth lower than $0.25 \times h$
- Stress lower than $0.4 \times f_k$
- No tensile forces under torsion and quasi-permanent loads

If proofs cannot be fulfilled for historical structures, in many cases it does not mean that the structure shows insufficient safety. One has to consider that the safety concepts as the basis of codes are mainly concerned with modern structures. Therefore, as already mentioned in Chapter 1, the safety concept may be altered, for existing structures. This statement does not mean that historical structures or bridges can be less safe, however the applied tools can differ.

References

Alfes C (1992) Spannungs-Dehnungsverhalten, Schwinden und Kriechen von Sandsteinen. Jahresberichte Steinzerfall – Steinkonservierung 1992, Verlag Ernst & Sohn, Berlin

Armaly M, Blasi C & Hannah L (2004) Stari Most: rebuilding more than a historic bridge in Mostar. Museum International, Blackwell Publishing, No. 224, Vol. 56, No. 4, pp 6–17

Atkinson RH, Noland JL & Abrams DP (1985) A Deformation Failure Theory for Stack-Bond Brick Masonry Prisms in Compression. Proceedings of the 7th International Brick/Block Masonry Conference, Melbourne

Baier G (1999) Statistische Auswertung von Schubversuchen an Sandsteinmauerwerk. Diplomarbeit, Technische Universität Dresden, Lehrstuhl für Massivbau

Baronio G & Binda L (1991) Experimental approach to a procedure for the investigation of historic mortars. Proceedings of the 9th International Brick/Block Masonry Conference, 13th–16th October 1991, Vol. 1, Berlin, Germany, pp 1397–1405

Becker G & Bernard R (1991) Nichtlineare Spannungs-Dehnungs-Verläufe bei Mauerwerk. Bautechnik 68, Heft 5, pp 5–15

Berndt E & Schöne I (1991) Tragverhalten von Natursteinmauerwerk aus Elbesandstein. Sonderforschungsbereich 315, Universität Karlsruhe, Jahrbuch 1990

Berndt, E (1996) Zur Druck- und Schubfestigkeit von Mauerwerk – experimentell nachgewiesen an Strukturen aus Elbsandstein. Bautechnik 73, Heft 4, pp 222–234

Bökea H, Akkurtb S, İpekoğlua B & Uğurlua E (2006) Characteristics of brick used as aggregate in historic brick-lime mortars and plasters. Cement and Concrete Research, 36 (6), June 2006, pp 1115–1122

Börner K & Hill D (1999) Lexikon der Natursteine, CD-ROM, Abraxas Verlag

Boye A (1998) Einfluss der Steinbreite auf die Tragfähigkeit des einschaligen Mauerwerks. Diplomarbeit, Technische Universität Dresden, Lehrstuhl für Massivbau

COST 345 (2006) European Commission Directorate General Transport and Energy: COST 345 – Procedures Required for Assessing Highway Structures: Working Group 6 Report on remedial measures for highway structures. http://cost345.zag.si/Reports/COST_345_WG6.pdf

Curbach M & Proske D (1998) Abschätzung des Verteilungstyps der Mauerwerksdruckfestigkeit bei Sandsteinmauerwerk. Jahresmitteilungen 1998, Schriftenreihe des Institutes für Tragwerke und Baustoffe, Heft 7, TU Dresden

Curbach M & Proske D (2003) Gutachten zur Ermittlung von Sicherheitselementen für die Biegetragfähigkeit von Granit-Steindeckern, Technische Universität Dresden

Curbach M, Günther L & Proske D (2004) Sicherheitskonzept für Brücken aus Granitsteindeckern. Mauerwerksbau, Heft 4, pp 138–141

Dienemann W & Burre O (1929) Die nutzbaren Gesteine Deutschlands und ihre Lagerstätten mit Ausnahme der Kohlen, Erze und Salze. II. Band: Feste Gesteine, Verlag von Ferdinand Enke, Stuttgart

DIN 1053-1 (1996) Mauerwerk – Teil 1: Berechnung und Ausführung. 11/1996

DIN 1053-100 (2004) Mauerwerk: Berechnung auf der Grundlage des semiprobabilistischen Sicherheitskonzeptes. NABau im DIN/Beuth Verlag: Berlin, 7/2004

DIN 18555 1-9 (1982) Prüfung von Mörteln mit mineralischen Bindemitteln. 12/1982

Domède N, Pons G, Sellier A & Fritih Y (2008) Mechanical behaviour of ancient masonry. Materials and Structures. DOI 10.1617/s11527-008-9372-z

Ebner B (1996) Das Tragverhalten von mehrschaligem Bruchsteinmauerwerk im regelmäßigen Schichtenverband. Berichte aus dem konstruktiven Ingenieurbau, Heft 24, TU Berlin

Eggermann R (1995) Tragverhalten mehrschaliger Mauerwerkskonstruktionen. Dissertation, Aus Forschung und Lehre 29, Institut für Tragkonstruktionen, Universität Karlsruhe

Francis AJ, Horman CB & Jerrems LE (1970) The Effect of Joint Thickness Other Factors on the Compressive Strength of Brickwork, 2. International Brick Masonry Conference, Stoke-on-Trent, England

Franken S & Müller HS (2001) Historische Mörtel und Reparaturmörtel. Untersuchen, Bewerten, Einsetzen. Empfehlungen für die Praxis, Fritz Wenzel und Joachim Kleinmanns (Hrsg) "Erhalten historisch bedeutsamer Bauwerke". Karlsruhe

Franken S (1995) Die Frauenkirche zu Dresden. Zur Entwicklung geeigneter Mörtel für den Wiederaufbau eines historischen Bauwerkes. In: Jahrbuch (1993) des Sonderforschungsbereiches 315 "Erhalten historisch bedeutsamer Bauwerke", 1995, pp 93–110

Frenzel M (2004) Untersuchung zur Quertragfähigkeit von Bogenbrücken aus Natursteinmauerwerk. Diplomarbeit Technische Universität Dresden

Freyburg E (1994) Bausteine und Mörtel historischer Eisenbahnbrücken zwischen Weißenfels und Weimar. Wissenschaftliche Zeitschrift der Hochschule für Architektur und Bauwesen Weimar, 40, 5/6/7, pp 75–77

Gäbert C, Steuer A & Weiss K (1915) Die Nutzbaren Gesteinsvorkommen Deutschlands. Verwitterung und Erhalt der Gesteine. Berlin Union Deutsche Verlagsgesellschaft

Garbrecht G (1995) Meisterwerke antiker Hydrotechnik. Einblicke in die Wissenschaft. B.G. Teubner Verlagsgesellschaft, Stuttgart

Geological map of the Free state Saxony (1992) 1:400,000. Sächsisches Landesamt für Umwelt und Geologie, 3. Auflage, Freiberg

Glock C (2004) Traglast unbewehrter Beton- und Mauerwerkswände. Nichtlineares Berechnungsmodell und konsistentes Bemessungskonzept für schlanke Wände unter Druckbeanspruchung. Institut für Massivbau, Technische Universität Darmstadt, Heft 9, Darmstadt

Gucci N & Barsotti R (1995) A non-destructive technique for the determination of mortar load capacity in situ. Materials and Structures, 28, pp 276–283

Herrbach S (1996) Sicherung historischen Mauerwerks am Beispiel des Schlosses in Schönfeld. Diplomarbeit, Technische Universität Dresden, Institut für Massivbau

Hilsdorf H (1969) Investigation into the Failure Mechanism of Brick Masonry Loaded in Axial Compression. Proceedings of International Conference on Masonry Structural Systems, Houston, Texas, pp 34–41

Huesmann M & Knöfel D (1991) Mörteluntersuchungen von historischem Mauerwerk am Beispiel der Michaelskirche Fulda. In: Proceedings of the 9th International Brick/Block Masonry Conference, 13th–16th October 1991, Vol. 1, Berlin, Germany, pp 1406–1411

Huster U (2000) Tragverhalten von einschaligem Natursteinmauerwerk unter zentrischer Druckbeanspruchung: Entwicklung und Anwendung eines Finite-Elemente-Programmes. Kassel University Press, Kassel

Jackson P (2004) Highways Agency BD on New Masonry Arch Bridges. 27th February 2004. http://www.smart.salford.ac.uk/February_2004_Meeting.php

Khoo CL & Hendry AW (1972) A Failure Criterion for Brickwork in Axial Compression. The British Ceramic Research Association, Technical Note, No. 179

Knöfel D & Middendorf B (1991) Chemisch-mineralogische Mörteluntersuchungen an historischen Ziegelgebäuden in Norddeutschland. In: Proceedings of the 9th International Brick/Block Masonry Conference, 13.–16. October 1991, Vol. 1, Berlin, Germany, pp 1420–1427

Knöfel D & Schubert P (1991) Zur Beurteilung von Mörtel für historische Bauwerke. In: Proceedings of the 9th International Brick/Block Masonry Conference, 13.–16. October 1991, Vol. 1, Berlin, Germany, pp 1412–1419

Lissai, F (1986) Elastisch-plastisches Verhalten von Mauerwerks-Wandscheiben, die in ihrer Ebene vertikal und horizontal belastet sind. Bautechnik, Heft 1, pp 7–17

Mann W (1983) Druckfestigkeit von Mauerwerk: Eine statistische Auswertung von Versuchsergebnissen in geschlossener Darstellung mit Hilfe von Potenzfunktionen, Mauerwerk-Kalender 1983, pp 687–699

Mann W (1991) Grundlagen der Bemessung von Mauerwerk unter vertikaler Belastung und Knicken nach dem Entwurf zu Eurocode EC 6 und Vergleich mit Versuchsergebnissen. Proceedings of the 9th International Brick/Block Masonry Conference, 13–16 Oct. 1991, Vol 1, Berlin, Germany, pp 1281–1291

Meskouris K, Butenweg C, Mistler M & Kuhlmann W (2004) Seismic Behaviour of Historic Masonry Buildings. In Proceeding of 7th National Congress on Mechanics of HSTAM, Chania, Crete, Greece

Müller B (1977) Beiträge zur Geschichte der Natursteinindustrie in der Sächsischen Oberlausitz. Abhandlungen des Staatlichen Museums für Mineralogie und Geologie zu Dresden. Band 27, Dresden, pp 111–142

Ohler A (1986) Zur Berechnung der Druckfestigkeit von Mauerwerk unter Berücksichtigung der mehrachsigen Spannungszustände in Stein und Mörtel. Bautechnik, Heft 5, pp 163–169

Papayianni I & Stefanidou M (2003) Mortars for intervention in monuments and historical buildings. Structural Studies, Repairs and Maintenance of Heritage Architecture VIII, CA Brebbia (Ed), WIT-Press, Southampton, pp 57–64

Pauser A (2005) Ingenieurbüro: Guide to the high-level assessment of masonry arch bridges. UIC Report (draft)

Peschel A (1984) Natursteine der DDR – Kompendium petrographischer, petrochemischer, petrophysikalischer und gesteinstechnischer Eigenschaften. Beilage zur Dissertation (B) an der Bergakademie Freiberg, Januar 1984

Proske D (2003) Ein Beitrag zur Risikobeurteilung alter Brücken unter Schiffsanprall. Dissertation, Technische Universität Dresden

Purtak F (2001) Tragfähigkeit von schlankem Mauerwerk. Dissertation. Technische Universität Dresden, Fakultät Architektur

Purtak F, Geißler K & Lieberwirth P (2007) Bewertung bestehender Natursteinbogenbrücken. Bautechnik 8, Heft 8, pp 525–543

Ril 805 (1999) Richtlinie 805: Tragsicherheit bestehender Eisenbahnbrücken

Rustmeier HG (1982) Untersuchungen über Einflüsse auf die Drucktragfähigkeit von Bruchsteinmauerwerk. Dissertation, Technische Hochschule Darmstadt, Darmstadt

Sabha A & Schöne I (1994) Untersuchungen zum Tragverhalten von Mauerwerk aus Elbsandstein. Bautechnik 71, Heft 3, pp 161–166

Sabha A & Weigert A (1996) Einfluss der Steinhöhe auf das Tragverhalten einschaligen Mauerwerkes. Sonderdruck SFB 315, Erhalten historisch bedeutsamer Bauwerke, Jahrbuch 1995, Verlag für Architektur und technische Wissenschaften, Berlin

Schnackers PJH (1973) Mauerwerk und seine Berechnung. Dissertation, TH Aachen

Schubert P (2004) Eigenschaftswerte von Mauerwerk, Mauersteinen und Mauermörtel. Mauerwerk-Kalender 1994, Verlag Ernst und Sohn, Berlin, pp 129–141

Schulenberg W (1982) Theoretische Untersuchungen zum Tragverhalten von zentrisch gedrücktem Mauerwerk aus künstlichen Steinen unter besonderer Berücksichtigung der Qualität der Lagerfugen, Dissertation, TH Darmstadt Fachbereich Architektur

Simon E (2002) Schubtragverhalten von Mauerwerk aus großformatigen Steinen. Dissertation, Technische Universität Darmstadt, Darmstadt

Stein A (1993) Bemessung von Natursteinfassaden. Rudolf Müller Verlag, Köln

Stiglat K (1984) Zur Tragfähigkeit von Mauerwerk aus Sandstein. Bautechnik 2, pp 54–59 and Bautechnik 3, pp 94–100

Straub H (1992) Die Geschichte der Bauingenieurkunst. Ein Überblick von der Antike bis in die Neuzeit. Birkhäuser Verlag Basel

UIC-Codex (1995) Empfehlungen für die Bewertung des Tragvermögens bestehender Gewölbebrücken aus Mauerwerk und Beton. Internationaler Eisenbahnverband. 1. Ausgabe 1.7.1995

Van der Vlist AA, Bakker MM, Heijnen WJ, Roelofs HJJ & Oosterhoff J (1998) Bruggen in Nederland 1800–1940. Bruggen van beton, steen en hout. Nedelandse Bruggen Stichting – Uitgeverij Matrijs

van Hees RPJ, Binda L, Papayianni I & Toumbakari E (2004) Characterisation and damage analysis of old mortars. Materials and Structures, 37, November 2004, pp 644–648

Verstrynge E, Schueremans L & Van Gemert D (2008) Life time expectancy of historical masonry structures subjected to creep – a probabilistic approach. Proceedings of the 6th International Probabilistic Workshop, CA Graubner, H Schmidt & D Proske (Eds), Darmstadt 2008

Walthelm U (1990) Rissbildungen und Bruchmechanismen in freistehenden Wänden aus Beton und Mauerwerk. Bautechnik 67, Heft 1, pp 21–26

Warnecke P (1995) Tragverhalten und Konsolidierung von historischem Natursteinmauerwerk. Dissertation, Technische Universität Braunschweig

Warnecke P (1997) Biegedruckfähigkeit von historischem Mauerwerk – Modelle und Anwendung. Konsolidierung von historischem Natursteinmauerwerk, 6.

und 7. Nov. 1997, Technische Universität Braunschweig, Institut für Baustoffe, Massivbau und Brandschutz Heft 135, Braunschweig, pp 13–28

Warnecke P, Rostasy FS & Budelmann H (1995) Tragverhalten und Konsolidierung von Wänden und Stützen aus historischem Natursteinmauerwerk. Mauerwerkskalender 1995, Ernst & Sohn, Berlin, pp 623–685

Wedler B (1997) Einfluss der Füllungsschichtung auf das Tragverhalten des mehrschaligen Mauerwerks in historischen Bauwerken. Diplomarbeit. Technische Universität Dresden. Lehrstuhl für Massivbau

Weigert A (1996) Untersuchung zum Einfluss der Steinhöhe auf das Tragverhalten des einschaligen Mauerwerks aus Elbsandstein. Diplomarbeit. Lehrstuhl für Massivbau. Technische Universität Dresden

Wenzel F (Ed) (1997) Mauerwerk – Untersuchen und Instandsetzen durch Injizieren, Vernadeln und Vorspannen. Erhalten historisch bedeutsamer Bauwerke. Empfehlungen für die Praxis. Sonderforschungsbereich 315, Universität Karlsruhe

Wisser S & Knöfel D (1988) Untersuchungen an historischen Putz- und Mauermörteln: Bautenschutz und Bausanierung 11, Nr 5, pp 163–171

5 Investigation Techniques

*"Here and elsewhere, we shall not obtain the best insight
into things until we see them growing from the beginning."*

Aristotle, taken from Vanmarcke (1997)

5.1 Introduction

The preparation of input data for the numerical modelling of the arch
bridges is a major part of the investigation of arch bridges: a sophisticated
numerical model is without much worth if the quality of the input data is
rather low. Therefore, the observation of the existing structure is of utmost
importance to understand such a structure. Even further, numerical models
and structural observation interact with each other. Whereas for new
structures the choice of the static system is part of the design process, for
existing structures the situation is different: here the static system has to be
identified. Usually in the beginning of such an observation, only very
limited knowledge is accessible. However, with the first numerical models
the required input data and the zones of interest for observation are
identified and investigated. This should improve the numerical model
yielding to further refinement of the interesting parameters. Lourenço
(2001) has described the initial state with the following:

- Important geometrical properties are unknown
- Information about the internal construction is unknown or limited
- The material properties are often unknown or difficult to identify
- The type of construction is unknown
- Possible damages are unknown
- The design basis of the structure is unknown and modern codes are
  usually not applicable
- The parameters show significant deviations, either due to the former
  construction type or due to the use of natural construction materials
  (Goretzky 2000, Franke, Deckelmann and Goretzky 1991 and Kirtschig
  1991)

D. Proske, P. van Gelder, *Safety of Historical Stone Arch Bridges*, DOI 10.1007/978-3-540-77618-5_5,
© Springer-Verlag Berlin Heidelberg 2009

Therefore, to provide a realistic description of the structure, a structural observation is compulsive. Different technologies can be used for such an observation and are listed in Table 5-1. Further techniques are summarized in Schueremans and Van Gemert (2001), Wenzel and Kahle (1993), Kaplan (1997), Colla (1997), Silman and Ennis (1993), Prieto et al. (2006), and Orbán et al. (2008). In general, the observation techniques are classified into destructive, semidestructive and nondestructive identification methods. Bién and Kamiński (2007) relate certain damages to certain investigation types (Table 5-2).

Table 5-1. Listing of certain techniques to evaluate historical masonry structures (Schueremans et al. 2003, Wenzel and Kahle 1993, Kaplan 1997, Silman and Ennis 1993, Prieto et al. 2006, and Orbán et al. 2008)

Technique	Degree of destruction	Location	General principle and application field
Historic research	NDT	IS and IL	Historical documents often include worthwhile information about the construction technology, used materials, and the geometry. Very often it is useful not only to investigate files but also to contact local history association
Visual inspection	NDT	IS	Visual inspection is compelling since it is cheap and the most efficient nondestructive test method. It can be extended by monitoring systems or geomarking tools
Photogrammetry	NDT	IS	Photogrammetry can be used to identify any type of deformations including discontinuities. In recent years, not only photogrammetry but also other techniques such as laser scanning and laser interference techniques have been applied and they have yielded extraordinary results
Electric resistivity	NDT	IS	This method can be applied to achieve information about the overall conditions of the masonry such as cavities and layering
Radiography	NDT	IS	The application of strong ionizing radiation, mainly gamma rays, can be used to identify not only the steel elements in the structure, but also the cavities and other types of discontinuities. However, safety considerations limit the applicability under practical conditions

Technique	Degree of destruction	Location	General principle and application field
Infrared thermography	NDT	IS	Identification of layering of the structure and further discontinuities
Magnetic methods	NDT	IS	Identification of steel or iron elements inside the masonry blocks
Radar	NDT	IS	Radar can give indications about certain types of discontinuities such as cavities
Mechanical pulse velocity	NDT	IS	Waves are introduced to the material. The wave velocity gives information about the integrity and density of the material
Ultrasonic	NDT	IS	Waves are introduced to the material. Again, information about not only the density, but also the humidity and discontinuities can be gained. Limited application of masonry
Vibration tests	NDT	IS	Investigation of the stiffness of the structural elements
Endoscopy	SDT and NDT	IS	After drilling, endoscopy can be used to investigate the internal structure. In most cases, it is combined with video taping
Flat jack	SDT	IS	Determination of the stress-strain relationship, also sometimes used for the identification of the maximum compressive strength
Proof loading	NDT	IS	Testing of load of some structural parts. It increases the certainty about the applied numerical models but may cause some slight damage
Monitoring	NDT	IS	Permanent measurement of certain structural parameters

Amount of destruction: DT – destructive test; SDT – semi-destructive test; NDT – non-destructive test
Location of test: IS – in situ; IL – in labo.

Table 5-2. Application of nondestructive tests (NDT) and minor-destructive tests (MDT) for the investigation of damages to masonry bridges (Bién and Kamiński 2007)

Test	Degradation mechanism	contamination	deformation	destruction	discontinuity	Displacement	Loos of material
	Basic methods						
	Visual inspections	●	●	●	●	●	●
	Direct geometric measurements	●	●		●	●	●
	Sclerometric test			●			
NDT	**Acoustic and stress wave methods**						
	Acoustic emission measurement			●	●		
	Impact echo test			●	●		●
	Parallel seismic method			●	●		●
	Ultrasonic echo test			●	●		●
	Electrical and electromagnetic methods						
	Electrical conductivity measurement			●	●		●
	Ground penetrating radar			●	●		●
	Thermal heat transfer methods						
	Pulse-phase thermography	●		●	●		●
	Transient thermography	●		●	●		●
	Proof load tests						
	Dynamic test		●	●	●	●	●
	Static tests		●	●	●	●	●
	Boroscopy	●		●	●		●
	Flat-jack test		●	●		●	
MDT	Pull out test			●			
	Specimen test – chemical	●					
	Specimen test – mechanical			●			

5.2 Destructive Tests

Besides geometrical parameters, the numerical models require data from the material properties. Such material properties are often measured on test specimens. The specimens have to be separated from the original structures. A wide literature for the extraction of test specimens from historical masonry propagating many different techniques can be found.

For example, Stiglat (1984) recommends the extraction of unnecessary stones from the structure. However, under realistic conditions, it is difficult to identify elements of the structure that can be confiscated without restriction of the functionality of the structure. Furthermore, only visible alterations of a historical structure are often prohibited due to conservation regulations of monuments and historic buildings (Budelmann 1997, Wenzel 1997a, b).

Therefore, in many cases, drilling of cores has to be carried out. Although this type of material extraction is often criticized due to uncertainties about the process (Stiglat 1984 and Berndt and Schöne 1990), the technology is simple to apply even under difficult conditions, for example under water. Furthermore, the drillings not only allow the extraction of material but also provide geometrical data about the internal structure. Additionally, the visual disturbance of the structure due to drillings is rather limited since only the diameter of the drilling machine has to be substituted on the structure surface.

Besides the material excavation, drilling also provides data from the process alone. For example, the volume of cooling water during the drilling process permits conclusions about the pore volume in the masonry. Finally, drilling can be combined with endoscopy, which allows visual views inside the structure. After the material has been used for material testing, the data can be compared with the drilling protocols.

The diameter of the core drilling depends on different conditions. In general, from the material investigation point of view, drilling diameters should be rather great to achieve characteristic material property data. For example, Stiglat (1984) recommends a minimum diameter of 20 cm for natural stone masonry walls. However, such big diameters are often not applicable due to visual disturbance of the structure, increased breaking rate of the cores to the high friction, and simply drilling costs or drilling conditions. Therefore, under practical conditions, often diameters in the range of 10–15 cm are used. Under specific conditions, diameters of 5 cm are also used; however, the measured material data may then be of restricted use. Wenzel (1997b) recommends minimum diameters of 3 cm for brick stones and 5 cm for natural stones.

After the drilling process, the drilling cores are visually observed (Fig. 5-1). The observation is used to create a drilling profile, for example, identifying layer thickness, rough material estimation, and identification of hollow sections. The layers are then classified according to the size and number of material pieces (lumpy, small sized) (DIN 4022 1987). Based on the classification, possible test specimens for material testing can be marked on the cores. The specimens are then sawed out. However, sawing requires cooling water and the water can influence some material properties. It can even prevent the production of test specimens. The material can also be used for chemical, petrographic (Fig. 5-2), spectrographic, and microscopic investigations (Fig. 5-3).

A description of drilling investigations at a historical arch bridge can be found in Aoki et al. (2004) or Proske (2003).

Fig. 5-1. Example of drilling cores. See also the marking of test specimen for material property tests

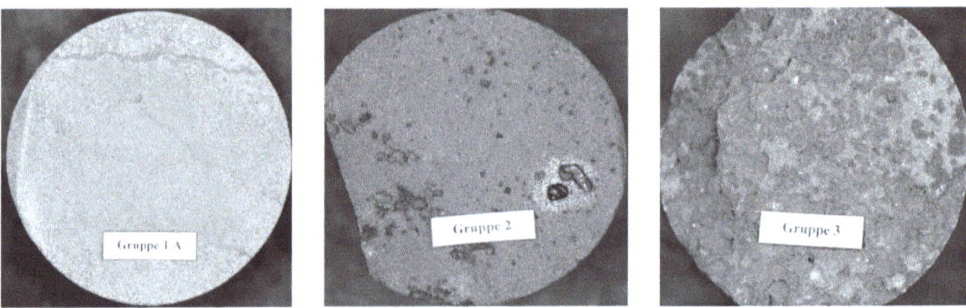

Fig. 5-2. Different sandstone varieties taken from one historical arch bridge

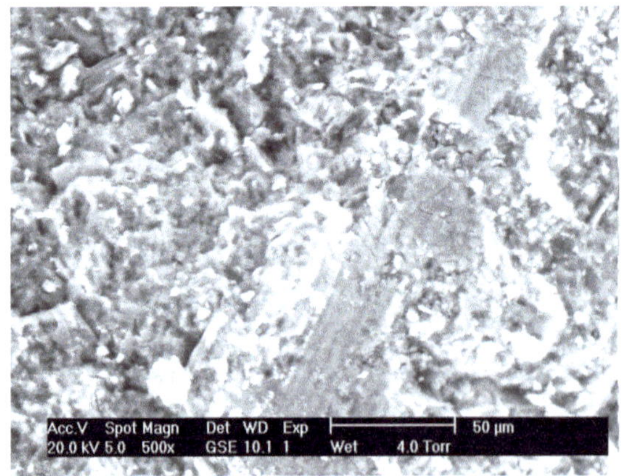

Fig. 5-3. X-ray microscope picture of historical masonry material

5.3 Semi-destructive Test Methods

According to Forde (1996), the semidestructive test methods, sometimes also called minor-destructive test methods, can be distinguished into

- Pull-out tests
- Pull-off tests
- Penetration tests (Windsor-Probe, Schmidt hammer)

  For further information, refer Corps of Engineers (2002).

5.4 Non-destructive Test Methods

A classification of nondestructive tests can be found in Corps of Engineers (2002), Orbán et al. (2008), and Bungey (1997). Nondestructive test methods can provide the following information about arch bridges (Forde 1996):

- Type and construction of the springing
- Thickness of the arch
- Type of backfill and existence of vaults inside the backfill
- Density of the backfill

Astudillo (1996) presents nondestructive investigation methods for the examination of bridges. Colla et al. (1997) report about nondestructive techniques on stone masonry bridges. Bensalem et al. (1998) use nondestructive investigation methods for the estimation of the safety factor decrease. Aoki et al. (2004) and Binda and Saisi (2001) have used nondestructive techniques for the investigation of masonry. With regard to masonry, the European research project ONSITEMASONRY should be mentioned, which focused on development of nondestructive and semidestructive investigation techniques for masonry (Wendrich et al. 2004, Maierhofer et al. 2003, and Köpp et al. 2005). During the sustainable bridge project of the EU, nondestructive tests for the assessment of bridges were also evaluated (Helmerich and Niederleithinger 2006 and Niederleithinger et al. 2006). Recently, Orbán et al. (2008) showed some application for historical arch bridges.

5.4.1 Ultrasound

The application of ultrasound for the investigation of historical masonry or stone structures has found wide application. As examples, the works by Schubert et al. (2002) and Müller and Garke (2005) should be mentioned.

The general idea is simple: the denser a material is or better the material joints are, the better mechanical waves can spread out in the material. The application of ultrasound uses this effect. For the investigation, sound waves in a frequency range between 46 and 350 kHz are introduced in the material investigated. The waves are then registered at a different location. The time difference between the sending and the reception of the waves is measured and the velocity of the ultrasonic waves is then computed. The sound velocity depends on certain material parameters, such as the material structure and the amount of pores. Furthermore, humidity or watering conditions of the material can influence the ultrasonic velocity. Due to these effects, a drop in the wave velocity may have different causes, which under practical conditions may restrict the application of ultrasonic techniques.

5.4.2 Impact-echo

The impact-echo technique is strongly related to the application of ultrasonic waves. In contrast, here a hammer is used to apply shock waves to the material. The echo is again measured with receivers. If material defects occur in the structural material, then the echo-impulses are decreased in comparison to homogenous materials (Leaird 1984).

Values for average ultrasonic velocities are given in Table 5-3.

According to Forde (1996), the frequency range lies between 1 kHz and 300 Hz. Forde (1996) has already used the impact-echo technique for investigations of the springings and abutments of historical arch bridges.

Table 5-3. Examples of ultrasonic velocity in different materials according to Forde (1996)

Material	Average ultrasonic velocity in m/s
High quality brick masonry	3,100
Low quality brick masonry	2,500–2,700
Structural concrete	>4,500
Granite masonry piers	3,300–3,500
Red sandstone piers	1,970
Yellow sandstone piers	2,040
White sandstone piers	1,700
Steel bar	5,100
Steel body	6,100
Dry sandy ground	200–300
Dry sandy clay	400–600
Water saturated clay	1,300–2,400
Water	1,430–1,680
Limestone and dolomite	4,000–6,000

5.4.3 Radar

The application of radar technology for masonry and masonry arch bridges is strongly related to the development of ground penetration radar. This radar has been widely applied for the investigation of ground structures. Based on the experience gained there, the technique has also been applied for archeological investigations, for the investigation of structures in the ground, and for investigation of buried vaults and foundation rests. Examples are given in Kahle and Illich (1992), Illich (1999), BAM (2006a, b), Wiggenhauser and Maierhofer (2002), Cameron et al. (2008), and Wenzel (1997b).

During a radar investigation, waves (now radar waves) are also entered into the structure. If the waves hit irregularities such as separated surfaces, change in the salt content and humidity, hollow cavities, or metal elements inside the element, then the waves are reflected. Since the sender and receiver are usually assembled jointly into one case, the intensity and the running time of the waves can be used to compute the depth of the reflection zone.

The radar case is moved slowly over the surface of the structural element to achieve not only results at some certain points, but also to cover surfaces. Using the depth information, three-dimensional maps can be prepared.

Very often 1 GHz antennas are used. Such antennas reach a penetration depth of 1.0–1.5 m. The resolution depends on the frequency and reaches 3 cm for 1 GHz. For masonry arch bridges, lower frequencies in the range of 100 MHz are often used according to Forde (1996). Here, radar investigations are mainly used to identify hollow sections inside piers or walls, to find cracks, and to estimate the salt content and the water content (Forde 1996). Clark et al. (2003a, b) describe the application of radar and infrared for the humidity investigation.

To give an impression about the quality of radar and ultrasonic investigations, Table 5-4 lists the results of such an investigation. The table also includes the achieved maximum flexural bending stress of the granite stone material investigated. The measured wave velocities and found irregularities are already transferred into qualitative statements about the flexural bending strength. All methods find the one weak stone (Nr. 1), however for the other stones the results are inconsistent.

Table 5-4. Comparison of the flexural bending strength of granite stone material and the results from radar and ultrasonic investigation

Test number	Qualitative statement about the strength according to ultrasonic test 1	Qualitative statement about the strength according to ultrasonic test 2	Qualitative statement about the strength according to radar test	Maximum flexural bending strength in MPa
1	Low	Low	Low	4.23
2	Average to good	–	Good to very good	9.66
3	Average to good	Good to very good	Good to very good	12.70
4	Good to very good	Low to average	Average to good	9.53
5	Good to very good	–	Low strength	10.21

5.4.4 Tomography

Cote (1996) presents the tomographical investigation of a pier of the Bridge Le Pont-Neuf in Paris. At first, 20 single-spot tomographical measurements were taken on the pregrouted pier. Based on that data, a spatial picture of the velocity distribution of ultrasonic waves inside the pier was constructed. Then the velocity was translated into the density of the material. The results showed a very inhomogeneous limestone inside the pier. In the next step, the pier was grouted and again investigated tomographically. As the result of the grouting, the more homogeneous density field was found. Computer tomography as presented by Schulze and Hampel (2004) may only be used for some small structural elements.

5.4.5 Thermography

Thermography can be used to identify hollow sections and humidity inside masonry walls (Orbán et al. 2008). However, the technique is used more for buildings than for bridges.

5.4.6 Electrical Conductivity

Conductivity tests can be used for salinity and stone thickness investigations (Forde 1996 and Helmerich et al. 2008).

5.4.7 Experimental Tests on Bridges on Site

Besides the introduced techniques, there exists finally the possibility to apply a load test on the arch bridge on site. Basic works for such experimental ultimate load-bearing tests can be found in Optiz and Steffens. However, some examples of application for historical arch bridges should be mentioned here.

Mildner (1996) describes load tests and deformation measurements of the Schrote Bridge and the Anna-Ebert Bridge in Magdeburg. Further tests by Milder can be found in Mildner and Mildner (2001). Vockrodt and Schwesinger (2002) have also carried out experimental load-bearing tests on historical arch bridges. Steffens (2001), Gutermann and Steffens (2005), Burkert and Steffens (2008), and Gutermann (2002) give recommendations about the application of in situ load tests and show examples. Fanning and Boothby (2003) also report about load tests on arch bridges. Bolle (2005) describes the permanent observation of a viaduct. Slowik et al. (2005) and Slowik (2004) also report on load testing on vault bridges and the development of the numerical model based on the data. Domède and Sellier (2008) have also carried out measurements and have used it to create an FEM model. Hughes and Pritchard (1998) have published on in situ measurement of masonry arch bridges. Armstrong et al. (1995a) have carried out dynamic measurements on arch bridges and have continued with modal analyses (Armstrong et al. 1995b). Measurement under load was also done by Bién et al. (2008), Prader et al. (2008), and Rücker et al. (2006). Measurement of fill pressure was done by Ponniah and Prentice (1999).

Fibre optic sensors can also be applied for strain measurements on arch bridges (Inaudi and Glisic 2008).

5.4.8 Photogrammetry and Lasercanning

During the observation of arch bridges under load, certain different methods can be applied to receive the deformations data. Photogrammetry is one of the contact-free techniques (Hampel and Maas 2003). Albert and Seyler (2004), for example, have used photogrammetry for the investigation of arch bridge deformations.

A second technique is laser scanning. Laser scanning is not only used to scan snow covers in the mountains now (Prokop 2007) but also widely applied since a few years for the capture of structural geometries (Ehmann 2000, Mönicke 2003).

References

Albert J & Seyler S (2004) Verformungsmessungen an Brückenbauwerken aus Bildsequenzen, 14th International Conference on Engineering Surveying Zürich, 15.–19. März 2004

Aoki T, Komiyama D, Sabia D & Rivella R (2004) Theoretical and Experimental Dynamic Analysis of Rakanji Stone Arch Bridge, Honyabakei, Oita, Japan, Proceedings of 7th International Conference on Motion and Vibration Control MOvIC'04, St. Louis, 8.-11. August, 2004, pp 9 (CD-ROM)

Aoki T, Komiyama T, Tanigawa Y, Hatanaka S, Yuasa N, Hamasaki H, Chiorino MA & Roccati R (2004) Non-destructive Testing of the Sanctuary of Vicoforte, Proceedings of 13th International Brick and Block Masonry Conference, Amsterdam, July 4–7, 2004, Vol. 4, pp 1109–1118

Armstrong DM, Sibbald A & Forde MC (1995a) Integrity assessment of masonry arch bridges using the dynamic stiffness technique. NDT & E International, 28 (6), pp 367–375

Armstrong DM, Sibbald A, Fairfield CA & Forde MC (1995b) Modal analysis for masonry arch bridge spandrel wall separation identification. NDT & E International, 28 (6), pp 377–386

Astudillo R (1996) Use of nondestructive methods for bridge auscultation in Spain. Recent Advances in Bridge Engineering. Evaluation, Management and Repair. Proceedings of the US-Europe Workshop on Bridge Engineering, organized by the Technical University of Catalonia and the Iowa State University, Barcelona, 15–17 July 1996, International Center for Numerical Methods in Engineering CIMNE, pp 202–213

BAM (2006a) http://www.bam.de/service/publikationen/zfp_kompendium/geraete/g050/g050_L.html 2006

BAM (2006b) http://www.ndt.net/article/dgzfp01/papers/p45/fig4_6.htm

Bensalem A, Ali-Ahmed H, Fairfield CA & Sibbald A (1998) NDT as a tool for detection of gradual safety factor deterioration in loaded arches. Arch Bridges: History, Analysis, Assessment, Maintenance and Repair. Proceedings of the Second International Arch Bridge Conference, A. Sinopoli (Ed), Venice 6.–9. October 1998, Balkema, Rotterdam, pp 271–277

Berndt E & Schöne I (1990) Tragverhalten von Natursteinmauerwerk aus Elbesandstein. Sonderforschungsbereich 315, Universität Karlsruhe, Jahrbuch 1990

Bień J & Kamiński T (2007) Damages to masonry arch bridges – proposal for terminology unification. Proceedings of the Arch 07 – 5th International Conference on Arch Bridges, PB Lourenço, DV Oliveira & A Portela (Eds), 12–14 September 2007, Madeira, pp 341–348

Bién J, Kamiński T & Trela C (2008) Numerical analysis of old masonry bridges supported by field tests. Bridge Maintenance, Safety, Management, Health Monitoring and Informatics, H Koh & DM Frangopol (Eds), Taylor & Francis Group, London, pp 2384–2391

Binda L & Saisi A (2001) Non destructive testing applied to historic buildings: The case of some Sicilian Churches. Historical Constructions, PB Lourenço & P Roca (Eds), University of Minho, Guimarães, pp 29–46

Bolle G (2005) Dauerüberwachung am Chemnitztalviadukt – eine innovative Lösung zur Datenanalyse und Zustandsbewertung. 3. Symposium Experimentelle Untersuchung von Baukonstruktionen, Technische Universität Dresden

Budelmann H (1997) Zustandserfassung und -beurteilung historischer Mauerwerke. Konsolidierung von historischem Natursteinmauerwerk, 6–7. November 1997, Technische Universität Braunschweig, Institut für Baustoffe, Massivbau und Brandschutz Heft 135, Braunschweig, pp 53–66

Bungey, J (1997) Non-destructive testing in civil engineering. In: Proceedings of the International Conference in Civil Engineering, JH Bungey (Ed), Liverpool, 08.–11. April 1997, The British Institute of NDT, Northampton 1997, Vol. 1 and 2

Burkert T & Steffens K (2008) Instandsetzung und Ertüchtigung von Mauerwerk, Teil 7: Experimentelle Bestimmung der Tragfähigkeit von Mauerwerk, Belastungsversuche an Mauerwerksbauten in situ, Mauerwerk-Kalender 2009, W Jäger (Ed), Ernst und Sohn, Berlin

Cameron J, Demoulin C, Diamanti N, Giannopoulos A & Forde MC (2008) Aspects of GPR testing of masonry arch bridges. Structural Faults + Repair-2008: 12th International Conference, MC Forde (Ed), Edinburgh, 10th–12th June 2008

Clark MR, Parsons P, Hull T & Forde MC (2003a) Case study of radar laboratory work on the masonry arch bridge. Extending the Life of Bridges, Concrete and Composites Buildings, Masonry and Civil Structures, MC Forde (Ed), The Commonwealth Institute, London, 1st–3rd July 2003

Clark MR, Parsons P, Hull T & Forde MC (2003b) Laboratory study of infrared thermography on masonry arch bridges. Extending the Life of Bridges, Concrete and Composites Buildings, Masonry and Civil Structures, MC Forde (Ed), The Commonwealth Institute, London, 1st–3rd July 2003

Colla C (1997) NDT of masonry arch bridges, PhD Thesis, The University of Edinburgh, Department of Civil and Environmental Engineering, Edinburgh

Colla C, Das, PC, McCann D & Forde MC (1997) Sonic, electromagnetic and impulse radar investigation of stone masonry bridges. NDT&E International, 30 (4), pp 249–254

Corps of Engineers (2002). Nondestructive Evaluation Techniques, Chapter 10, TM 5-809-3/NAVFAC DM-2.9/AFM 88-3

Cote P (1996) NDT applied to concrete and masonry bridges. Recent Advances in Bridge Engineering. Evaluation, Management and Repair. Proceedings of the US-Europe Workshop on Bridge Engineering, JR Casas, FW Klaiber & AR Marí (Eds), Organized by the Technical University of Catalonia and the Iowa State University, Barcelona, 15–17 July 1996, International Center for Numerical Methods in Engineering CIMNE, pp 192–201

DIN 4022 (1987) Part 1–3: Baugrund und Grundwasser; Benennen und Beschreiben von Boden und Fels; Schichtenverzeichnis für Bohrungen mit/ohne durchgehende Gewinnung von gekernten Proben im Boden und Fels (Fest- und Lockergestein), September 1987

Domède N & Sellier A (2008) Assessment method of masonry arch bridges. Structural faults + Repair-2008: 12th International Conference, MC Forde (Ed), 10th–12th June 2008, Edinburgh, UK

Ehmann D (2000) Untersuchung und Erprobung der zielmarkierungsfreien Deformationsmessung an Bauwerken im Nahbereich. Diplomarbeit, HfT Stuttgart, Fb. Vermessung und Geoinformatik

Fanning PJ & Boothby TE (2003) Experimentally-based assessment of masonry arch bridges. Bridge Engineering, 156 (3), September 2003, pp 109–116

Forde MC (1996) Non-destructive evaluation of bridges: research in progress. Recent Advances in Bridge Engineering. Evaluation, management and repair. Proceedings of the US-Europe Workshop on Bridge Engineering, JR Casas, FW Klaiber, AR Marí (Eds), Organized by the Technical University of Catalonia and the Iowa State University, Barcelona, 15–17 July 1996, International Center for Numerical Methods in Engineering CIMNE, pp 127–152

Franke I, Deckelmann G & Goretzky W (1991) Einfluss der Streubreite der Materialeigenschaften auf die Tragfähigkeit von Mauerwerk – Theoretische und Experimentelle Untersuchungen. In: Proceedings of the 9th International Brick/Block Masonry Conference, 13.–16. October 1991, Vol. 1, Berlin, Germany, pp 180–187

Goretzky W (2000) Tragfähigkeit druckbeanspruchten Mauerwerks aus festigkeits- und verformungsstreuenden Mauersteinen und -mörteln. Dissertation GCA Verlag Herdecke

Gutermann M & Steffens K (2005) Belastungsversuche zur Bausubstanzerhaltung. 3. Symposium Experimentelle Untersuchung von Baukonstruktionen, Technische Universität Dresden

Gutermann M (2002) Ein Beitrag zur experimentell gestützten Tragsicherheitsbewertung von Massivbrücken. Dissertation, Technische Universität Dresden

Hampel U & Maas H-G (2003) Application of digital photogrammetry for measuring deformation and cracks during load tests in civil engineering material testing. In: 6th Conference on Optical 3-D Measurement Techniques. Prof. A. Grün, Institute of Geodesy and Photogrammetry, Swiss Federal Institute of Technology, Zürich, 22–25.9.2003, Vol. II, pp 80–88

Helmerich R & Niederleithinger E (2006) NDT-toolbox for the Inspection of Railway Bridges. Sustainable Bridges Project – Assessment for Future Traffic Demands and Longer Lives.

Helmerich R, Niederleithinger E, Trela C, Bien J & Bernardini G (2008) Complex multi-tool inspection of a masonry arch bridge using non-destructive testing. Bridge Maintenance, Safety, Management, Health Monitoring and Informatics, H Koh & DM Frangopol (Eds), Taylor & Francis Group, London, pp 3716–3723

Hughes TG & Pritchard R (1998) In situ measurement of dead and live load stresses in a masonry arch. Engineering Structures, 20 (1–2), pp 5–13

Illich B (1999) Zerstörungsfreie Zustandserfassung an Mauer- und Betonbauteilen mit Radar. DGZfP-Berichtsband 66-CD, Vortrag 9, 21.–22. Januar 1999 in der Neuen Messe München. pp 89–93

Inaudi D & Glisic B (2008) Overview of 40 bridge monitoring projects using fiber optic sensors. Bridge Maintenance, Safety, Management, Health Monitoring and Informatics, H Koh & DM Frangopol (Eds), Taylor & Francis Group, London, pp 2514–2521

Kahle M & Illich B (1992) Erkundung historischen Mauerwerks mit Radar. 3. Internationales Kolloquium für Werkstoffwissenschaften und Bausanierung, Ehningen, pp 379–393

Kaplan ME (1997) Non-destructive evaluation techniques for masonry construction (Masonry), U.S. Dept. of the Interior, National Park Service, Cultural Resources, Heritage Preservation Services

Kirtschig K (1991) Zur Beschreibung der Mauerwerksdruckfestigkeit über die mittlere oder charakteristische Steindruckfestigkeit. In: Proceedings of the 9th International Brick/Block Masonry Conference, 13.–16. October 1991, Vol. 1, Berlin, Germany, pp 196–201

Köpp C, Maierhofer C & Wendrich A (2005) Entwicklung zerstörungs-freier und zerstörungsarmer Prüfverfahren zur Untersuchung historischen Mauerwerks. Mauerwerk 9, Heft 3, pp 102–107

Leaird JD (1984) A report on the pulsed acoustic emission technique applied to masonry, Journal of Acoustic Emission, 3 (4), pp 204–210

Lourenço PB (2001) Analysis of historical constructions: From thrust-lines to advance simulations. Historical Constructions, PB Lourenço & P Roca (Eds), University of Minho, Guimarães, pp 91–116

Maierhofer C, Wendrich A & Köpp C (2003) ONSITEFORMASONRY – A European Research Project: On-site investigation techniques for the structural evaluation of historic masonry, DGZfP (Ed) International Symposium Non-Destructive Testing in Civil Engineering (NDT-CE) in Berlin, Germany, September 16–19, 2003, Proceedings on BB 85-CD, V70, Berlin

Mildner K & Mildner A (2001) Bauwerksmessungen vor und nach der Instandsetzung einer Eisenbahn-Gewölbebrücke und statische Modellbildung. Ehrenkolloquium Prof. Dr.-Ing. Möller zum 60. Geburtstag, Lehrstuhl für Statik und Dynamik der Tragwerke, Technische Universität Dresden, 2001, Dresden

Mildner K (1996) Tragfähigkeitsermittlung von Gewölbebrücken auf der Grundlage von Bauwerksmessungen. 6. Dresdner Brückenbausymposium, 14. März 1996, Fakultät Bauingenieurwesen, Technische Universität Dresden, pp 149–162

Mönicke H-J (2003) Profil- und flächenhafte Objektvermessung mit Tachymeter oder 3D-Laser-Scanner. In: Ingenieurblatt für Baden-Württemberg, Heft 4/2003, pp 102–106

Müller K-H & Garke R (2005) Auswertung von Ultraschallmessungen unter Berücksichtigung von Vorinformationen. Revitalisierung von Bauwerken. Schriften der Bauhaus-Universität, Heft 117, pp 33–40

Niederleithinger R et al. (2006) Guideline for Condition Assessment and Inspection of Railway Bridges. Sustainable Bridges Project – Assessment for Future Traffic Demands and Longer Lives

O'Connor A, Pedersen C, Enevoldsen I, Gustavsson L & Hammarbäck J (2008) Probability based assessment of a large riveted truss railway bridge. Bridge

Maintenance, Safety, Management, Health Monitoring and Informatics, H Koh & DM Frangopol (Eds), Taylor & Francis Group, London, pp 1812–1819

Orbán Z, Yakovlev G & Pervushin G (2008) Non-Destructive Testing of masonry arch bridges – an overview. Bautechnik (85), Heft 10, pp 711–717

Ponniah DA & Prentice DJ (1999) Long term monitoring of fill pressure in a new brickwork arch bridge. Construction and Building Materials 13, pp 159–167

Prader J, Weidner JW, Hassanain H, Moon F, Aktan E, Jalinoos F, Buchanan B & Ghasemi H (2008) Load testing and analysis of bridges missing critical documentation. Bridge Maintenance, Safety, Management, Health Monitoring and Informatics, H Koh & DM Frangopol (Eds), Taylor & Francis Group, London, pp 1877–1885

Prieto F, Martin M & Viola M (2006) Application of laser vibromety analysis to damage detection in masonry structures. Heritage, Weathering and Conservation, R Fort, M de Buergo, M Gomez-Heras & C Vazquez-Calvo (Eds), Taylor & Francis Group, London, Volume 2, pp 627–632

Prokop A (2007) The application of terrestrial laser scanning for snow and avalanche research. Universität für Bodenkultur Wien

Proske D (2003) Ein Beitrag zur Risikobeurteilung alter Brücken unter Schiffsanprall. Dissertation, Technische Universität Dresden

Rücker W, Hille F & Rohrmann R (2006) Guideline for the Assessment of Existing Structures. SAMCO Final Report 2006

Schubert F, Köhler B & Lausch R (2002): Zerstörungsfreie Ultraschallprüfung von Betonbauteilen – Impulsecho und Impactechomethode in Theorie und Praxis. 2. Symposium Experimentelle Untersuchungen von Baukonstruktionen – Ehrenkolloquium Prof. Dr.-Ing. habil. Heinz Opitz. Heft 17 der Schriftenreihe des Institutes für Tragwerke und Baustoffe, Fakultät Bauingenieurwesen, Technische Universität Dresden, pp 25–34

Schueremans L & Van Gemert D (2001) Assessment of existing masonry structures using probabilistic methods – state of the art and new approaches. STRUMAS 2001 18.–20. April 2001, Fifth International Symposium on Computer Methods in Structural Masonry

Schueremans L, Van Rickstall F, Ignoul S, Brosens K, Van Balen K & Van Gemert D (2003) Continuous Assessment of Historic Structures – A State of the Art of applied Research and Practice in Belgium. KULeuven, Department of Civil Engineering

Schulze M & Hampel U (2004) Erfassung und Analyse von Prüfkörpern in der Materialprüfung durch Computertomographie. In: Publikationen der Deutschen Gesellschaft für Photogrammetrie, Fernerkundung und Geoinformation. Band 13, Halle (Saale), Germany, 2004, pp 317–324

Silman R & Ennis, M (1993) Non-destructive evaluation to document historic Structures, International Association for Bridge and Structural Engineering (IABSE), Zurich, pp 195–203

Slowik V (2004) Combining experimental and numerical methods for the safety evaluation of existing concrete structures, Fracture Mechanics of Concrete Structures, VC Li, CKY Leung, KJ Willam & SL Billington (Ed), Proceedings of Framcos-5, April 12–16, 2004, Vail, USA, IA-FraMCoS, Volume 2, pp 701–707

Slowik V, Fiedler LD & Kapphahn G (2005) Zur kombinierten Anwendung experimenteller und rechnerischer Methoden bei der Tragsicherheitsbewertung bestehender Massivbauwerke, 3. Symposium Experimentelle Untersuchungen von Betonkonstruktionen, Technische Universität Dresden, Institut für Massivbau, 23. Juni 2005, Tagungsband, pp 157–168

Steffens K (2001) Experimentelle Tragsicherheitsbewertung von Bauwerken. Grundlagen und Anwendungsbeispiele. Ernst und Sohn Verlag

Stiglat K (1984) Zur Tragfähigkeit von Mauerwerk aus Sandstein. Bautechnik 2/1984, pp 54–59 and Bautechnik 3/1984, pp 94–100

VanMarcke E (1997) Quantum Origins of Cosmic Structure. Balkema, Rotterdam

Vockrodt H-J & Schwesinger P (2002) Experimentelle Tragfähigkeitsanalyse einer historischen Bogenbrücke in Erfurt, Bautechnik 79, Heft 6, pp 355–367

Wendrich A, Maierhofer C, Kuritz M & Köpp C (2004) Aufbau eines historischen Mauerwerkskörpers zur Bewertung von zerstörungsfreien Verfahren. Jahrestagung 2004 Zerstörungsfreie Materialprüfung. Kurzfassungen der Vorträge und Plakatbeiträge. Salzburg, 17.–19. Mai 2004. Deutsche, Österreichische und Schweizerische Gesellschaft für Zerstörungsfreie Prüfung

Wenzel F & Kahle M (1993) Indirect methods of investigation for evaluating historic masonry, International Association for Bridge and Structural Engineering (IABSE), Zürich 1993, pp 75–90

Wenzel F (1997a) Behutsame Vorgehensweise bei der Mauerwerksinstandsetzung. Tagung Konsolidierung von historischem Natursteinmauerwerk, 6.–7. November 1997, Technische Universität Braunschweig, Institut für Baustoffe, Massivbau und Brandschutz Heft 135, Braunschweig, pp 1–11

Wenzel F (1997b) (Ed) Mauerwerk – Untersuchen und Instandsetzen durch Injizieren, Vernadeln und Vorspannen. Erhalten historisch bedeutsamer Bauwerke. Empfehlungen für die Praxis. Sonderforschungsbereich Karlsruhe. Universität Karlsruhe

Wiggenhauser H & Maierhofer C (2002) Zerstörungsfreie Bauwerksuntersuchungen mit Radar. 2. Symposium Experimentelle Untersuchungen von Baukonstruktionen – Ehrenkolloquium Prof. Dr.-Ing. habil. Heinz Opitz. Heft 17 der Schriftenreihe des Institutes für Tragwerke und Baustoffe, Fakultät Bauingenieurwesen, Technische Universität Dresden, pp 71–78

6 Damages and Repair

6.1 Introduction

Aging is a common phenomenon in biological systems. We do not consider it as an illness, but as a normal process. It is defined as a decrease of the capability of the organism to cope with the requirements of the environment with increasing age. Aging can also be related to decrease of reserves (Strasser 2006).

Such a process can also be found in the case of technical products. The engineer has to consider the aging of structures during the design process, if the safety requirements are to be fulfilled over the entire lifetime of the structure. Design life spans for certain structures are shown in Table 6-1. So, the structure should function well, not only if it is new but also at the end of its lifetime. This is sometimes called "graceful degradation". We do not only desire it for humans, it is also a desired criterion for technical products.

Table 6-1. Design work time of concrete structures according to the Eurocode 1

Design life in year	Example
1 ... 10	Temporary structures
10 ... 25	Replaceable structural parts, e.g. gantry girders and bearings
15 ... 30	Agricultural used structures
50	Buildings and other common used structures
100	Monument building structures, bridges and other structures

D. Proske, P. van Gelder, *Safety of Historical Stone Arch Bridges*, DOI 10.1007/978-3-540-77618-5_6,
© Springer-Verlag Berlin Heidelberg 2009

6.2 Damages on Historical Arch Bridges

6.2.1 Overview

For reinforced concrete structures, lifetime restricting loads are clearly identified such as carbonation, chloride attack, sulphate attack, and Frost-Thaw (DIN 1045-1 2001). Also for historical masonry arch bridges, such durability loads have been identified. They usually yield to changes of properties of the structure and furthermore to damages. A definition of the term "damage" can be found in Chapter 7. Such changes and damages can be found on many historical arch bridges. However, that is not mainly because these types of structure have been designed so weak, but because many of these bridges are quite old.

A rough list of damages was given by Bién and Kamiński (2004). They list the following damages:

- Incompatible deformations (deformations which yield to changes of the initial geometry)
- Destruction of material caused either by chemical or by physical processes
- Material discontinuities (cracks)
- Loss of material (falling stones)
- Damage on auxiliary elements (damaged sealing)
- Deformation damages (deformation on the structure that does not yield to a change of the initial geometry, for example sliding spandrel walls)
- Contamination (natural cover, besmirch)

Typical damage patterns for arch bridges were shown by Angeles-Yáñez and Alonso (1996), as shown in Fig. 6-1. A classification of damage patterns for historical stone arch bridges of the European railway organizations was also given by Orbán (2004) and is summarized in Table 6-2. The latest catalogue of damages was presented by Bień and Kamiński (2007) in relation to degradation processes and damages (Table 6-3). Mildner (1996) has also mentioned typical damages on masonry arch bridges.

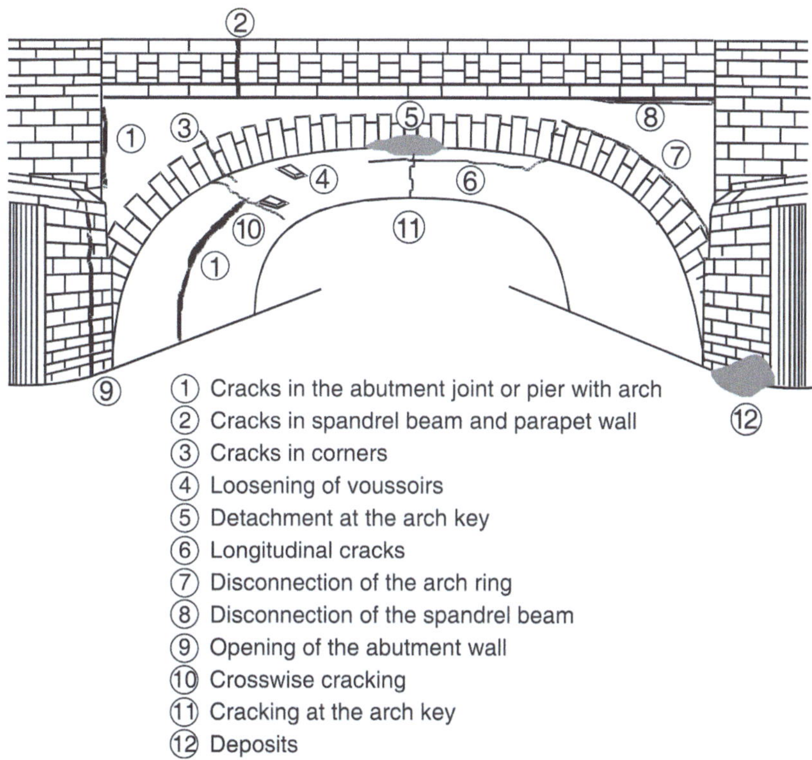

Fig. 6-1. Most frequent damages found on arch bridges according to Angeles-Yáñez and Alonso (1996)

Table 6-2. Types of damages at arch bridges of railways organizations and their frequency according to Orbán (2004)

Nr.	Type of damage[1]	Frequency[2]
1	Damage at sealing[3]	2.1
2	Deterioration of material	2.4
3	Separation and movement of wing wall	3.0
4	Separation and movement of spandrel wall	3.5
5	Damages at piers, foundation and skewback	4.0
6	Geometrical problems with the structure	4.0
7	Other problems[4]	4.0
8	Cracks in arch caused by settlement	4.2
9	Damages at the road crossing construction	4.3
10	Damages caused by overload	4.3

Nr.	Type of damage[1]	Frequency[2]
11	Deformation	4.4
12	Cracks in arch caused by overload	4.5
13	Damages at the parapet caused by single loads	4.6

[1] In general, in many cases, the cause of the damage cannot be identified.
[2] Calculated as mean value based on information provided by the different railway organizations. The numbers represent the following:
1 = Very frequent = about 50% of all bridges
2 = Frequent   = about 25% of all bridges
3 = Occasional   = about 10% of all bridges
4 = Rare   = about   5% of all bridges
5 = Exceptional = less than 5% of all bridges
[3] Many historical arch bridges were built without sealing. But of course, damages caused by water can be found there. These bridges have therefore been added to this statistic.
[4] Other problems include damages caused by plants, damages caused by earthquakes, impacts and wrong maintenance.

Table 6-3. Degradation mechanisms subject to the damages to masonry bridges (Bién and Kamiński 2007)

Degradation mechanism	contamination	deformation	destruction	discontinuity	displacement	Loos of material
Physical						
Effects of high temperature	X	X	X	X		X
Fatigue		X	X	X	X	
Freeze-Thaw		X	X	X	X	
Change of foundation conditions		X		X	X	
Overloading		X	X	X	X	
Shrinkage		X	X			
Water penetration	X		X			X
Chemical						
Carbonation	X		X			X
Crystallization	X		X	X		X
Leaching	X		X	X		X
Salt and acid actions	X		X	X	X	X
Biological						
Accumulation of contamination	X		X	X		X
Living organisms activities	X		X	X	X	X

6.2.2 Recent Collapses of Historical Arch Bridges

Recent failures of arch bridges were often related to accidental loads. For example, the historic Pöppelmann arch bridge in Grimma was heavily damaged during the 2002 flooding of the river Mulde (Fig. 6-2). The bridge had to be blasted afterwards since reconstruction using the remaining parts was not possible (Curbach et al. 2003a). Another example was the failure of the arch bridge in Benairbeig over the Rio Girona in Spain due to a flashflood called Gota Fría in October 2007 (Meyer 2007). The failure was actually filmed because television was reporting about the flashflood onsite. The movies are visible on YouTube. A further example was the flood-related failure of a farm track and public footpath masonry arch bridge over the river Devon in 2007 (Bottesford Living History 2007). Ural et al. (2008) also mentions the failure of Turkish arch bridges by floods.

Fig. 6-2. Pöppelmann Bridge in Grimma after the flooding in 2002

Flooding and ice loads often caused failures of historical arch bridges. Drdácký and Slízková (2007) report on the repeated damages and partial failures of the historical Charles Bridge in Prague caused by flooding in 1359, 1367, 1370, 1373, 1374, 1432, 1496, 1503, 1655, 1784, 1890, and in 2002.

Furthermore, not only are historical bridges exposed to flooding but also to all gravity-driven mass movements such as debris flows, rock falls, and avalanches.

Examples of debris flow impacts against historical arch bridges can be found in Proske (2009). For example, in Log Pod Mangartom in Slovenia, a huge debris flow killed several people, and destroyed houses and also one historical arch bridge. The arch bridge over the Lattenbach in Austria is regularly exposed to debris flow impacts. The last overflow occurred in September 2008. Ural et al. (2008) mention fluvial mass transport including dead wood as cause of an arch bridge failure. A research project to develop load design procedures for bridges under debris flow impacts has been submitted by the first author of this book. Explicit examples of arch bridge damages or failures either due to rock falls or due to avalanches are not known, however in general the failure of bridges due to such loads is well-known (Proske 2009).

Besides natural accidental loads, technical load may also be applied to arch bridges such as car, railway or ship impacts, or bombardment. The problem of ship impacts against arch bridges has been intensively discussed in Proske (2003). Further discussion of arch bridge failures can be found in Ural et al. (2008).

Some further examples from the last few decades are also mentioned. The first example is the partial failure of the Molins de Rei bridge close to Barcelona, Spain on 7th February 1971 and on 1st January 1972 (Troyano 2003). Pictures of the structure after the failure are shown in Troyano (2003). On 9th April 1978, 6 of the 15 arches of the Wilson Bridge in Tours, France collapsed (Troyano 2003, Rombock 1994). A further example was the failure of the Westminster Bridge in Humberside Country 1983 (Tingle and Heelbeck 1995).

Besides structural damages, the building material itself can also be damaged. Masonry, as already mentioned, is a multi component material. The damages can therefore effect single elements alone, such as the mortar or the stone, or they can effect the masonry. Translation of terms related to masonry can be found on the ICOMOS (2008) web page or in Bau.de (2008).

6.2.3 Weathering of the Mortar

Mortar, the joint material of masonry, usually exhibits a much lower lifetime and strength than the natural stones or masonry bricks. Natural stones can reach hundreds or thousands of years of lifetime still keeping their strength and showing only minor weathering effects.

However, the low weathering resistance of some mortar types can also influence the stone material. The breakout of the mortar material enables the penetration of humidity and acceleration of a further weathering of the remaining mortar and stone or brick material. Figure 6-3 shows the principal consequences of wrong pointing application and mortar weathering.

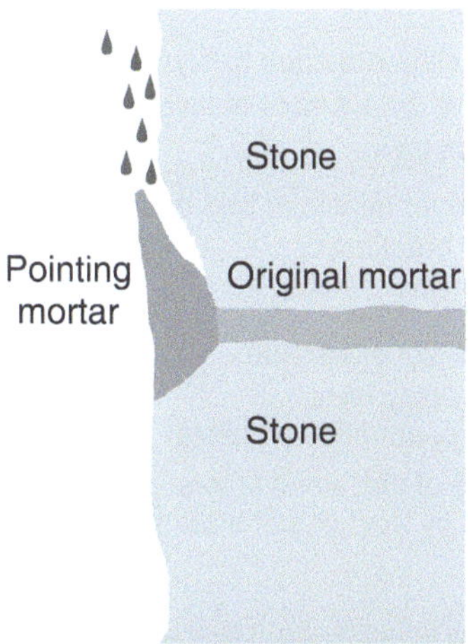

Fig. 6-3. Example of wrong application of pointing mortar in joints (Bartuschka 1995)

6.2.4 Spalling and Contour Scaling

Near-to-surface damages on natural stones are usually spallings and contour scaling (Fig. 6-4). Based on the penetration depth, such damages are classified into the following areas:

- Chipping of stone material either in convex or in concave shape
- Flaking or exfoliation in thin layers
- Spalling or detachment of crusts with stone layers of more than 10 mm

The cause of such damages is manifold. Classical weathering, loss of binding agent, cracks caused by frost, bacteria as nitrificants, or salt attack. These issues are widely discussed in literature such as Sauder and Wiesen (1993), Beeger (1992), Poschlod (1990), Weiss (1992), or Bläuer (1992).

A very interesting example of spalling was presented by Mann. He reported on spalling of masonry in a tunnel. The spalling was caused by the smoke gas of the steam locomotive, which caused a chemical reaction of the mortar towards gypsum.

A further example on weathered masonry surfaces can be found in Garrecht (1997).

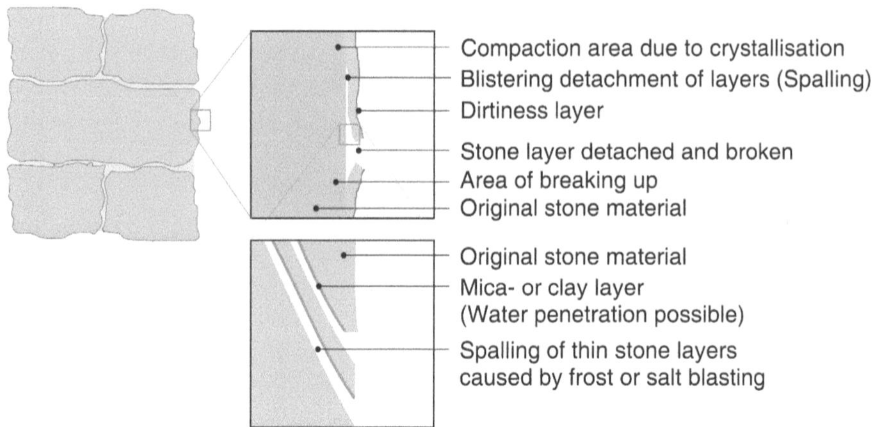

Fig. 6-4. Spalling and contour scaling (Bartuschka 1995)

6.2.5 Salt Attack

Salts that are able to damage structural material can be characterized in many cases by their water solubility. Besides the solubility, salts can also damage by hygroscopic water absorption. Here, salts feature a blasting effect. This blasting effect is caused by an increase in volume during the changing of moist and dry crystalline phases of the salt. If the pore system is already saturated, then the crystallization pressure can damage the structural material. Besides the crystallization pressure, hydration pressure can also be observed. Water is then chemically attached to the salt in certain temperature regions. Also, this process is characterized by a volume increase. Detailed information about certain crystallization and hydration pressures subject to saturation grade can be found in Weber (1993). A short summary of damaging salts is given in Table 6-4.

Table 6-4. Summary of different building material damaging salts according to Weber (1993)

Class of chemical compound		Name
Sulphate compounds	$MgSO_4 \cdot 7\,H_2O$	Acrid salt
	$CaSO_4 \cdot 2\,H_2O$	Gypsum, calcium sulphate
	$Na_2SO_4 \cdot 10\,H_2O$	Sodium sulphate
Nitrate compounds	$Mg(NO_3)_2 \cdot 6\,H_2O$	Magnesium nitrate
	$Ca(NO_3)_2 \cdot 4\,H_2O$	Calcium nitrate
	$5\,Ca(NO_3)_2 \cdot 4\,NH_3NO_3 \cdot 10\,H_2O$	
Chloride compounds	$CaCl_2 \cdot 6\,H_2O$	Calcium chloride
	$NaCl$	Common salt, sodium chloride
Carbonate compounds	$Na_2CO_3 \cdot 10\,H_2O$	Sodium carbonate
	K_2CO_3	Potash, Calcium carbonate

In comparison to damage caused only by humidity and wetness, usually the damage by the salts is greater. However, the salts require humidity and water as a transport medium and therefore mixed damages are common (Weber 1993).

Damages related to humidity and salt are the following:

- Frost damage
- Spalling caused by hydraulic swelling and shrinkage
- Crystallization damage by salts
- Hydration damage by salts
- Frost-thaw damage
- Binding material reaction caused by acid exhausts
- Damage caused by microorganisms

Because damage by salts can be simply avoided by water penetration exception, hydrophobicity of stones not only prevents water damage but also salt damage. Hydrophobicity decreases the capillary suction capability of materials. Most construction materials such as masonry or concrete suck water on the surface. The wetting angle of contact is zero. Hydrophobicity increases the wetting angle up to 90° or 180°. Since the capillary suction is proportional to the cosine of the wetting angle, this yields to a cancellation of capillary suction. However, this does not mean that the material is sealed. If water with pressure is applied, this water can penetrate the material.

6.2.6 Chemical Weathering

Chemical weathering describes the natural transformation process of masonry material subject to different chemical reactions. Usually, moisture and humidity are common requirements for such chemical reactions.

With the assimilation of moisture from the environment, usually other chemicals are assimilated such as sulphur dioxide, nitric oxide, or carbon dioxide. These elements then form acids and bases that solve the binding material of the stones and mortar. The loss of binding material yields to spalling, contour scaling, chipping, and flaking (Bartuschka 1995).

6.2.7 Biological Weathering

Biological weathering describes the damaging of structural materials or structures by biological processes. Such biological processes can be microorganisms, moss growth, or the growth of plants and trees. Such organisms either cause some chemical reactions or are able to introduce stresses and forces inside the structural elements.

Mattheck et al. (1993) have measured the compression and tensile strength of tree roots and found maximum compression stresses of up to 0.7 MPa, and tensile stresses of up to 50 MPa in longitudinal direction of the tree root. Müller (2005) and Bauriegel (2004) investigate the strength of trees under certain types of loadings on normal land. Although such tests do not reflect in detail the conditions discussed here, they give a good impression about the load transfer into roots. Garston (1985) has investigated the influence of old trees to houses. Mattheck et al. (1993) give a good example about the load-bearing capabilities of tree roots: in 1993 in a northern German city, a tree root lifted up a gas pipe. This caused a gas explosion.

6.2.8 Mechanical and Physical Weathering

Physical weathering is based on some physical properties of the construction material. For example, the coefficient of thermal expansion can yield to different strains, causing different compression or tensile stresses inside the material. If the stresses exceed the strength, then the material will crush or crack. Such cracks can accelerate the weathering in combination with water and salt penetration. However, physically caused damages are generally of minor importance for historical arch bridges (Bartuschka 1995).

6.2.9 Deformations

Deformation of elements or the entire structure is required for structures to perform. However, if the magnitude of the deformations is too high, deformations can be considered as damage. The line between common deformations and damage is difficult to find under practical conditions. For example, arch bridges have already shown significant deformations after the destruction of the falsework. Early bridges in the 18th century showed vertical deformations in the crown of more than a 1 cm per m span. The spandrel walls were often completed half a year after the finishing of the arch. At that time, nearly 1/3 of the creep deformation of the arch had already occurred. Bridges constructed in the second part of the 19th century did not reach such high vertical deformations (usually between 0.1 and 0.4 mm/m). This can probably be related to an increased mortar quality (Brencich and Colla 2002).

A good example to illustrate deformations in arch bridges is the Syratal Bridge in Plauen, Germany. The deformation at the crown reached 55.57 cm in 1995. The bridge has a span of 90 m. It has been assumed that the crown deformation will reach 56.90 cm by 2070. This represents a ratio of nearly 0.57/90 = 6/1,000. The temporal development of the deformation is shown in Fig. 6-5. The cause of the high deformation is manifold. Figure 6-6 tries to relate certain causes to certain deformation values.

Perronet already knew about the great deformation of arch bridges. At the Neuilli Bridge in Paris (1782–1783), he measured a deformation of 0.7 cm per m after the destruction of the falsework. He assumed that this value represented about 60% of the overall assumed deformation. In the next 12 months, a further 30% of the overall deformation was observed and the final deformation was found after five years (Brencich and Colla 2002).

Weber (1999) gives a deformation of 66 mm for the Lavour Bridge after removal of the falsework. Harvey (2006) mentions deformation at the springing of 0.1 mm under traffic load. He indicates that stones can come off in the section of maximum traffic load.

According to Brencich and Colla (2002), high deformations of the arch can cause cracks in hidden vaults and in the spandrel walls. Such cracks yield to a separation, and therefore the ultimate load-bearing capacity of the bridge may be changed due to the changed interaction of the single structural elements.

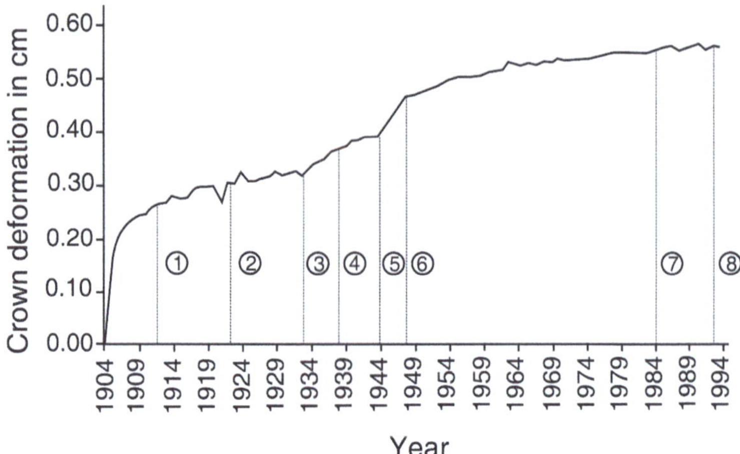

Fig. 6-5. Crown deformation of the Syratal Bridge, Plauen over time (Span 90 m) (Bartuschka 1995)

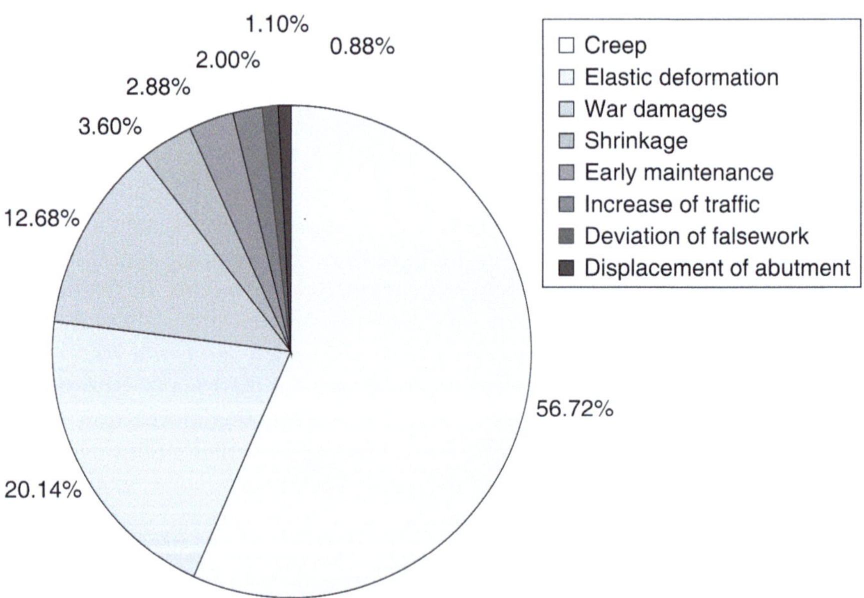

Fig. 6-6. Contribution of different causes to the crown deformation of the Syratal Bridge, Plauen (Bartuschka 1995)

6.2.10 Cracks

A crack is the linear physical disconnection of former homogenous body. Cracking is caused by the exceedance of the tensile strength of a material. Cracks are common phenomena in brittle, low tensile strength materials. Such materials are, for example, glass, natural stones or concrete. Single cracks are not necessarily a damage as seen with reinforced concrete. Here, the concrete has to crack to permit the steel reinforcement contribution in the load bearing. In such cases, the cracking is considered during the design process and only the crack size has to be limited.

Certain types of cracks in masonry are shown in Fig. 6-7. Such crack patterns are strongly related to the failure surfaces as discussed in Chapter 3. However, looking more towards masonry arch bridges, further classification of cracks seems to be useful.

Cracks in the arch or vault can be an indication of overload on the structure. Since the geometrical location of the crack permits further interpretation, certain types of arch bridge cracks are introduced in Fig. 6-8 and classified. The comments of the cracks are mainly taken from Bienert (1976), Bartuschka (1995), and the UIC-Codex (1995).

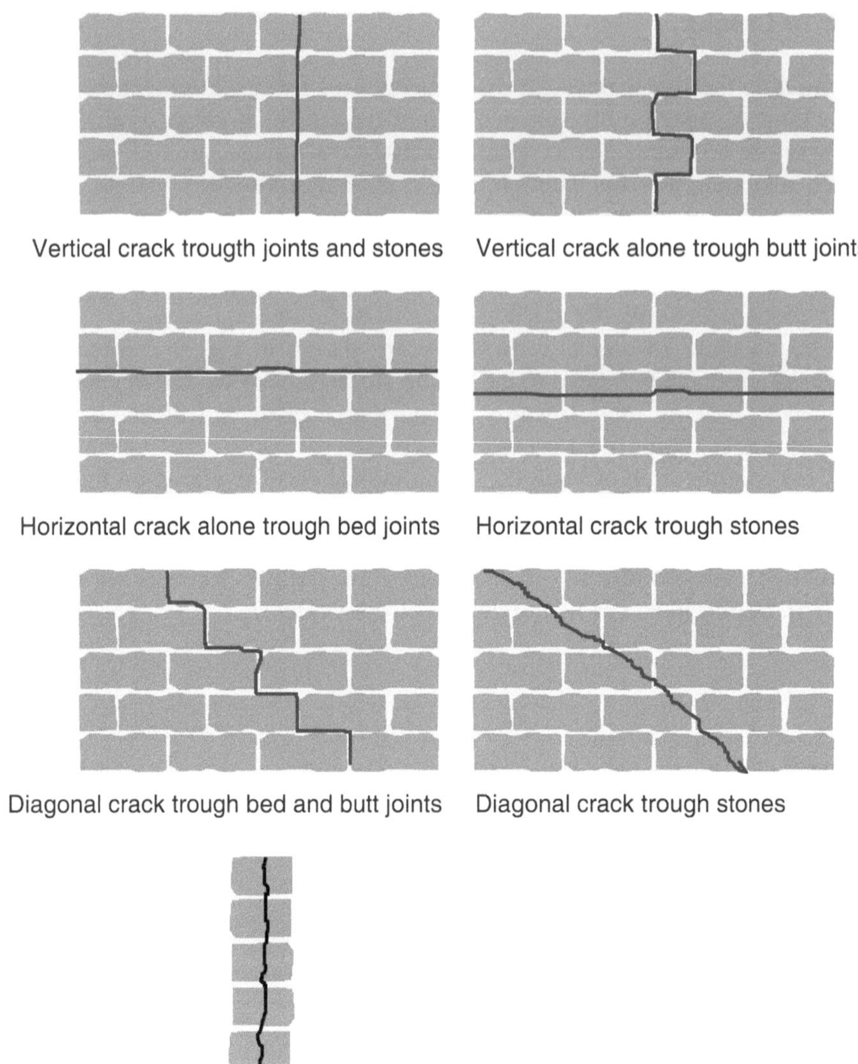

Fig. 6-7. Typical crack patterns in masonry walls according to Al Bosta (1999), Jäger (2006), and Walthelm (1990, 1991)

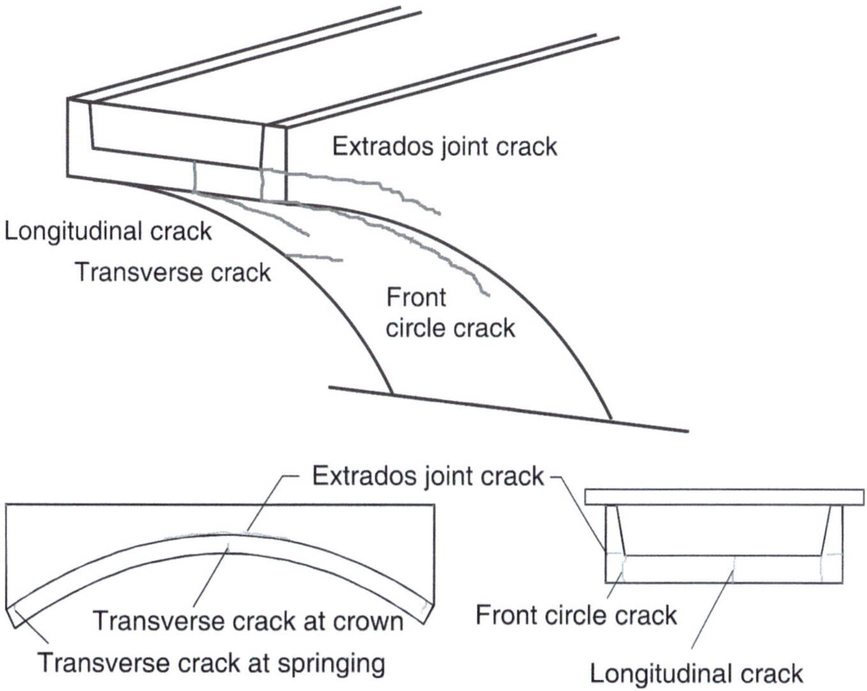

Fig. 6-8. Typical crack types in masonry arch bridges according to Bienert (1976)

6.2.10.1 Longitudinal cracks

Longitudinal cracks run parallel with the span of the bridge. They can either run over the entire span of the bridge or cover parts of the bridge. Since cracks always indicate tensile forces rectangular to the crack direction, longitudinal cracks indicate tensile forces in transversal direction of the bridge. Such tensile forces can be caused by one-sided settlement of the bridge, which is the main cause, while transversal bending can be caused by one-sided traffic load on wide arches or vaults with several lanes or high shrinkage stresses on wide arches. Longitudinal cracks often indicate damage to the sealing. In general, longitudinal cracks do not indicate an immediate threat to the ultimate load-bearing capacity of the bridge. Longitudinal cracks are, for example, mentioned in Boothby et al. (2004).

6.2.10.2 Front circle cracks

Front circle cracks or front ring cracks are a special type of longitudinal cracks in arch bridges. Front circle cracks are located directly behind the spandrel wall in the arch or vault. Sometimes they reach down to the piers or abutment. Usually, the causes are different: stiffnesses of the spandrel wall and the backfill, high traffic loads which cause movements of the spandrel wall, or moisture penetration caused by damaged sealing. In contrast to the good-natured longitudinal cracks, front circle cracks may cause a distinct change of the load-bearing behaviour of the arch bridges because the spandrel wall separates from the arch. Usually, the spandrel walls contribute strongly to the load bearing of the arch and therefore the loss of this contribution has strong effects on load-bearing behaviour. The question then arises whether the spandrel wall has been considered in the static computation of the arch bridge (Bartuschka 1995).

6.2.10.3 Extrados joint crack

Extrados joint cracks or arch back cracks are longitudinal cracks located at the back side of the arch. They also yield to a separation of the spandrel wall from the arch. However, the consequences are lower compared to the front circle cracks since they indicate an original weak interlook between the spandrel wall and the arch. Under such conditions, the load capacity of the spandrel wall cannot be considered anyway since the interaction between the spandrel wall and the arch was not realized during design and construction. Causes of such cracks can be differences of stiffness between arch and spandrel wall, differences of stiffness between arch and the backfill, shrinkage deformations, or high horizontal loading inside the backfill caused by traffic load or frost (Bartuschka 1995).

6.2.10.4 Transversal cracks

Transversal cracks run rectangular to the span of the arch. They occur mainly at the springing, at the quarter point of the arch or at the crown, and indicate the development of hinges in the arch (Fig. 6-9). Therefore, they are a serious sign of overloading of the arch. Further causes besides the overloading of the arch can be settlement, introduction of high single loads with low coverage, improper shape of the arch, or high shear stresses in the horizontal working joints of the backfill.

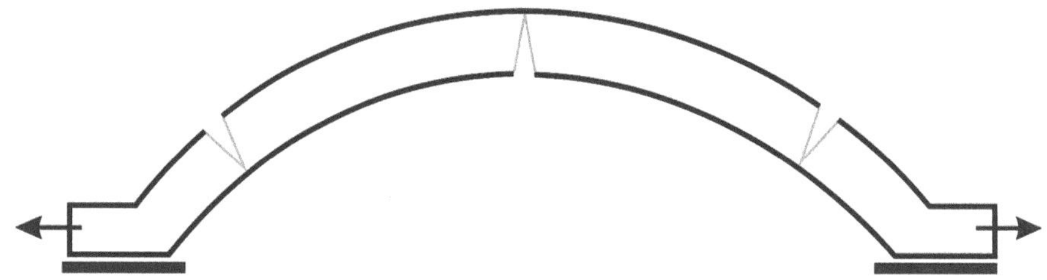

Fig. 6-9. Transversal cracks caused by horizontal movements of the support according to Como (1998)

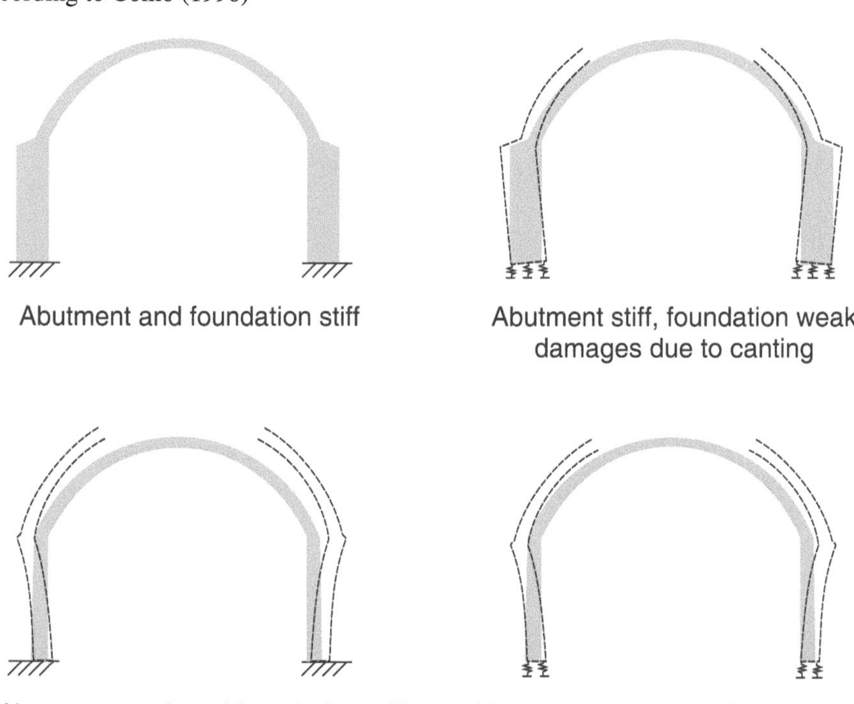

Fig. 6-10. Damages to vault constructions caused by abutment and foundation weakness (Bauriegel 2004)

An intensive discussion of loading of arches by horizontal displacement of the support can be found in Ochsendorf (2002) and Ochsendorf et al. (2004). A more general introduction to the consequences of settlement for historical structures was given by Bauriegel (2004). As an example, damages to vault constructions are shown in Fig. 6-10.

6.2.10.5 Diagonal cracks

Diagonal cracks appear rather seldom in arches and vaults. If such cracks are found, they have to be inspected. Causes for the development of the cracks can be local weakness of the masonry or unequal load distribution. The unintended load bearing of the spandrel wall at the springing can cause diagonal cracks in the spandrel wall (Bartuschka 1995).

6.2.10.6 Displaced stones

Sometimes displaced stones can be found in the arch. The reason why the stones have moved has to be investigated very carefully. In most cases, such moved stones occur on arch bridges with low coverage, high single loads, and low bound between the stones. Such stones can endanger traffic and humans under the bridges (Bartuschka 1995, Harvey 2006).

6.2.10.7 Special damage on spandrel walls

Spandrel walls and parapets can show some special types of damage. Melbourne (1991) has classified the damage as shown in Fig. 6-11. An example of sliding is shown in Fig. 6-12. Several examples of the complete failure of spandrel walls by overturning after earthquakes can be found in Rota (2004).

Como (1998) describes crack patterns in spandrel walls of arch bridges caused by settlement of the middle pier (Fig. 6-13). Fauchoux and Abdunur (1998) have indeed found such crack patterns at a bridge with pier settlement. They have repaired the bridge by removal of the backfill, needling of the arch masonry, and inserting a new concrete backfill.

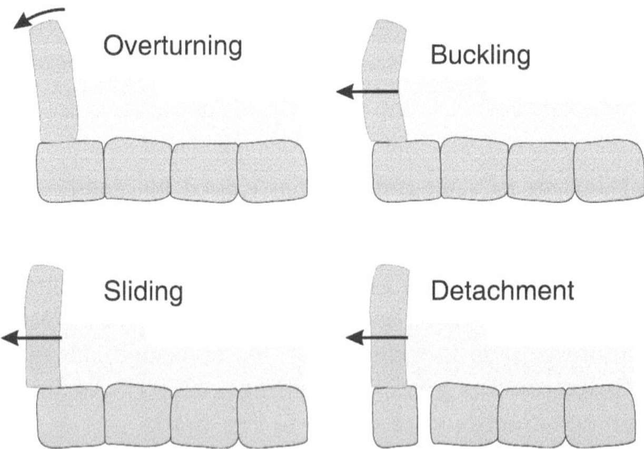

Fig. 6-11. Different damage types on spandrel walls and parapets according to Melbourne (1991)

Fig. 6-12. Sliding spandrel wall

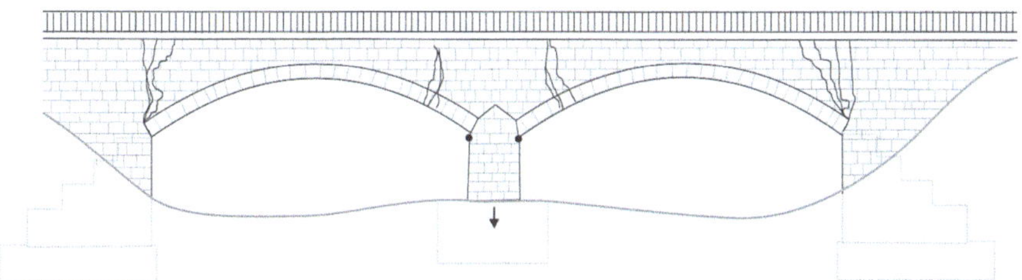

Fig. 6-13. Crack pattern on spandrel walls caused by vertical settlement of the middle pier according to Como (1998)

6.3 Repair and Strengthening

6.3.1 Introduction

If damages are found at structures, they are usually repaired or refurbished. Repair is a part of the maintenance of structures. According to DIN 31 051 (2003), all actions to conserve and restore the normal conditions and to investigate and assess the actual conditions are integral parts of what is called "maintenance." Therefore, maintenance includes all types of inspection, servicing, and repair. For example, it includes damage and failure investigation, undertaking mitigation measures, repairing and mending, replacement and assembly, testing, and clearance. Inspection itself includes all means to investigate and assess the actual conditions of a system. That includes testing, measuring, assessment, and documentation. Servicing includes all actions to keep a system in normal conditions. For example, testing, adjusting, exchanging, supplementing, preserving, and cleaning are parts of servicing. It seems to be meaningful to link inspections and servicing. Therefore, often the costs for both are given together (Curbach et al. 2003b).

Table 6-5 gives some German rules for investigation periods of bridges. In contrast, Table 6-6 gives some indications for the maintenance planning for arch bridges.

Table 6-5. Inspection intervals of bridges according to Switaiski (2006)

DIN 1076	RiL 804	Time distance
Ongoing observation	Observation	Permanent, half-yearly
Observation	-------------------	Yearly
Simple inspection	Inspection	3 years
Main inspection	Expertise	6 years
Inspection caused by event	Special inspection	

Table 6-6. Return period of different maintenance actions on arch bridges according to Steele et al. (2006)

Maintenance activity	Every (years)
Vegetation removal	5
Coping stone replacement/realignment	10
Brickwork maintenance – repoint/renewal	15
Parapet repairs/replacement	15
Invert clearance	20
Cutwaters replaced	40
First refurbishment scheme	120
Second refurbishment scheme	200

To describe the degree of maintenance, Melchers and Faber (2001) introduce a maintenance factor:

$$R_{0,R} = M \cdot R_0.$$

(6-1)

The value M depends on the quality of maintenance (Table 6-7).

Table 6-7. Statistical properties of factor M according to Melchers and Faber (2001)

Quality of maintenance	Mean value	Standard deviation
Low	0.90	0.10
Good	0.95	0.05
Excellent	1.00	0.02

An excellent maintenance restores the original load-bearing capacity of a structure. In contrast, a low quality of maintenance restores the load-bearing capacity only to 90%. However, this model is rather simple. In many cases, some maintenance actions restore the capacity, whereas other actions decrease the capacity by structural work. In many cases the load-bearing capacity of historical arch bridges also has to be increased.

Orbán (2004) has published an investigation about common repair works on masonry railway arch bridges (Table 6-8). Not only does it list the different methods but it also indicates the frequency of the certain methods. Rombock (1994) mentions the following maintenance works for natural stone material:

- Cleaning and hydrophobicity of natural stones
- Treatment of salt damages on natural stone masonry
- Conservation of natural stones
- Chemical stone cleaning
- Drying of masonry
- Mechanical cleaning
- Strengthening of natural stones
- Supplement of natural stones

The list of refurbishment techniques suggested by Page (1996) is listed in Table 6-9.

Table 6-8. Repair techniques for masonry railway arch bridges according to Orbán (2004)

Repair technique	Percent of railway organizations with experience in the repair technique (%)
Maintenance of sealing:	
Drainage pipes re-positioned and put through the arch	58
New backfill and roadway slab with sealing	42
Sealing without bond on the arch	33
Grouting of cement and micro-cement into the arch or vault	25
Grouting of gel through the arch	17

Repair technique	Percent of railway organizations with experience in the repair technique (%)
Increase of load-bearing capacity	–
Injection into the arch or vault	83
Shotcrete at the introdos of the arch	58
New backfill and roadway slab on the vault	42
Nailing of cracks with grouting of the nails	33
Support of the vault by steel arches	25
Increase of load-bearing capacity of the abutments, foundations and piers	–
Piles through the abutment	67
Nailing and grouting	50
Protection against erosion (sheet pile, concrete cover, stone plasters around the pier)	42
Addition of reinforced concrete elements	33
Injection into the ground	33
Introduction of load-bearing capacity into the width of the arch	
Tie road and anchor plates	67
Connection between spandrel wall and arch	17
Reinforced concrete slab on the arch	25
Shotcrete at the intrados and connection of the spandrel walls on the concrete by tensile elements	8

Table 6-9. Maintenance on masonry arch bridges according to Page (1996) and COST 345 (2006)

Fault	Measure
Deteriorated pointing	Repoint
Deteriorated arch ring	Repair masonry Install saddle Apply sprayed concrete to intrados Install pre-fabricated liner Grout arch ring Apply proprietary repair technique
Arch ring inadequate to carry in-service loads	Install saddle Apply sprayed concrete to intrados Install pre-fabricated liner Replace fill with concrete Install steel beam-relieving arches Install relieving slab Apply proprietary repair technique
Internal deterioration of mortar, which could lead to ring separation, for example	Grout arch ring Stitch (using tie bars spanning across a crack)

Fault	Measure
Foundation movement	Install mini-piles or underpin
	Grout piers and abutments
Outward movement of spandrel walls	Install tie bars
	Install spreader beams
	Replace fill with concrete
	Demolish walls and rebuild
	Grout fill
Separation of arch ring beneath spandrel wall from remainder of arch ring	Stitch together
Weak fill	Replace fill with concrete
	Grout fill
	Reinforce fill
Water leakage through arch ring	Make road surfacing water resistant
	Install waterproofing
	Waterproof extrados and improve drainage
Scour, or damage to scour protection works	Install protection measures
	Repair or enhance protection system, for example by placing riprap or concrete around substructures at risk

According to Bartuschka (1995), the following actions can be used to restore the load-bearing capacity of the arch:

- Replacement of the backfill
- Application of shotcrete shell
- Construction of a bridge inside the bridge (see for example Notkus and Dulinskas 2002 and Stritzke 2007 and Fig. 6-14)
- Construction of a railway reinforced concrete slab

Fig. 6-14. Example of a bridge in bridge construction taken from Stritzke (2007)

Bartuschka (1995) distinguishes between an investigation and a construction phase. Both sequences are shown in Figs. 6-15 and 6-16.

Furthermore, a summary of certain reconstruction techniques can be found in Hamid et al. (1994) or at Mathur et al. (2006).

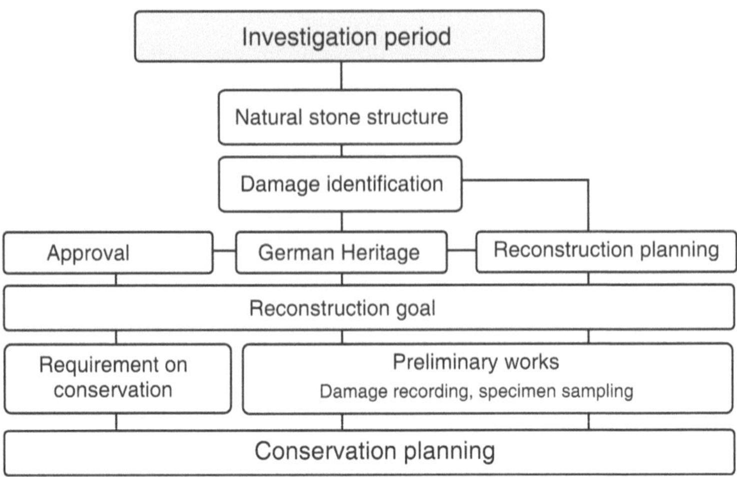

Fig. 6-15. Investigation phase (Bartuschka 1995)

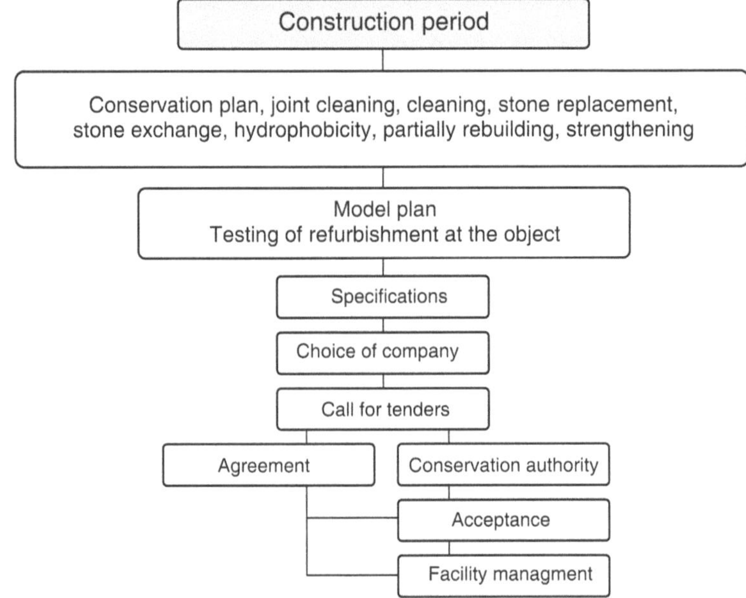

Fig. 6-16. Construction phase (Bartuschka 1995)

Rombock (1994) gives a comprehensive list of damage and refurbishment cases on historical arch bridges built with natural stones. Standfuß and Thomass (1987) state that the view of historical arch bridges should be changed by only minor refurbishment. For example, shotcrete and concrete layers such as shown in Knoblauch et al. (2008) should only be applied if no other strengthening measures can be used. However, they indicate that under nearly all conditions, the so-called replacement of lay bricks can be applied to re-establish the original conditions. Replacement of lay bricks considers the replacement of damaged masonry parts by constructing a falsework that supports the arch and lay bricks in the damaged regions after replacing the damaged parts. For this technology, comparable natural stones should be applied. Furthermore, sealing of the historical arch bridges is a must according to Standfuß and Thomass (1987).

Mabon (2002) considers the construction of a concrete backfill and the reinforced concrete railway slab as the most popular techniques for the strengthening of historical natural stone arch bridges. Mabon (2002) furthermore mentions the application of reinforcement in cut slots and of a shotcrete layer.

Vockrodt (2005) and Vockrodt et al. (2003) also mention some restoration examples in detail. Witzany et al. (2008) describe experimental studies in strengthening techniques.

6.3.2 Strengthening Techniques

6.3.2.1 Stone refurbishment

Certain types of damage mechanisms were given in this chapter. One major type was chemical attacks, such as salt penetration. Measures against salt can be classified into chemical and physical ones. The chemical measures mainly use the idea to transform aggressive salts into non-aggressive salts. However, since different types of salts can be found inside the stone under realistic conditions, it seems to be improbable to transform all aggressive salts into harmless ones.

The physical salt decontamination uses intermediate plasters. Unfortunately, such a technology seems to be impracticable for bridges. Additionally, electro-physical technologies based on electro-osmosis are known.

Under all conditions, a cleaning of the natural stones should be carried out. Table 6-10 lists and relates certain types of cleaning technologies to certain types of stones. Local damages of stones can be repaired by reimbursement. For example, for stone damage zones smaller than 200 cm^2, restoration mortar can be used. If the damage zones are greater, then for the reimbursement natural stone parts should be used. Bartuschka (1995) gives the following working steps:

1. Pick out at least 2 cm deep and dovetail shaped (Fig. 6-17).
2. If the damage zones are greater, stainless steel reinforcement should be built in. The reinforcement should not run over joints.
3. The surface of the reimbursement should be prepared by a stone cutter.

Table 6-10. Cleaning technologies for certain natural stones according to Bartuschka (1995)

		Cold water cleaning without pressure	Press water cleaning (cold/hot)	Steam jet cleaning	Sand blasting	Cleaning compound
Sandstone	Beebly	O	+	+	–	O
	Limy	O	O	+	–	+
	Clayey	O	+	+	–	+
Lime stone	Absorbent, soft	O	O	+	–	O
	coarsely porous	+	O	+	O	O
	coarsely porous, buffed	+	O	+	–	O
	Dense	+	O	+	O	O
	Dense, buffed	+	O	+	–	+
Granite, Diorite		+	O	+	O	+
Syenite, Labradorite (buffed)		+	O	+	–	+
Tuff		O	O	+	–	+
Marble	Not buffed	+	O	+	–	+
	Buffed	+	O	+	–	+
crystalline schist (not buffed)		O	O	+	O	O
Phyllite, Serpentinite (buffed)		O	O	+	–	+
Brick	Not glazed	O	O	+	–	+
	Glazed	O	O	+	–	+

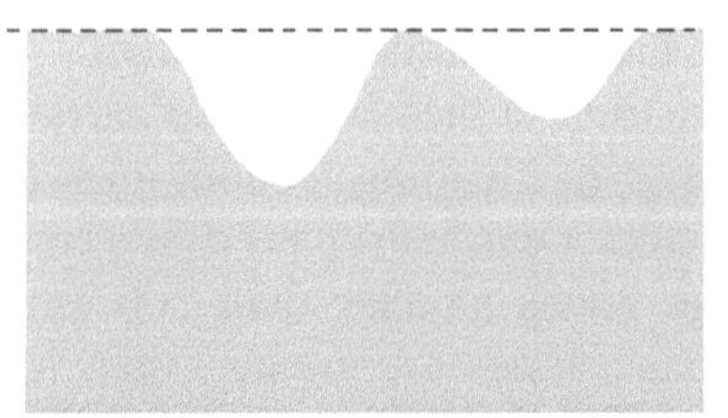

Wrong picking out of a stone

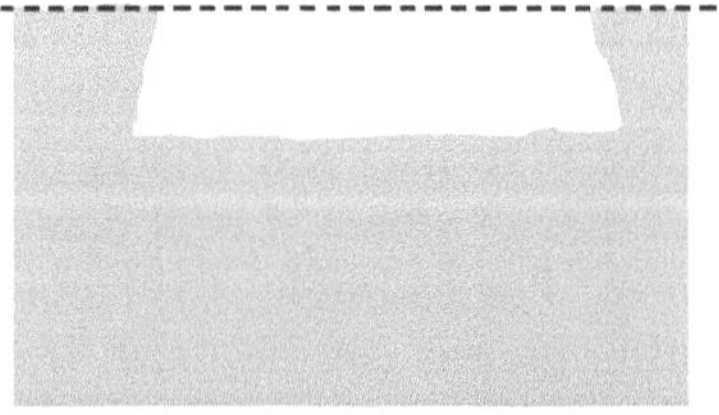

Correct picking out of a stone

Fig. 6-17. Examples of stone picking out according to Bartuschka (1995)

Many works have been published about the renovation of natural stones. Further references of Ruffert (1981), Nodoushani (1997, 1998), Wihr (1980), Reul (1994), Bienert (1976), and Sauder and Wiesen (1993) are given.

The renovation of the stone material should always further include a restoration of the joint material since both act together. Furthermore, the joint mortar often damages the stone material.

6.3.2.2 Mortar refurbishement

Mortar refurbishment mainly includes the removal of loose mortar and the building of new joint material. This can be done either by hand or by pressing by force. Several different techniques are shown in Bartuschka (1995) and Jäger (2006). For example, the steps for the dry injection method are shown in Fig. 6-18.

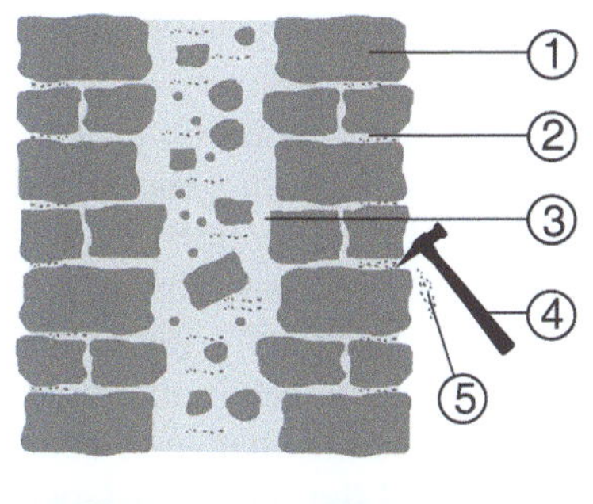

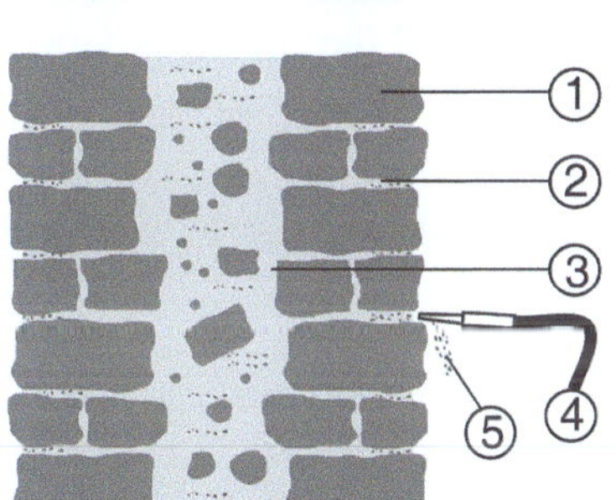

Fig. 6-18. (Continued)

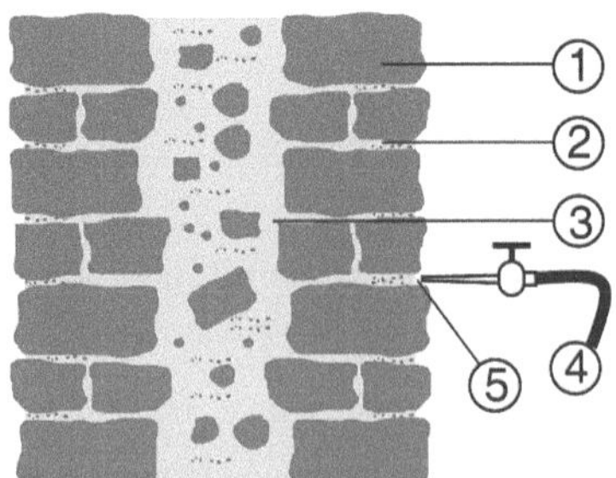

Fig. 6-18. Sequence of dry spraying method (Bartuschka 1995)

6.3.2.3 Reinforced concrete slab

Reinforced concrete slabs are a common load-bearing capacity-increasing strengthening technology (Fig. 6-19). Miri and Hughes (2004) have carried out tests on a scale of 1:12 which prove the increase impressively. The application of a concrete slab increases the load-bearing capacity by a factor between 3.2 and 3.7, depending on the ratio rise to span. In general, the tests by Miri and Hughes (2004) have shown the load-deflection curve in Fig. 6-20. It should be mentioned here that a former German railway recommendation gave an increase in the load-bearing capacity without any detailed computation of a factor 1.2. The increase can be easily explained by the change of the loading mechanism in arches. See the change of beam models in Chapter 3 from simple arch models toward the model of Gocht (1978).

Fig. 6-19. Example of arch bridge width increase by a concrete slab

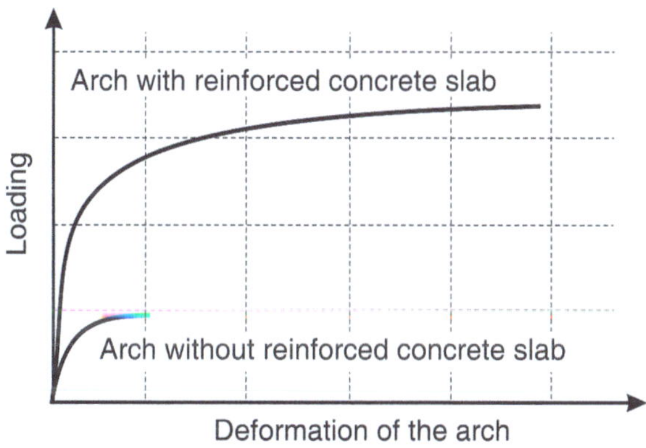

Fig. 6-20. Qualitative deformation of an arch with and without reinforced concrete slab according to Miri and Hughes (2004)

6.3.2.4 Injection and grouting

Injections and grouting can be applied to masonry bridges for different reasons. For example, the behaviour of the masonry can be homogenized, sealed, and hollow sections can be closed.

If sealing is reached by injection, then the injecting material does have to fulfill the requirements for not only the injecting process but also sealing properties. From a technological point of view, the injection material should have a high penetration capability. Therefore, usually a low molecular material is used. Additionally, mainly true fluid solutions are common. Emulsions and suspensions are only rarely used. Since the 1980s, besides cement slurry and cement suspensions, further injection material has become widespread. Such materials are

- Alkali silicate dissolutions
- Alkali methyl silicone dissolutions
- Combination of alkali silicate dissolution and alkali methyl silicone dissolutions
- Alkali propyl silicone dissolutions
- Silane and low-molecular oligomer siloxane in organic dissolvent
- Water soluble silicone microemulsions concentrate
- Bitumen solution and melting mass
- Bitumen emulsion
- Organic resin in organic dissolvent
- Alkanes

However, injections and grouting include a great uncertainty subject to the effectiveness and any long-term effects. Therefore, before the application is launched, the technology has to be evaluated regarding quality assurance. To provide that, usually test injections are carried out and can be used for the assessment of the injection value. Several months after the test injection, a destructive and nondestructive test should be applied to investigate the quality and effectiveness of the injection and grouting. That injection and grouting can be successfully applied, for example, has been shown by Schueremans et al. (2003).

6.3.2.5 Reinforcement

Already grouting and injection are technologies for improving the mechanical material properties of the construction material – here, masonry. To this class also belongs the technique of reinforcement. However, the volume or area ratio is rather low. The technique concentrates much more on a well-selected location of the reinforcement inside the masonry. This requires an understanding of the load path flow inside the masonry. Examples of the application of reinforcement, either for the strengthening of the pier or for the strengthening of the arch can be found in Figs. 6-21 and 6-22.

In general, the reinforcement can be distinguished either according to the material used in metallic and nonmetallic reinforcements, or according to the forces that should be partially covered by the reinforcement such as shear force reinforcement or bending moment reinforcement.

A special example of the application of reinforcement is given in the next section.

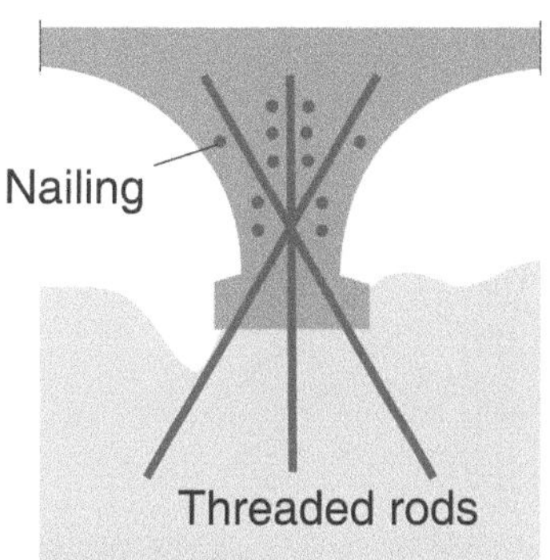

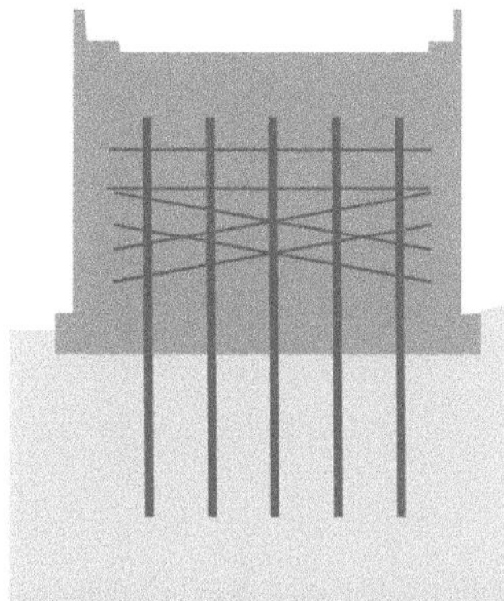

Fig. 6-21. Installation of threaded rods and nailing on a pier according to HA (1997) and COST 345 (2004)

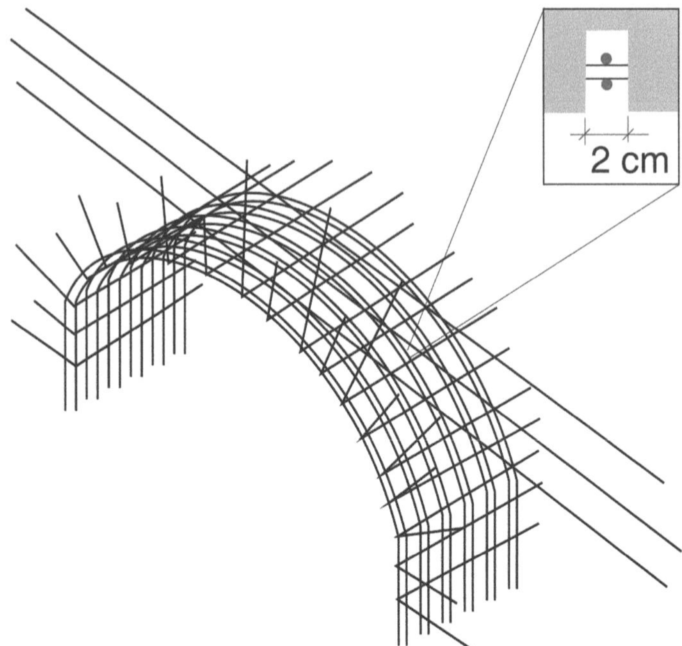

Fig. 6-22. Reinforcement concept for arches according to Woodward (1997) and COST 345 (2004)

6.3.2.6 Archtec techniques

The Archtec technique has been applied more than 130 times in Great Britain, the United States, and Australia since 1998 (Brookes and Mullet 2004). As a specific example, the Wisconsin Avenue Bridge in the United States is mentioned (Darden and Scott 2006)

The general idea of the technology is the strengthening of arch segments by additional reinforcement elements. Since the most frequent failure of arches is the development of mechanisms, including hinges at the quarter points, the application is applied there to increase the bending moment capacity in the cross sections. However, in contrast to other strengthening measures, such as replacement of the backfill by concrete, here no massive construction works have to be undertaken at the bridge. Rather, only drillings are carried out and then the reinforcement such as threaded rods is applied into the drilling holes. Figure 6-23 shows, as an example, the location of reinforcement elements with a length of 2.5 m. The effectiveness of this concept has been proven in tests where the reinforcement elements as well as the arch were equipped with strain-measuring devices. Of course, the strengthening can only be active for traffic load. The dead load

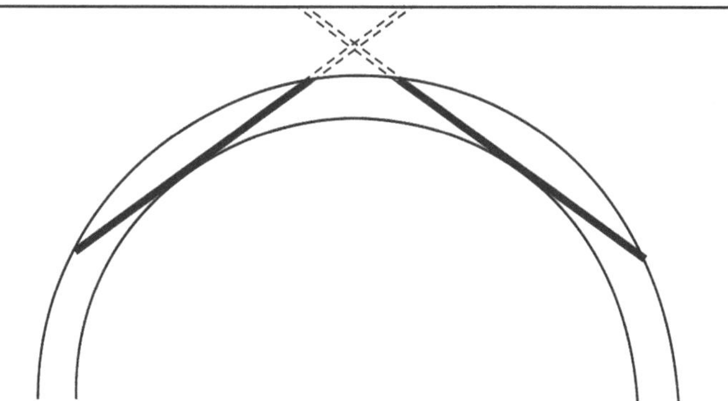

Fig. 6-23. Profile of an arch bridge with the position of the reinforcement elements

still has to be taken entirely by the original arch (Brookes and Mullet 2004, Mabon 2002, Owen et al. 2005 and Tilly and Brookes 2005).

6.3.2.7 Profile bars and threaded rods

Oliveira and Lourenço (2004) introduced the installation of transversal reinforcement elements into arches (Fig. 6-24) or above the extrados. The elements are designed to overtake tensile forces in transversal direction. The reinforcement elements are profile bars. A comparable solution has been introduced by Falconer (1999). He anchored the reinforcement above the arch stones. In Germany, model drawings and recommendations for the anchoring of reinforcement elements in the spandrel walls exist (BMVBW 1993). A further example can be found in Welch (1995).

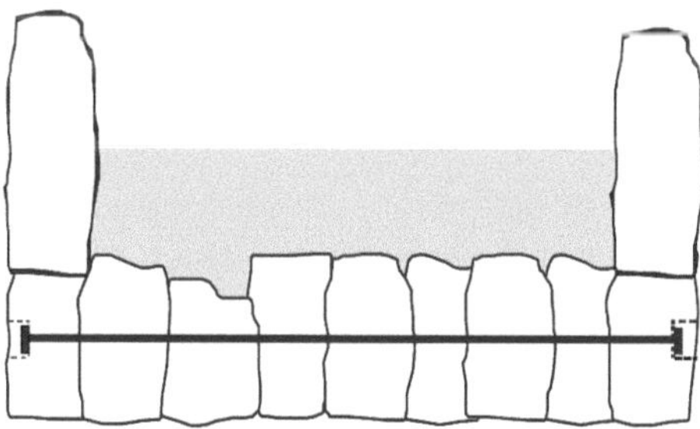

Fig. 6-24. Transversal reinforcement elements in an arch taken from Oliveira and Lourenço (2004)

6.3.2.8 Non-metallic reinforcement

Besides the application of classical steel reinforcement, nonmetallic reinforcement elements can also be applied. The major advantage is the exclusion of corrosion. Such techniques are applied, not only to natural stone bridges, but also to classical steel reinforced concrete elements. The first author has used textiles for reinforcement for concrete slabs (Proske 1997). Other materials have also been applied to arch bridges.

For example, Modena et al. (2004) report about the installation of carbon fibre-reinforced polymer (CFRP) elements transversally and longitudinally into arches. Melbourne and Tomor (2004) and Hodgson (2003) have also used CFRP elements for arch strengthening.

Bergmeister (2003) introduces the strengthening of a concrete arch bridge in transversal direction by CFRP elements.

Fiber-reinforced polymers (FRP) have been applied for the strengthening of historical arch bridges by De Lorenzis and Nanni (2004), Creazza and Saetta (2001), Valluzzi and Modena (2001), Borri et al. (2002) – Fig. 6-25, Bati and Rovero (2008), Ricamato (2007), Drosopoulos et al. (2007), and Panizza et al. (2008).

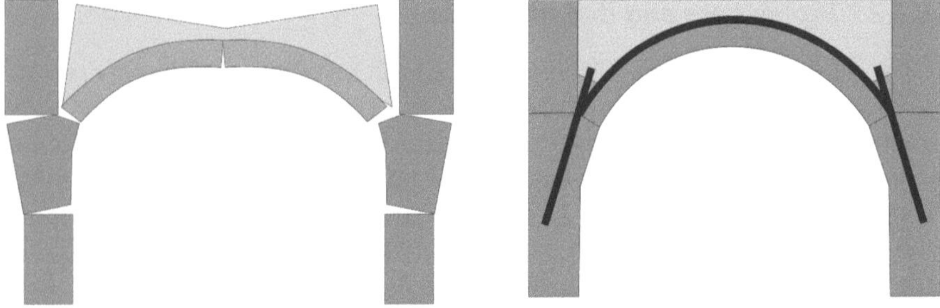

Fig. 6-25. Strengthening example of arch bridge using FRP by Borri et al. (2002)

6.3.2.9 Pre stressing

Pre stressed and nailed masonry applied to historical arch bridges will not be discussed here in detail. For details, see Ganz (1990), Ullrich (1989), and Wenzel (1997).

6.3.2.10 Shotcrete

The application of shotcrete with historical masonry arch bridges is not recommended due to conservation criteria for monuments and historical structures. However, under some conditions, it cannot be avoided, or it can be applied to unimportant structures. Especially in Germany, model drawings and recommendations exist about the application of shotcrete at the intrados (BMVBW 1993).

6.3.2.11 Restoration of parapet

In the section on damages, damages on spandrel walls and parapets was also mentioned. Here, restoration techniques have been developed with either using some hidden application of strengthening material, such as reinforced concrete, or using a decoupling of the backfill and the spandrel walls. Examples are shown in Figs. 6-26 and 6-27. Further details can be found in the COST 345 report (2006).

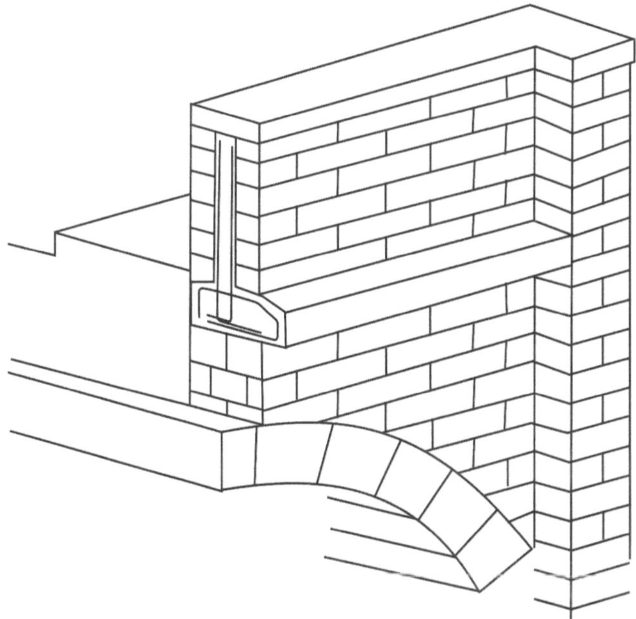

Fig. 6-26. Reinforced parapet according to Welch (1995) and COST 345 (2006)

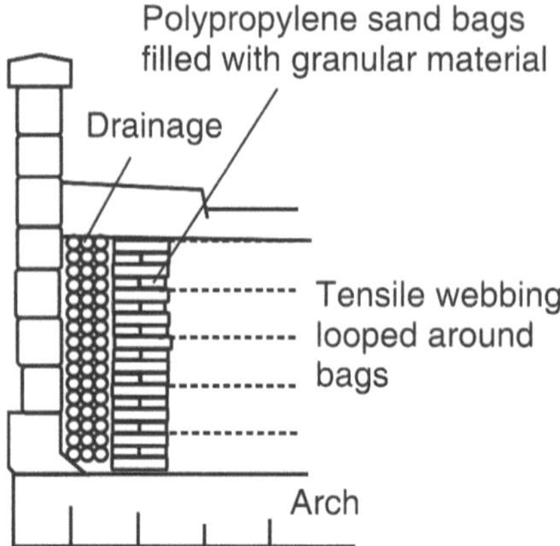

Fig. 6-27. Decoupling of the spandrel wall according to Welch (1995) and COST 345 (2006)

6.3.2.12 Increase of width

An insufficient width of historical bridges is a major problem for the adaptation of such structures to modern traffic requirements. Therefore, widening of the bridges is common (Fig. 6-28). For example, in the last few years in Spain, one fifth of all highway bridges have been widened (Angeles-Yáñez and Alonso 1996). According to Harrison (2004), the major cause for destruction of historical bridges in Great Britain was insufficient width of the road track.

An example of the successful widening of a historical arch bridge can be found in Troyano (2003). Here, the widening of the Pont Vieux de Albin over the river Tarn and Burgo in Spain was carried out using a low-pitched arch. Another example was given by Parikh and Patwardhan (1999). The widening of the Marienbrigde in Dresden was described by Koettnitz and Schwenke (1998). See also Scheidler (1992), Sobrino (2007), Boronczyk-Plaska and Radomski (2008), and Vockrodt et al. (2003).

A very special example is the Chemnitz Viaduct in Germany, where the span of the arch was too low for the highway traffic under the bridge. Therefore, one pier was supported by a bridge under the historical arch bridge (Fig. 6-29). Also, the New Saale Bridge South Jena on German highway A4 is mentioned here (Martin and Becker 2005). Although the bridge is a new reinforced bridge to provide sufficient overall width for the highway, it uses the shape of the historical arch bridge in which neighbourhood it is built (Fig. 6-30).

It is virtually impossible to present here all maintenance and repair strategies applied to historical arch bridges. Therefore, the following section summarizes some refurbishment examples for historical arch bridges but does not intend to be a complete list. However, it should give an impression about the problems faced under practical conditions.

Fig. 6-28. Example of arch bridge width increase by an attached beam

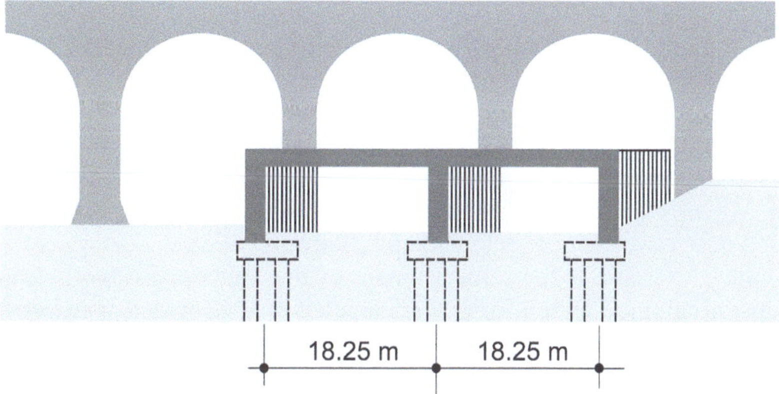

Fig. 6-29. Chemnitz Viaduct after removal of the piers (Germany) according to Reintjes (2002)

Fig. 6-30. New Saale Bridge South Jena

6.3.3 Examples

The historical bridge over the river Werra in the city of Münden in Lower
Saxony, Germany was found to show an insufficient load-bearing capacity
after a major bridge investigation. This assumption was based on erosion
found at the piers, heavy curvatures and shifts of the pier and arch masonry, efflorescence and strong weathering of the masonry joints, and finally strong corrosions of the steel anchors, iron clips, and the iron parapet
(Schwartz 1988).

Due to the historical importance of the bridge, a major maintenance action was launched. The maintenance plan included, for example, the substitution of the iron parapet by a massive parapet. Furthermore, the arch was
cleaned by high-pressure cold water. Damage to the stones was repaired by
substitution using natural stone material and small damage was repaired
using Mineros stone mass. This material includes stone flour, which is embedded into a two-component synthetic (Schwartz 1988).

Another example is the refurbishment of the Taubern Bridge in Lauda.
The historical three-arch bridges were erected in 1512 with a span between
6 and 7 m. Due to the increasing traffic load, spandrel walls and wing
walls were moved and the masonry arch showed wide cracks. To provide
safety for the bridge, anchors and clips were built in. In the 1930s, a
steel jacket was installed at the extrados. However, the amount of damage increased, and in the 1960s the weight restriction of the bridge was
intensified from 16 to 9 tonnes. Finally, the bridge was demolished and

reconstructed with reinforced concrete. The foundation using reinforced concrete piles, the span and, very importantly, the width of the reconstructed bridge were changed in relation to the original one. To conform at least partially to conservation rules, parts of the original bridge were used for the reconstruction. For example, the coverage of the bridge was carried out using natural stone, sometimes even original parts. Also, for the bridge platforms, the original bridge crucifixes were used (BMV 1988).

The third example is the Hoch Bridge Dingolfing. The bridge was originally constructed in 1612 as a five-span brick masonry bridge with single spans between 5.40 and 6.35 m. The piers reach a width of 1.20 and 1.35 m. The bridge reaches an overall length of 54 m and a maximum height of 5.6 m. The brick material showed heavy damages due to long moisture penetration.

At several locations, stones had separated from the masonry. Settlement of the piers had yielded to cracks in the spandrel walls. The bridge was maintained in 1750, 1850, and 1890. The latest refurbishment was carried out in 1966. The refurbishment had contained a complete disassembly and reconstruction of the spandrel walls, clearage of the backfill and refilling with lean concrete, and installation of a sealing and drainage system. Furthermore, special bricks were produced for the reconstruction, the remaining elements were cleaned by sandblasting, and afterwards the joints were filled again. For this filling, a special mortar was designed using pit lime mortar with tuff additive. Additionally, the bridge-in-bridge system was applied because the arch was separated from the reinforced concrete slab by a reinforced concrete structure that carried the load directly from the railway slab towards the piers. The arch crown was separated from the concrete slab by a 5 cm strong polystyrene layer (BMV 1988).

The old Dreisam Bridge Eichstetten is a five-span basket arch bridge with single spans between 4.70 and 5.00 m. The piers have a width of 1.35 m. The width of the bridge reaches 4.6 m. The arches are covered with natural stone ashlar masonry, however the inner parts are only built with quarries. The spandrel walls are also made of ashlar masonry. The parapet again is built with quarries. In 1950, the bridge was refurbished by building in a backfill with lean concrete and adding a reinforced concrete railway slab. The spandrel walls were secured by tendons with steel anchors. Such steel anchors were also used at the Kocherbridge Griesbach (BMV 1988).

The Wurm Bridge in Hessia, constructed in 1777, was maintained in 1979 only by adding a reinforced concrete slab and a new sealing to the structure. This five-span bridge built of new red stone masonry, and with spans between 3.0 and 4.5 m, is one of the few bridges in Germany not blasted at the end of World War II (BMV 1988).

The Nagol Bridge in Hirsau was constructed in 1560 as an arch bridge using new red sandstone. The bridge reached an overall length of 51.30 m with four segmental arches spanning between 7.30 and 13.00 m. The piers reached a width between 3.00 and 5.40 m. In 1852 and 1855, the bridge was strongly maintained. In 1914, the width of the bridge was extended from 5.0 to 12 m. This was done by constructing reinforced concrete arches covered with natural stones (BMV 1988).

Many further examples can be found in the literature. As mentioned in Chapter 1, the number of arch bridges still functioning in the infrastructure is overwhelming. The efforts to keep such an essential piece are represented in many papers dealing with the strengthening of arch bridges. Some examples are mentioned in Koettnitz and Schwenke (1998), Günther et al. (1999), Vockrodt (2005), Patzschke (1996), and Zahn (1999). Recent examples can be found in Fotheringham (2008), Asmar et al. (2008), Beben and Manko (2008), and Siwowski and Sobala (2008).

6.4 Arch Bridges of the Second Generation

Although the maintenance efforts for arch bridges are low if the lifetime is considered, Weber (1999) suggests an improvement of arch bridges and call these bridges second generation stone arch bridges. Such new arch bridges feature the following:

- Abandonment of back and lining masonry above the extrados. This would yield to an improved numerical description of the load-bearing behaviour of the arch with lower construction costs
- Backfill material used should be cohesionless and coarse grained. Geotextiles or steel ribbons should be applied to limit the compression on the spandrel walls. Sealing should be built in above the backfill
- Drainage is very important and therefore should be long lasting and controllable
- Abandonment of spandrel walls depending on the conditions
- Application of new technologies to decrease falsework costs
- Usage of new developed natural stone stocks in Europe
- Application of automatic stone cutting techniques

Examples of new stone arch bridges are double-curved arch bridges in China (1964). Another example is the Kimbolton Butts Bridge, a brick stone arch bridge constructed in the 1990s in Great Britain.

Such new arch bridges have to fully comply with the safety requirements for modern structures.

References

Al Bosta S (1999) Risse im Mauerwerk. Verformungen infolge von Temperatur und Schwinden. Baupraktische Beispiele. Werner Verlag 2. Auflage

Angeles-Yáñez M & Alonso AJ (1996) The actual state of the bridge management system in the state national highway network of Spain. Recent Advances in Bridge Engineering. Evaluation, management and repair. Proceedings of the US-Europe Workshop on Bridge Engineering, organized by the Technical University of Catalonia and the Iowa State University, JR Casas, FW Klaiber & AR Marí (Eds), Barcelona, 15–17 July 1996, International Center for Numerical Methods in Engineering CIMNE, pp 99–114

Asmar R, Buys JP, Deschamps D, Kassis B, Salame A, Slaha W, Bouchon E & Vion P (2008) Expertise and repair of Maameltein arch bridge in Lebanon. In: Forde: Structural Faults and Repair 2008. 12th International conference, Edinburgh, 10–12th June 2008, Structural faults and repair 2008

Bartuschka P (1995) Instandsetzungsmaßnahmen an hochwertiger Bausubstanz von Gewölbebrücken. Diplomarbeit. Technische Universität Dresden, Institut für Tragwerke und Baustoffe

Bati SB & Rovero L (2008) Towards a methodology for estimating strength and collapse mechanism in masonry arches strengthened with fibre reinforced polymer applied on external surfaces. Materials and Structures, 41 (7), pp 1291–1306

Bau.de (2008) Bauen und Wohnen im Internet. www.bau.net/woerterbuch/

Bauriegel A (2004) Empfehlung Nr. 3 des Arbeitskreises Geotechnik historischer Bauwerke und Naturdenkmäler der DGGT: Geotechnische Schadensfeststellung und -behandlung an historischen Bauwerken. Bautechnik, 81 (9), pp 760–765

Beben D & Manko Z (2008) Modernization of old arch bridges by steel shell structure made from corrugated plates. In: Forde: Structural faults and repair 2008. 12th International conference, 10–12th June 2008, Edinburgh

Beeger D (1992) Naturstein in Dresden, Taschenbuch, Staatliche Naturhistorische Sammlungen Dresden, Museum für Mineralogie und Geologie

Bergmeister K (2003) Kohlenstofffasern im Konstruktiven Ingenieurbau. Verlag Ernst und Sohn, Berlin

Bién J & Kamiński T (2004) Masonry arch bridges in Poland. Arch Bridges IV – Advances in Assessment, Structural Design and Construction, P Roca & C Molins (Eds), CIMNE, Barcelona, pp 183–191

Bień J & Kamiński T (2007) Damages to masonry arch bridges – proposal for terminology unification. Proceedings of the Arch 07 – 5th International Conference on Arch Bridges, PB Lourenço, DV Oliveira & A Portela (Eds), 12–14 September 2007, Madeira, pp 341–348

Bienert G (1976) Brückenerhaltung. 2. Auflage Transpress Verlag

Bläuer C (1992) Mineralogische Schadenaufnahme und Interpretation der Schadenursachen an der Nydeggbrücke in Bern. Straße und Verkehr. Zeitschrift des Vereins Schweizerischer Straßenfachleute 2, pp 88–89

BMV (1988) Bundesminister für Verkehr: Steinbrücken in Deutschland. Beton-Verlag Düsseldorf

BMVBW (1993) Bundesministeriums für Verkehr, Bau- und Wohnungswesen: Sofortinstandsetzungsmaßnahmen an Brücken und anderen Ingenieurbauwerken der Bundesfernstraßen in den neuen Bundesländern; Sammlung von Arbeitshilfen für die Planung und Vergabe, Verkehrsblatt-Verlag

Boothby TE, Hulet HM & Stanton TR (2004) Inspection, assessment and monitoring of Railroad arch bridges in Sothwestern Pennsylvania. Arch Bridges IV – Advances in Assessment, Structural Design and Construction, P Roca & C Molins (Eds), CIMNE, Barcelona, pp 144–151

Boronczyk-Plaska G & Radomski W (2008) Bridge widening – technical, economical and aesthetical aspects. Bridge Maintenance, Safety, Management, Health Monitoring and Informatics, H Koh & DM Frangopol (Eds), Taylor & Francis Group, London, pp 2850–2857

Borri A, Corradi M & Vignoli A (2002) Seismic upgrading of masonry structures with FRP. Universita' Degli Studi di Perugia, Facoltà di Ingegneria

Bottesford Living History (2007) www.bottesfordhistory.org/page_id__242_path__0p25p.aspx

Brencich A & Colla C (2002) The influence of construction technology on the mechanics of masonry railway bridges, Railway Engineering 2002, 5th International Conference, 3–4 July 2002, London

Brookes CL & Mullet PJ (2004) Service load testing, numerical simulations and strengthening of masonry arch bridges. Arch Bridges IV – Advances in Assessment, Structural Design and Construction, P Roca and C Molins (Eds), CIMNE, Barcelona, pp 489–498

Como M (1998) Minimum and maximum thrust state in statics of ancient masonry bridges. Arch Bridges: History, Analysis, Assessment, Maintenance and Repair. Proceedings of the Second International Arch Bridge Conference, A. Sinopoli (Ed), Venice 6.–9. October 1998, Balkema, Rotterdam, pp 133–137

COST 345 (2006) European Commission Directorate General Transport and Energy: Procedures Required for Assessing Highway Structures: Working Group 6 Report on remedial measures for highway structures. http://cost345.zag.si/Reports/COST_345_WG6.pdf

Creazza G & Saetta AV (2001) Analysis of masonry structures reinforced by FRP. Historical Constructions, PB Lourenço & P Roca (Eds), University of Minho, Guimarães, pp 539–545

Curbach M, Bösche T, Scheerer S, Michler H, Proske D (2003a) Wiederaufbau der Pöppelmannbrücke Grimma

Curbach M, Köster T., Proske D, Schmohl L, Taferner J & Ehmann J (2003b) Parkhäuser, Betonkalender 2004, Teil II, Ernst & Sohn, pp 3–153

Darden C & Scott TJ (2006) Strengthening from Within. Federal Highway Administration, Public Roads, http://www.tfhrc.gov/pubrds/05mar/07.htm

De Lorenzis L & Nanni A (2004) International Workshop on Preservation of Historical Structures with FRP Composites. Final Report. National Science Foundation, USA, July 2004

DIN 1045-1 (2001) Tragwerke aus Beton, Stahlbeton und Spannbeton, Teil 1: Bemessung und Konstruktion, Juli 2001

DIN 31051 (2003) Maßnahmen zur Verzögerung des Abbaus des vorhandenen Abnutzungsvorrates

Drdácký MF & Slízková Z (2007) Flood and post-flood performance of historic stone arch bridges. Proceedings of the Arch 07 – 5th International Conference on Arch Bridges, Lourenço, Oliveira & Portela (Eds), 12-14 September 2007, Madeira. pp 163-170

Drosopoulos GA, Stavroulakis GE & Massalas CV (2007) FRP reinforcement of stone arch bridges: Unilateral contact models and limit analysis. Composites: Part B 38, pp 144–151

Falconer RE (1999) Test on masonry arch bridges strengthened using stainless steel reinforcement. Current and future trends in bridge design, construction and maintenance in Bridge design construction and maintenance, PC Das, DM Frangopol & AS Nowak (Eds), Thomas Telford, London, pp 415–423

Fauchoux G & Abdunur C (1998) Strengthening masonry arch bridges through backfill replacement by concrete. Arch Bridges: History, Analysis, Assessment, Maintenance and Repair. Proceedings of the Second International Arch Bridge Conference, A Sinopoli (Ed), Venice 6.–9. October 1998, Balkema Rotterdam, pp 417–422

Fotheringham N (2008) B9136 Ruthven bridge: Stone masonry refurbishment. In: Forde: Structural Faults and Repair 2008. 12th International conference, 10–12th June 2008, Edinburgh

Ganz H-R (1990) Vorgespanntes Mauerwerk. Schweizer Ingenieur und Architekt, Nr. 8. 22. Februar 1990, pp 177–182

Garrecht H (1997) Zum Verhalten bewitterter Mauerwerksoberflächen. In: Universität Karlsruhe, Sonderforschungsbereich 315 "Erhalten historisch bedeutsamer Bauwerke", Jahrbuch 1994, pp 187–198

Garston (1985) The influence of trees on house foundations in clay soils, Building Research Establishment Digest; Building Research Establishment, 298, pp 13–75

Gocht R (1978) Untersuchungen zum Tragverhalten rekonstruierter Eisenbahngewölbebrücken, Dissertation Hochschule für Verkehrswesen "Friedrich List" Dresden

Günther L, Schäfer J & Schwenke F (1999) Instandsetzung der Friedensbrücke Bautzen. 9. Dresdner Brückenbausymposium, Technische Universität Dresden, pp 123–150

HA (1997) Highway Agency. BA 16: The assessment of highway bridges. Design Manual for Roads and Bridges, HMSO, London

Hamid AA, Mahmoud ADS & El Magd SA (1994) Strengthening and repair of unreinforced masonry structures: State of the Art. Proceedings of the 10th IB2 MaC, Calgary, Canada, 5.–7.July 1994, pp 485–497

Harrison D (2004) The Bridges in Medieval England – Transport and Society 400–800. Clarendon Press, Oxford

Harvey B (2006) Some problems with arch bridge assessment and potential solutions. Structural Engineer, 84 (3), 7th February 2006, pp 45–50

Hodgson JA (2003) The Use of CFRP Plates to strengthen Masonry Arch Bridges. Extending the Life of Bridges, Concrete and Composites Buildings, Masonry and Civil Structures, MC Forde (Ed), The Commonwealth Institute, 1st–3rd July 2003, London

ICOMOS (2008) International Scientific Committee for Stone (ICOMOS-ISCS) http://lrmh-ext.fr/icomos/consult/consultation.php?lang=1

Jäger W (2006) Rissbilder und Putzabplatzungen bei Mauerwerksbauten – Ursachen und Strategien zur Schadensbeseitigung, Lehrstuhl Tragwerksplanung, Fakultät Architektur der TU Dresden

Knoblauch F-J, Laubach A, Sauer W, Schmidt M & Sprinke P (2008) Erneuerung des Burtscheider Viadukts in Aachen. Beton- und Stahlbetonbau 103, Heft 3, pp 195–203

Koettnitz R & Schwenke F (1998) Umbau und Instandsetzung der Marienbrücke in Dresden. 8. Dresdner Brückenbausymposium, Technische Universität Dresden, pp 139–156

Mabon L (2002) Assessment, strengthening and preservation of masonry structures for continued use in today's infrastructures. IABSE symposium, Melbourne

Martin R & Becker M (2005) Neubau Saalebrücke Süd bei Jena im Zuge der BAB A 4 – Umsetzung der gestalterischen Vorgaben. 15. Dresdner Brückenbausymposium, 15. März 2005, TU Dresden, pp 115–136

Mathur M, Kumar SV & Singh K (2006) Assessment and retrofitting of arch bridges. www.iricen.gov.in/projects/622/ARCH%20BRIDGE.pdf

Mattheck C, Tesari I & Bethge K (1993) Roots and Buildings. Structural Repair and Maintenance of Historical Building III (STREMAH), CA Brebbia & RJB Frewer. Computational Mechanics Publication, Glasgow, pp 751–760

Melbourne C & Tomor AK (2004) Fatigue Performance of Composite and Radial-Pin Reinforcement on Multi-Ring Masonry Arches. Arch Bridges IV - Ad-vances in Assessment, Structural Design and Construction, Roca and Molins (Eds), CIMNE, Barcelona, pp 428–434

Melbourne C (1991) Conservation of masonry arch bridges. In: Proceedings of the 9th International Brick/Block Masonry Conference, 13.–16. October 1991, Volume 1, Berlin, Germany, pp 1563–1570

Melchers RE & Faber MH (2001) Aspects of Safety in Design and Assessment of Deteriorating Structures, Proceedings to the International IABSE Conference on Safety, Risk and Reliability - Trends in Engineering, March 21-23, Malta 2001, pp 161–166

Meyer MG (2007) www.spanien-bilder.com/aktuelles_aus_spanien_details3328.htm

Mildner K (1996) Tragfähigkeitsermittlung von Gewölbebrücken auf der Grundlage von Bauwerksmessungen. 6th Dresdner Brückenbausymposium, 14th March 1996, Fakultät Bauingenieurwesen, Technische Universität Dresden, pp 149–162

Miri M & Hughes TG (2004) Increased load capacity of arch bridges using slab reinforced concrete. Arch Bridges IV – Advances in Assessment, Structural Design and Construction, P Roca & C Molins (Eds), pp 469–478

Modena C, Valluzzi MR, da Porto F, Casarin F & Bettio C (2004) Structural up-grading of a brick masonry arch bridge at the Lido (Venice). Arch Bridges IV –

Advances in Assessment, Structural Design and Construction, P Roca & C Molins (Eds), pp 435–443

Müller P (2005) Biomechanische Beschreibung der Baumwurzel und ihre Verankerung im Erdreich. Dissertation. Fakultät für Maschinenbau, Universität Karlsruhe

Nodoushani M (1997) Instandsetzung von Natursteinbrücken. Beton-Verlag. Düsseldorf

Nodoushani M (1998) Natursteingewölbebrücke (... über Fluss Sauer in Grundsteinheim, Ortsausgang Richtung Igenhausen/Stadt Lichtenau...). Bausanierung, Gütersloh 9, 4, pp 45–48

Notkus AJ & Dulinskas E (2002) Spatial non-linear analysis of complex structure "bridge-in-bridge". Journal of Civil Engineering and Management. Vilnius, Technika, Volume VIII, Suppl 1, pp 29–34

Ochsendorf JA (2002): Collapse of Masonry Structures. University of Cambridge, Dissertation, Cambridge

Ochsendorf JA, Hernando JI & Huerta S (2004) Collapse of Masonry Buttresses. Journal of Architectural Engineering, ASCE, September 2004, pp 88–97

Oliveira DV & Lourenço PB (2004) Repair of Stone Masonry Arch Bridges. Arch Bridges IV – Advances in Assessment, Structural Design and Construction, Roca & Molins (Eds), CIMNE, Barcelona, pp 451–458

Orbán Z (2004) Assessment, reliability and maintenance of masonry arch railway bridges in Europe. Arch Bridges IV – Advances in Assessment, Structural Design and Construction, P Roca & C Molins (Eds), CIMNE, Barcelona, pp 152–161

Owen DR, Peric D, Petrinic N, Smokes C & James PJ (2005) Finite/Discrete Element Models for Assessment and Repair of Masonry Structures. www.cintec.com

Page J (1996) A guide to repair and strengthening of masonry arch highway bridges. TRL Report 204, TRL Limited, Crowthorne

Panizza M, Garbin E, Valluzzi MR & Modena C (2008) FRP Strengthening – shear mechanism of brick masonry vaults. In: Forde: Structural Faults and Repair 2008. 12th International conference, 10–12th June 2008, Edinburgh

Parikh SK & Patwardhan YG (1999) Widening of an existing old masonry arch bridge through RCC box girder deck. Current and Future Trends in Bridge Design, Construction and Maintenance in Bridge Design Construction and Maintenance, PC Das, DM Frangopol & AS Nowak (Eds), Thomas Telford, London, pp 218–226

Patzschke F (1996) Der Wahrener Viadukt in Leipzig – Rekonstruktion einer Gewölbebrücke der DB AG. 6. Dresdner Brückenbausymposium, 14.3.1996, Fakultät Bauingenieurwesen, Technische Universität Dresden, pp 135–148

Poschlod K (1990) Das Wasser im Porenraum kristalliner Naturwerksteine und sein Einfluss auf die Verwitterung. Dr. Friedrich Pfeil Verlag

Proske D (1997) Textilbewehrter Beton als Verstärkung von Platten, 34. Forschungskolloquium des Deutschen Ausschuss für Stahlbeton, Technische Universität Dresden, Institut für Tragwerke und Baustoffe, Dresden, pp 51–58

Proske D (2003) Ein Beitrag zur Risikobeurteilung alter Brücken unter Schiffsanprall. Dissertation, Technische Universität Dresden

Proske D (2009) Brücken unter alpinen Stoßeinwirkungen. Proceedings of the 19th Dresdner Brückenbausymposium, 9–10 March 2009, Dresden

Reintjes K-H (2002) Die Unterfangung des Bahrmühlenviadukts Erfahrungen aus der Bauabwicklung. 12. Dresdner Brückenbausymposium, 14. März 2002, TU Dresden, pp 145–163

Reul H (1994) Handbuch Bautenschutz Bausanierung, Verlag Müller

Ricamato M (2007) Numerical and experimental analysis of masonry arches strengthened with FRP materials. University of Cassino, Department of Engineering, Graduate School in Civil Engineering, PhD thesis

Rombock U (1994) Restaurierung von Steinbrücken. Informationszentrum Raum und Bau der Fraunhofer-Gesellschaft, 2., erweiterte Auflage, IRB-Verlag, Stuttgart

Rota M (2004) Seismic Vulnerability of Masonry Arch Bridge Walls. European School of Advance Studies in Reduction of Seismic Risk, Rose School, Dissertation

Ruffert G (1981) Sanierung von Baudenkmälern, Beton-Verlag

Sauder M & Wiesen H (1993) Sanierung von Bauwerken aus Naturstein, Taschenbuch, Landesinstitut für Bauwesen NRW

Scheidler J (1992) Umbau und Verbreiterung von Bogen- und Gewölbebrücken. Würzburger Betonseminare der LGA Bayern am 11.–12. März 1992, Oberste Baubehörde Bayerns

Schueremans L, Van Rickstall F, Ignoul S, Brosens K, Van Balen K & Van Gemert D (2003) Continuous Assessment of Historic Structures – A State of the Art of applied Research and Practice in Belgium, KULeuven, Department of Civil Engineering

Schwartz J (1988) Sanierung der historischen Werrabrücke 1986–1988: Schadensanalyse und Sanierungskonzept. Werrabrücken Münden – Geschichte und Sanierung 1986–1988, pp 25–33

Siwowski T & Sobala D (2008) The revitalization of XIX century masonry arch bridge. In: Forde: Structural Faults and Repair 2008. 12th International conference, 10–12th June 2008, Edinburgh

Sobrino JA (2007) Extending the life of masonry and concrete arch bridges. Structural Engineering International, 17 (4), pp 328–336

Standfuß F & Thomass S (1987) Grundsätze der baulichen Maßnahmen an alten Steinbrücken. Bausubstanz, Nr. 3, p 28

Steele K, Cole G, Parke G, Clarke B & Harding J (2006): Application of life cycle assessment technique in the investigation of brick arch highway bridges. Proceedings of the 2002 Conference for the Engineering Doctorate in Environmental Technology,
http://www.cintec.co.uk/en/applications/Archtec/documents/Chapter05_4.htm

Strasser A (2006) Vorlesung Altern, Institut für Physiologie, Veterinärmedizinische Universität Wien, Vienna

Stritzke J (2007) Bogen- und Stabbogenbrücken. In: Handbuch Brücken, Mehlhorn (Ed), Springer Verlag, Berlin – Heidelberg

Switaiski B (2006) Ein systematischer Ansatz für eine Strategie zur Gewährleistung der Sicherheit von Gebäuden. Proceedings of the 4th International Probabilistic Symposium 2006, Medianpour (Eds), Gucma & Proske, BAM Berlin.

Tilly GP & Brookes CL (2005) Archtec Strengthening of Masonry Arch Bridges. www.cintec.com

Tingle B & Heelbeck M (1995) The Management of a new County's Bridge Stock Humberside 20 Years on. Arch Bridges, C Melbourne (Ed), Proceedings of the 1st International Conference on Arch Bridges, Bolton, Thomas Telford, London, pp 37–46

Troyano LF (2003) Bridge Engineering – A Global Perspective. Thomas Telford

UIC-codex (1995) Empfehlungen für die Bewertung des Tragvermögens bestehender Gewölbebrücken aus Mauerwerk und Beton. Internationaler Eisenbahnverband. 1. Ausgabe 1.7.1995

Ullrich M (1989) Statusbericht über Forschungsarbeiten zur ingenieurmäßigen Sicherung von historischem Mauerwerk durch Verpressen, Vernadeln und Vorspannen. Bautenschutz + Bausanierung 12, pp 19–27

Ural A, Oruc S, Dogangün A & Iskender Tuluk Ö (2008) Turkish historical arch bridges and their deteriorations and failures. Engineering Failure Analysis 15, pp 43–53

Valluzzi MR & Modena C (2001) Experimental analysis and modelling of masonry vaults strengthened by FRP. Historical Constructions, PB Lourenço & P Roca (Eds), Guimarães, pp 627–635

Vockrodt HJ (2005) Instandsetzung historischer Bogenbrücken im Spannungsfeld von Denkmalschutz und modernen infrastrukturellen Anforderungen. 15. Dresdner Brückenbausymposium, 15. März 2005, Dresden, pp 221–241

Vockrodt HJ, Feistel D & Stubbe J (2003) Handbuch Instandsetzung von Massivbrücken. Untersuchungsmethoden und Instandsetzungsverfahren. Basel, Boston, Berlin: Birkhäuser Verlag

Walthelm U (1990) Rissbildungen und Bruchmechanismen in freistehenden Wänden aus Beton und Mauerwerk. Bautechnik 67, Heft 1, pp 21–26

Walthelm U (1991) Theoretische Spannungszustände, Rissbildungen und Traglasten von einachsig ausmittig gedrückten Rechteckquerschnitten. Bautechnik 68, Heft 6, pp 206–212

Weber H (1993) Instandsetzung von feuchte- und salzgeschädigtem Mauerwerk. expert-Verlag. Ehningen bei Böblingen

Weber WK (1999) Die gewölbte Eisenbahnbrücke mit einer Öffnung. Begriffserklärungen, analytische Fassung der Umrisslinien und ein erweitertes Hybridverfahren zur Berechnung der oberen Schranke ihrer Grenztragfähigkeit, validiert durch einen Großversuch. Dissertation, Lehrstuhl für Massivbau der Technischen Universität München

Weiss G (1992) Die Eis- und Salzkristallisation im Porenraum von Sandsteinen und ihre Auswirkungen auf das Gefüge unter besonderer Berücksichtigung gesteinsspezifischer Parameter. Dr. Friedrich Pfeil Verlag

Welch PJ (1995) Renovation of masonry bridges. Arch Bridges, C Melbourne (Ed), Proceedings of the 1st International Conference on Arch Bridges, Bolton, Thomas Telford, London, pp 601–611

Wenzel F (Ed) (1997): Mauerwerk – Untersuchen und Instandsetzen durch Injizieren, Vernadeln und Vorspannen. Erhalten historisch bedeutsamer

Bauwerke. Empfehlungen für die Praxis. Sonderforschungsbereich 315, Universität Karlsruhe

Wihr R (1980) Restaurierung von Steindenkmälern, Verlag Callwey

Witzany J, Čejka T & Zigler R (2008) Experimental research of masonry vaults strengthening. In: Forde: Structural Faults and Repair 2008. 12th International conference, 10–12th June 2008, Edinburgh

Woodward RJ (1997) Strengthening arch bridges. Annual Review 1997, TRL Limited, Crowthorne

Zahn A (1999) Wieder ganz tragfähig – Kulturhistorisch wertvoller Brückenbau gerettet und heutigen Verkehrsanforderungen angepasst. Bautenschutz und Bausanierung. Köln 22, pp 29–30

7 Safety Assessment

7.1 Definition of Safety and Safety Concepts

Structures have to be safe. However, there is no common understanding of the term "safety." Often the term "safety" is defined as a situation with a lower risk compared to an acceptable risk or as a situation "without any impending danger." Other definitions describe safety as "peace of mind." Whereas the first definition using the term "risk" is already based on a substitution, the later term using "peace of mind" is a better definition. The authors consider "safety" to be the result of an evaluation process of a certain situation. The evaluation can be carried out by every system that is able to perform a decision-making process, such as animals, humans, societies, or computers that use some algorithms. However, algorithms usually use some numerical representation, such as risk R, for the description of safety S:

$$\text{existing } R \leq \text{ permitted } R \rightarrow S \qquad (7\text{-}1)$$

$$\text{existing } R > \text{ permitted } R \rightarrow \cancel{S}$$

In contrast, the authors consider not only numerical presentations as results of decision-making processes, but also human feelings. Therefore, safety is understood here as a feeling. The decision-making process focuses mainly on preservation. Furthermore, the decision-making process deals with whether some resources have to be spent to decrease hazards and danger to an acceptable level. In other terms, "safety" is a feeling, which describes that no further resources have to be spent to decrease any threats. If one considers the term "no further resources have to be spent" as a degree of freedom of resources, one can define "safety" as a value of a function that includes the degree of freedom of resources. Furthermore, one can assume that the degree of freedom is related to some degree of distress and relaxation. Whereas in safe conditions relaxation occurs, in dangerous situations a high degree of distress is clearly reached.

D. Proske, P. van Gelder, *Safety of Historical Stone Arch Bridges*, DOI 10.1007/978-3-540-77618-5_7,
© Springer-Verlag Berlin Heidelberg 2009

The possible shape of the function between degree of relaxation, which ranges from "danger" to "peace of mind," and the value of the function as degree of freedom of resources is shown in Fig. 7-1. It is assumed here that the relationship is nonlinear, with at least one region of over proportional growth of the relative freedom of resources. In Figure 7-1, this region of over proportional growth is defined as the starting point of the safety region:

$$S = \{x \mid f''(x) = 0\}$$

(7-2)

However, the question still remains: where does the region of safety start since other points are possible? Such further points can be located either at regions of maximum curvature or at the point of inflection.

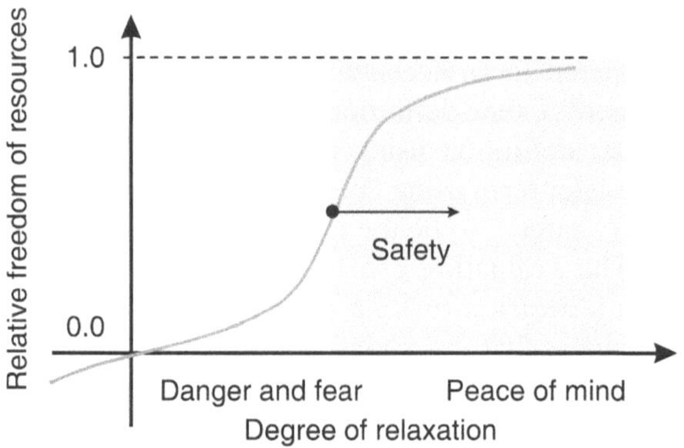

Fig. 7-1. Definition of "safety" (Proske 2008a)

In general, safety is a general requirement for humans. This claim is manifested in many laws, like the human rights of the United Nations, the German constitution with the right to life and personal integrity, the Product Liability Act, civil code, or some administrative fiats. Codes of practice are administrative fiats and state the requirement that structures have to be safe (Proske 2008b). Also, in the sense of building laws, structures have to be safe and should not endanger public safety, life, and health. Especially in the codes, safety is understood as capability of structures to resist loads. Reliability is then understood as a measure to provide this capability in different engineering fields. Here, a change from the general qualitative statement to a quantitative statement becomes obvious. This is very important for the engineers: now the engineer is enabled to prove safety by computation. The reliability is meanly understood as probability of failure (Fig. 7-2). Risk, which would be an alternative measure of (un)safety, is only considered for accidental loads. Then, a comparison between different accidental loads and emergency situations is possible (Proske 2008b).

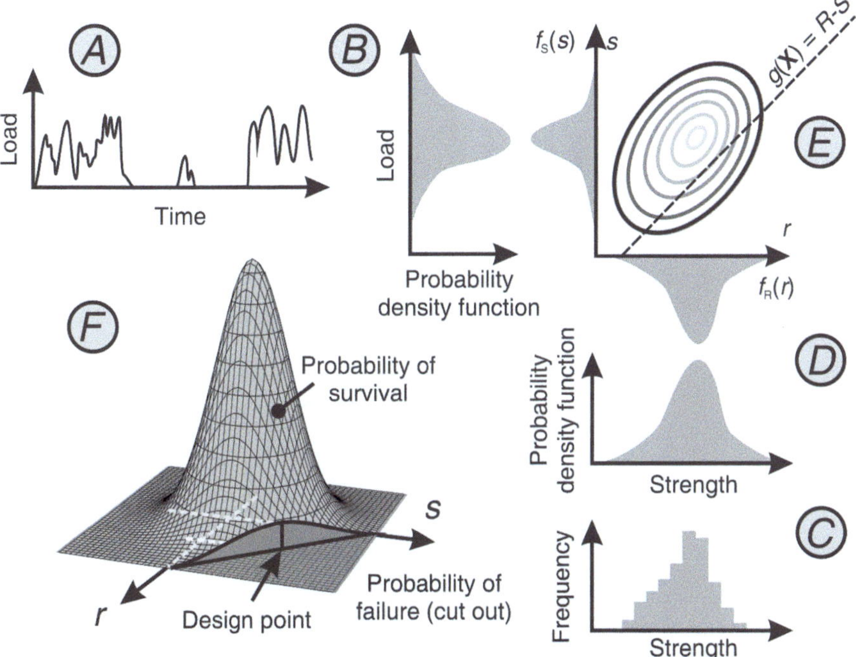

Fig. 7-2. Probability of failure for two random variables. First, (**A**) and (**C**) statistical data about the load and the strength are investigated. Then, a statistical investigation is carried out (**B**) and (**D**). Both distribution functions resulting from the statistical investigation are then merged to (**E**) further introducing a limit state function g(X). In (**F**), the two-dimensional distribution function is shown in three-dimensional illustrations of the probability of failure.

The choice of probability of failure determines stochasticity as the basis for the exposure of indeterminacy and uncertainty. Other concepts, like fuzzy-sets, rough-sets, Grey numbers, or further mathematical techniques are not considered. However, research is carried out in this field. Figure 7-3 shows the different safety concepts for structures over time.

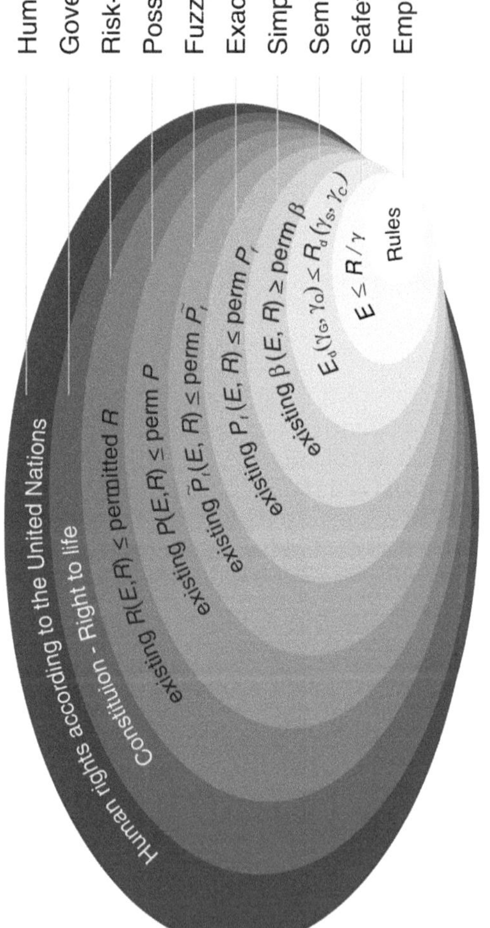

Fig. 7-3. Different safety concepts for structures

The current probabilistic or semiprobabilistic safety concept in structural engineering assumes that the exact value of many design variables is unknown. This uncertainty is based on

- Random aberrations of characteristic values of the structural resistance
- Random aberrations by transferring laboratory test results to the structure
- Random aberrations of cross-section sizes and other geometrical measures
- Geometric imperfections
- Random aberrations of internal forces like moments, shear forces, or axial forces
- Inherent uncertainties in the choice of characteristic value of loads
- Differences in the models for the loads

However, the stochastic models do not consider systematical errors like computational errors in structural design processes or bad workmanship. Such errors have to be avoided by control mechanisms (DIN 1055-100 1999).

7.2 Probabilistic Safety Concept

7.2.1 Introduction

First proposals about probabilistic-based safety concepts can be found by Mayer (1926) in Germany and Chocialov (1929) in the Soviet Union (Murzewski 1974). In the third decade of the 20th century, the number of people working in that field had already increased, just to mention Streleckij (1935) in the Soviet Union, Wierzbicki (1936) in Poland, and Prot (1936) in France (Murzewski 1974). Already in 1944 in the Soviet Union, the introduction of the probabilistic safety concept for structures had been forced by politicians (Tichý 1976). The development of probabilistic safety concepts in general experienced a strong impulse during and after World War II, not only in the field of structures but also in the field of aeronautics. In 1947, Freudenthal (1947) published his famous work about the safety of structures. Until now, a model code for the probabilistic safety concept of structures has been published by the JCSS (2004).

The probability of failure p_f as proof measure for safety is computed as function of the design values x. It can be referred to one year or the lifetime of the structures:

$$p_f = \int \cdots \int_{g(\mathbf{X}) \le 0} f_{\mathbf{X}}(x)dx \qquad (7\text{-}3)$$

$$p_f(n) = 1 - (1 - p_f)^n . \qquad (7\text{-}4)$$

The safety index is defined as the inverse Gauss standard distribution of the probability of failure:

$$\beta = -\Phi^{-1}(p_f). \qquad (7\text{-}5)$$

Results are given in Table 7-1. The integration of the probability of failure volume can then be transferred into an optimization task to determine the safety index. This is shown in Figs. 7-4 and 7-5.

The explained safety concept can be found in many regulations, such as Eurocode 1 (1994), DIN 1055-100 (1999), GruSiBau (1981), and JCSS Modelcode (2004). In these regulations, goal values for safety indexes can also be found. These values are then the basis for the estimation of safety factors, which are introduced for practical reasons.

Table 7-1. Conversion of probability of failure to safety index

Probability of failure	10^{-12}	10^{-11}	10^{-10}	10^{-9}	10^{-8}	10^{-7}	10^{-6}	10^{-5}	10^{-4}	10^{-3}	10^{-2}	10^{-1}	0.5
Safety index		7.03	6.71	6.36	5.99	5.61	5.19	4.75	4.26	3.72	3.09	2.33	1.28	0.0

7.2.2 First-order Reliability Method (FORM)

Since structures should feature a low probability of failure, the computation of the multidimensional probability may be simplified due to the low value. The simplification explained in this section increases the speed of the computation tremendously compared to a numerical integration of a multidimensional space.

In general, the simplification is based on the transfer of the integration of a multidimensional volume into an extreme value task. The result of this extreme value computation is the substitute measure safety index. The safety index itself describes the shortest distance between the origin in a standard normal distributed space and the limit state function, $g(\mathbf{X})$. A relationship between the probability of failure volume and the distance expressed by the safety index exists in this space (Table 7-1).

The standard normal distribution is characterized by a normal distribution with a mean value of 0 and a standard deviation of 1. The general assumption of this procedure is the transformability of all arbitrary random distribution functions into standard normal distribution functions. The second assumption is the linearization of the limit state function. The linearization gave the following name for the technique: first-order reliability method (FORM). The point of linearization on the limit state function is the so-called design point. This point is characterized by maximum probability of failure at the limit state function.

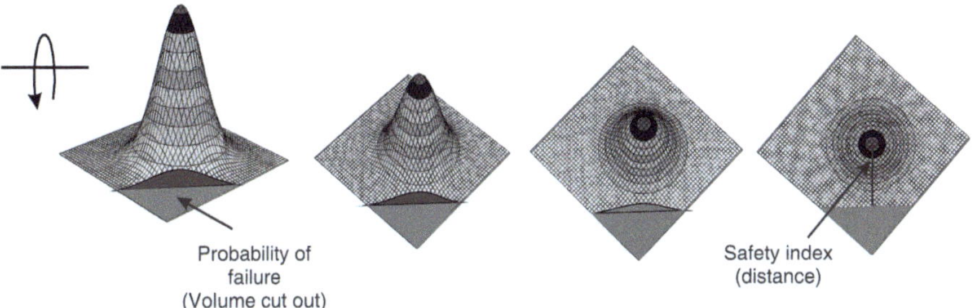

Fig. 7-4. Transfer of the probability volume into an extreme value computation

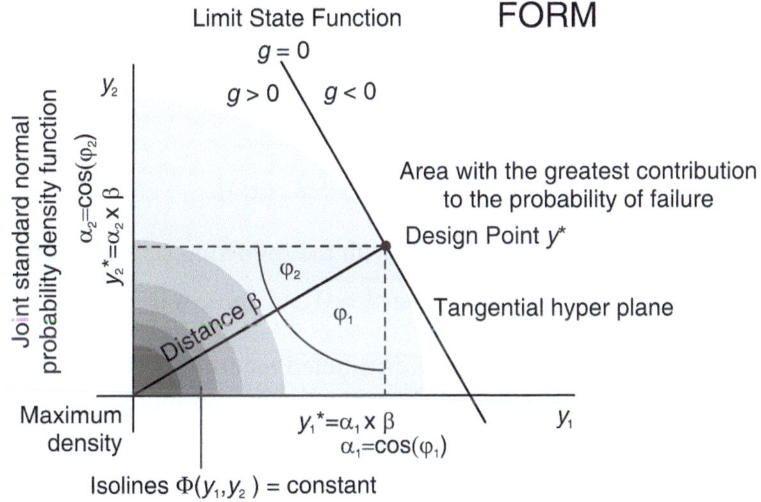

Fig. 7-5. Visualization of FORM

However, a FORM computation not only delivers the safety index as result, but also measures which may be useful to compute partial safety factors, characteristic values, and design values. These values can be found in many codes of practice and indicate the strong relationship between the codes and this probabilistic safety concept. Therefore, current safety concepts are called semiprobabilistic safety concepts.

In the following paragraphs, the FORM-methodology will be introduced in detail. The method is often called the Rackwitz-Fießler (Fießler et al. 1976) algorithm or normal tail approximation. In general, the concept is based on the fundamental work by Hasofer and Lind (1974).

In the procedure, first the non-normal distributed random variables have to be transferred into normal random variables. The following formulas will be used

$$f_{x_i}(x_i^*) = \frac{1}{\sigma_{x_i}^*} \varphi\left(\frac{x_i^* - m_{x_i}^*}{\sigma_{x_i}^*}\right) \tag{7-6}$$

$$F_{x_i}(x_i^*) = \Phi\left(\frac{x_i^* - m_{x_i}^*}{\sigma_{x_i}^*}\right) \tag{7-7}$$

with x_i^* as design point, $m_{x_i}^*$ as mean value, and $\sigma_{x_i}^*$ as standard deviation of the normal distribution. Since the normal distribution should be used as an approximation of the original distribution, mean value and standard deviation have to be computed by rearranging the formulas

$$\sigma_{x_i}^* = \frac{1}{f_{x_i}(x_i^*)} \varphi(\Phi^{-1}(F_{x_i}(x_i^*))) \tag{7-8}$$

$$m_{x_i}^* = x_i^* - \sigma_{x_i}^* \Phi^{-1}(F_{x_i}(x_i^*)) . \tag{7-9}$$

After that, an iteration cycle with the following steps is started:

1. Define an iteration counter $k = 0$ and chose a design point for the first iteration.
2. Transfer all non-normal distributed random variables into normal distributed random variables according to the following equations, with $i = 1, 2, \ldots, m$ and m as number of random variables considered:

$$\sigma_{x_i}^{*(k)} = \frac{1}{f_{x_i}(x_i^{(k)})} \varphi(\Phi^{-1}(F_{x_i}(x_i^{(k)}))) \tag{7-10}$$

$$m_{x_i}^{*(k)} = x_i^{(k)} - \sigma_{x_i}^{(k)} \Phi^{-1}(F_{x_i}(x_i^{(k)})) . \tag{7-11}$$

3. Compute the value of $x_i^{(k)}$ in the standardized space $y_i^{(k)}$:

$$y_i^{(k)} = \frac{x_i^{(k)} - m_{x_i}^{*(k)}}{\sigma_{x_i}^{*(k)}} . \qquad (7\text{-}12)$$

4. Compute the limit state function and the first derivative at $y_i^{(k)}$:

$$h(\mathbf{y}^{(k)}) = g(\mathbf{x}^{(k)}) \qquad (7\text{-}13)$$

$$\left.\frac{\partial h}{\partial y_i}\right|_{\mathbf{y}=\mathbf{y}^{(k)}} = \left.\frac{\partial g}{\partial x_i}\right|_{\mathbf{x}=\mathbf{x}^{(k)}} \cdot \frac{\partial x_i}{\partial y_i} = \left.\frac{\partial g}{\partial x_i}\right|_{\mathbf{x}=\mathbf{x}^{(k)}} \cdot \sigma_i^{*(k)} . \qquad (7\text{-}14)$$

5. Compute the coefficients of the tangential hyperplane at $h(y) = 0$ at point $y_i^{(k)}$:

$$\alpha_i^{(k)} = \frac{\left.\dfrac{\partial h}{\partial y_i}\right|_{\mathbf{y}=\mathbf{y}^{(k)}}}{\left(\displaystyle\sum_{j=1}^{m}\left(\left.\dfrac{\partial h}{\partial y_j}\right|_{\mathbf{y}=\mathbf{y}^{(k)}}\right)^2\right)^{1/2}} \qquad (7\text{-}15)$$

$$\delta^{(k)} = \frac{h(y^{(k)}) - \displaystyle\sum_{j=1}^{m} y_j^{(k)} \left.\dfrac{\partial h}{\partial y_j}\right|_{\mathbf{y}=\mathbf{y}^{(k)}}}{\left(\displaystyle\sum_{j=1}^{m}\left(\left.\dfrac{\partial h}{\partial y_j}\right|_{\mathbf{y}=\mathbf{y}^{(k)}}\right)^2\right)^{1/2}} . \qquad (7\text{-}16)$$

6. Compute a new estimation of the design point in the original space for $i = 1, 2, \ldots , m$:

$$x_i^{(k+1)} = m_{x_i}^{*(k)} - \alpha_i^{*(k)} \cdot \sigma_{x_i}^{*(k)} \cdot \delta^{(k)} . \qquad (7\text{-}17)$$

7. Verify if $x_i^{(k+1)} \approx x_i^{(k)}$. If it is fulfilled, then the design point has been found and the safety index is $\beta = \delta$ with $h(0) > 0$. If it is not fulfilled, then the iteration starts again with Step 2.

This method is very practicable and has experienced a major spreading. It gives fast and accurate results if the probability of failure is small, the random distribution functions do not diverge too strong from the normal distribution, and the limit state function does not show a strong curvature.

7.2.3 Second-order Reliability Method (SORM)

7.2.3.1 Breitung's method

If the limit state function shows a strong curvature, then the curvature has to be considered in the computation of the safety index (Fig. 7-6). This can be done by the second-order reliability method. Here, the curvature of the limit state function is approximated using

$$h(\mathbf{y}) = h(\mathbf{y}^*) + (\mathbf{y} - \mathbf{y}^*)^T \cdot \nabla h(\mathbf{y}^*) + \frac{1}{2}(\mathbf{y} - \mathbf{y}^*)^T \cdot \mathbf{B}_y \cdot (\mathbf{y} - \mathbf{y}^*) = 0 . \tag{7-18}$$

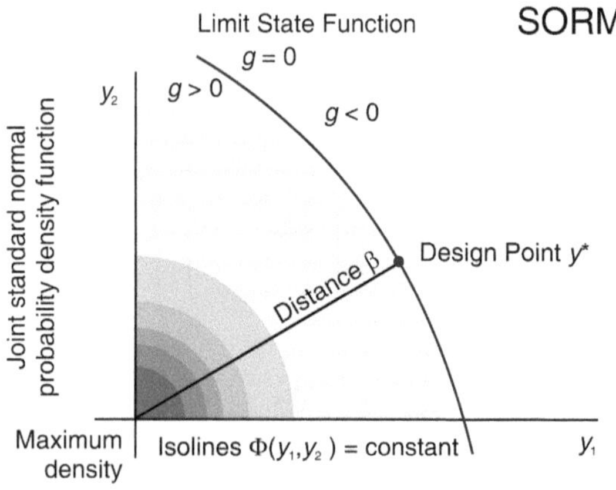

Fig. 7-6. Visualization of SORM

$\mathbf{B}_y$ is the matrix of the second and mixed derivates from $h(\mathbf{y})$ in the standardized space at the design point. Breitung (1984) has introduced the following equation with a_i and $i = 1, 2, \ldots , m-1$ as principal curvature of h at the design point in the standard normal space

$$P_f = \Phi(-\beta) \prod_{i=1}^{m-1} (1 - \beta \cdot a_i)^{-1/2} . \tag{7-19}$$

However, the computation of the principal curvatures represents the major part of this method. To compute the principal curvatures, a rotation of the coordinate system is required. The rotation requires orthogonalizing using the Schmidt process.

The old and new coordinates are linked as follows:

$$\mathbf{y} = \mathbf{D} \cdot \mathbf{u} . \tag{7-20}$$

After the rotation the new coordinates can be computed as

$$\mathbf{u} = \mathbf{D}^T \cdot \mathbf{y}. \tag{7-21}$$

The matrix $\mathbf{D}$ consists of

$$\mathbf{D} = (\mathbf{d}_1, \mathbf{d}_2, ..., \mathbf{d}_m)^T \tag{7-22}$$

with

$$\mathbf{d}_1 = \alpha \tag{7-23}$$

$$\mathbf{d}_k = \frac{f_k}{|f_k|}$$

and

$$f_k = \mathbf{e}_k - \sum_{l=1}^{k-1} (\mathbf{e}_k^T \mathbf{d}_l) \mathbf{d}_l \tag{7-24}$$

$$k = 2, 3, ..., m.$$

and $\mathbf{e}_k$ is the kth unit vector. In the new system,

$$\mathbf{u}^* = \mathbf{D}^T \mathbf{y}^* = (\beta, 0, 0, ..., 0)^T \tag{7-25}$$

and

$$\nabla g_u = \mathbf{D} \cdot \nabla h = \left(\frac{\partial g_u}{\partial u_1}, 0, 0, ..., 0 \right)^T. \tag{7-26}$$

$$g_u(\mathbf{u}^*) = 0 \tag{7-27}$$

$$\mathbf{B}_u = \mathbf{D}^T \mathbf{B}_y \mathbf{D} \tag{7-28}$$

Taylor's theorem becomes

$$(u_1 - u_1^*) \cdot \frac{\partial g_u}{\partial u_1} + \frac{1}{2} (\mathbf{u} - \mathbf{u}^*)^T \cdot \mathbf{B}_u \cdot (\mathbf{u} - \mathbf{u}^*) = 0 . \tag{7-29}$$

The principal curvatures are the roots of the following equation:

$$\det \left(\frac{\hat{\mathbf{B}}_u}{\partial g_u / \partial u_1} - a \cdot \mathbf{I} \right) = 0 , \tag{7-30}$$

When $\hat{\mathbf{B}}_u$ is the matrix of the second and mixed derivates, and can be derived by deleting the first line and first column from $\mathbf{B}_u$, I is the identity matrix.

The approximation by Breitung gives good results, if the curvature of the limit state function still is not too strong and the safety index is small.

7.2.3.2 Tvedt's method

Tvedt (1988) has introduced an extension of Breitung's formulae:

$$\gamma = 1 - P_f \tag{7-31}$$

$$\gamma = 1 - (A_1 + A_2 + A_3) \tag{7-32}$$

with

$$A_1 = \Phi(-\beta)\prod_{j=1}^{n-1}(1 + \beta \cdot a_j)^{-1/2} \tag{7-33}$$

$$A_2 = [\beta \cdot \Phi(-\beta) - \varphi(\beta)] \cdot \left\{ \prod_{j=1}^{n-1}(1 + \beta \cdot a_j)^{-1/2} - \prod_{j=1}^{n-1}(1 + (\beta+1)a_j)^{-1/2} \right\} \tag{7-34}$$

$$A_3 = (\beta+1)[\beta \cdot \Phi(-\beta) - \varphi(\beta)] \tag{7-35}$$
$$\cdot \left\{ \prod_{j=1}^{n-1}(1 + \beta \cdot a_j)^{-1/2} - \mathrm{Re}\left[\prod_{j=1}^{n-1}(1 + (\beta+\mathrm{i}) \cdot a_j)^{-1/2} \right] \right\} \cdot$$

The extension of Breitung's formula provides accurate results with low and semi-low probabilities of failure. High probabilities of failure and negative curvatures are not covered by the method. A further improvement was done by the so-called Tvedt's exact solution for parabolic limit state function. Here, the size of the probability of failure is not limited:

$$\gamma = 1 - P_f \tag{7-36}$$

$$\gamma = 0.5 + \frac{1}{\pi}\int_0^\infty \sin\left[\beta \cdot \theta + \frac{1}{2}\sum_{j=1}^{n-1}\arctan(a_j\theta) \right] \frac{\exp(-1/2\theta^2)}{\theta\prod_{j=1}^{n-1}(1 + a_j^2\theta^2)^{1/4}}\,\mathrm{d}\theta. \tag{7-37}$$

The disadvantage of this method is the requirement of a numerical integration.

7.2.3.3 Method by Köylüoglu and Nielsen

Additionally, Köylüoglu and Nielsen (1994) have tried to improve the accuracy of SORM methods for higher probabilities of failure. They also formulate

$$\gamma = 1 - P_f . \tag{7-38}$$

Assuming that all curvatures a_i are positive, then

$$\gamma = 1 - \Phi(-\beta) \prod_{j=1}^{n-1} \frac{1}{\sqrt{1 + a_j / c_{0,1}}}$$

$$\cdot \left\{ 1 + \frac{1}{2} c_{1,1} \sum_{k=1}^{n-1} \frac{a_k}{1 + a_k / c_{0,1}} + \frac{1}{4} c_{2,1} \left[\left(\sum_{k=1}^{n-1} \frac{a_k}{1 + a_k / c_{0,1}} \right)^2 + 2 \sum_{k=1}^{n-1} \left(\frac{a_k}{1 + a_k / c_{0,1}} \right)^2 \right] \right.$$

$$+ \frac{1}{8} c_{3,1} \left[\left(\sum_{k=1}^{n-1} \frac{a_k}{1 + a_k / c_{0,1}} \right)^3 + 2 \left(\sum_{k=1}^{n-1} \frac{a_k}{1 + a_k / c_{0,1}} \right) \left(\sum_{k=1}^{n-1} \left(\frac{a_k}{1 + a_k / c_{0,1}} \right)^2 \right) \right.$$

$$\left. \left. + 12 \sum_{k=1}^{n-1} \left(\frac{a_k}{1 + a_k / c_{0,1}} \right)^3 \right] + \dots \right\} .$$

$$\tag{7-39}$$

If all curvatures a_i are negative, then

$$\gamma = 1 - \Phi(+\beta) \prod_{j=1}^{n-1} \frac{1}{\sqrt{1 - a_j / c_{0,2}}}$$

$$\cdot \left\{ 1 + \frac{1}{2} c_{1,2} \sum_{k=1}^{n-1} \frac{a_k}{1 - a_k / c_{0,2}} + \frac{1}{4} c_{2,2} \left[\left(\sum_{k=1}^{n-1} \frac{a_k}{1 - a_k / c_{0,2}} \right)^2 + 2 \sum_{k=1}^{n-1} \left(\frac{a_k}{1 + a_k / c_{0,2}} \right)^2 \right] \right.$$

$$+ \frac{1}{8} c_{3,2} \left[\left(\sum_{k=1}^{n-1} \frac{a_k}{1 + a_k / c_{0,2}} \right)^3 + 2 \left(\sum_{k=1}^{n-1} \frac{a_k}{1 + a_k / c_{0,2}} \right) \left(\sum_{k=1}^{n-1} \left(\frac{a_k}{1 + a_k / c_{0,2}} \right)^2 \right) \right.$$

$$\left. \left. + 12 \sum_{k=1}^{n-1} \left(\frac{a_k}{1 + a_k / c_{0,2}} \right)^3 \right] + \dots \right\} .$$

$$\tag{7-40}$$

The generalized form is given as follows:

$$\gamma = \Phi(\beta) + \Phi(-\beta) \left[\prod_{j=m}^{n-1} \frac{1}{\sqrt{1 - a_j / d_{0,2}}} \left\{ 1 + \frac{1}{2} d_{1,2} \sum_{k=m}^{n-1} \frac{a_k}{1 - a_k / c_{0,2}} + \ldots \right\} \right. \tag{7-41}$$

$$\left. \cdot \left\{ 1 - \prod_{j=1}^{m-1} \frac{1}{\sqrt{1 + a_j / c_{0,1}}} \left\{ 1 + \frac{1}{2} c_{1,1} \sum_{k=m}^{m-1} \frac{a_k}{1 + a_k / c_{0,1}} + \ldots \right\} \right\} \right]$$

$$- \Phi(\beta) \left[\prod_{j=1}^{m-1} \frac{1}{\sqrt{1 + a_j / d_{0,1}}} \left\{ 1 + \frac{1}{2} d_{1,1} \sum_{k=1}^{m-1} \frac{a_k}{1 + a_k / c_{0,1}} + \ldots \right\} \right.$$

$$\left. \cdot \left\{ 1 - \prod_{j=m}^{n-1} \frac{1}{\sqrt{1 - a_j / c_{0,2}}} \left\{ 1 + \frac{1}{2} c_{1,2} \sum_{k=1}^{n-1} \frac{a_k}{1 - a_k / c_{0,2}} + \ldots \right\} \right\} \right].$$

Based on the cut-off of the terms, different approximation formulations can be derived. For one term with positive curvature, the coefficient becomes

$$c_{0,1} = \frac{\Phi(-\beta)}{\varphi(\beta)} \tag{7-42}$$

and $c_{1,1} = c_{2,1} = \ldots = 0$. If two terms are chosen, then the coefficients can be computed as

$$c_{0,1} = \frac{\Phi(-\beta)}{\varphi(\beta)} \left(\frac{1}{1 + \sqrt{1 - \beta \Phi(-\beta) / \varphi(\beta)}} \right) \tag{7-43}$$

$$c_{1,1} = \frac{\varphi(\beta)}{\Phi(-\beta)} \sqrt{1 - \frac{\beta \Phi(-\beta)}{\varphi(\beta)}} \tag{7-44}$$

and $c_{2,1} = c_{3,1} = \ldots = 0$. With three terms, one achieves

$$\frac{1}{c_{0,1}} - c_{1,1} = \frac{\varphi(\beta)}{\Phi(-\beta)} \tag{7-45}$$

$$\frac{1}{c_{0,1}^2} - 2\frac{c_{1,1}}{c_{0,1}} + 2c_{2,1} = \frac{\beta \varphi(\beta)}{\Phi(-\beta)} \tag{7-46}$$

$$\frac{1}{c_{0,1}^3} - 3\frac{c_{1,1}}{c_{0,1}^2} + 6\frac{c_{2,1}}{c_{0,1}} = \frac{(\beta^2 - 1)\varphi(\beta)}{\Phi(-\beta)}. \tag{7-47}$$

The three equations can be summarized to a cubic equation with at least one positive solution. The solution of $c_{0,1}$ less than the value assessed with the following formulae should be taken:

$$c_{0,1} = \frac{\Phi(-\beta)}{\varphi(\beta)}\left(\frac{1}{1+\sqrt{1-\beta\Phi(-\beta)/\varphi(\beta)}}\right) \tag{7-48}$$

Based on the cut off of the terms, different approximation formulations can be derived. For one term with negative curvature, the coefficient becomes

$$c_{0,2} = \frac{\Phi(\beta)}{\varphi(\beta)} \tag{7-49}$$

and $c_{1,2} = c_{2,2} = \ldots = 0$. If two terms are chosen, then the coefficients can be computed as

$$c_{0,2} = \frac{\Phi(\beta)}{\varphi(\beta)}\left(\frac{1}{1+\sqrt{1+\beta\Phi(\beta)/\varphi(\beta)}}\right) \tag{7-50}$$

$$c_{1,2} = \frac{-\varphi(\beta)}{\Phi(\beta)}\sqrt{1+\frac{\beta\Phi(\beta)}{\varphi(\beta)}} \tag{7-51}$$

and $c_{2,2} = c_{3,2} = \ldots = 0$. With three terms, one achieves

$$\frac{1}{c_{0,2}} + c_{1,2} = \frac{\varphi(\beta)}{\Phi(\beta)} \tag{7-52}$$

$$\frac{1}{c_{0,2}^2} + 2\frac{c_{1,2}}{c_{0,2}} + 2c_{2,2} = -\frac{\beta\varphi(\beta)}{\Phi(\beta)} \tag{7-53}$$

$$\frac{1}{c_{0,2}^3} + 3\frac{c_{1,2}}{c_{0,2}^2} + 6\frac{c_{2,2}}{c_{0,2}} = \frac{(\beta^2 - 1)\varphi(\beta)}{\Phi(-\beta)} \tag{7-54}$$

Again, the three equations can be summarized in to a cubic equation with at least one positive solution. The solution of $c_{0,2}$ less than the value assessed with the following formulae should be taken.

$$c_{0,2} = \frac{\Phi(\beta)}{\varphi(\beta)} \left(\frac{1}{1 + \sqrt{1 - \beta\Phi(\beta)/\varphi(\beta)}} \right) \qquad (7\text{-}55)$$

If the d-terms are approximated, one achieves for a $d_{0,1} = 2c_{0,1}$, $d_{0,2} = 2c_{0,2}$ $d_{1,1} = d_{2,1} = \ldots = 0$ and $d_{1,2} = d_{2,2} = \ldots = 0$ one-term formulation.

7.2.3.4 Method by Cai and Elishakoff

Cai and Elishakoff (1994) also try to improve Breitung's method by extending the original formulae into a Taylor theorem. The application is quite simple:

$$P_f = \Phi(\beta) + \frac{1}{\sqrt{2\pi}} \exp\left(-\frac{\beta^2}{2} \right) (D_1 + D_2 + D_3 + \ldots) . \qquad (7\text{-}56)$$

The single elements of the Taylor theorem are

$$D_1 = \sum_j \lambda_j \qquad (7\text{-}57)$$

$$D_2 = -\frac{1}{2}\beta \left(3 \cdot \sum_j \lambda_j^2 + \sum_{j \neq k} \lambda_j \lambda_k \right) \qquad (7\text{-}58)$$

$$D_3 = \frac{1}{6}(\beta^2 - 1) \left(15 \cdot \sum_j \lambda_j^3 + 9 \cdot \sum_{j \neq k} \lambda_j^2 \lambda_k + \sum_{j \neq k \neq l} \lambda_j \lambda_k \lambda_l \right) . \qquad (7\text{-}59)$$

The basis for the computation of the elements is again the principal curvatures. However, these values have already been computed, if Breitung's formulae were employed:

$$a_j = -2\lambda_j . \qquad (7\text{-}60)$$

7.2.3.5 Further developments

The latest SORM method was introduced by Zhao and Ono (1999a, b) and by Polidori et al. (1999). The first method is based on the inverse fast Fourier transformation (IFFT). These methods will not be explained here. However, it should be noted that Zhao and Ono (1999a) not only give a summary about the conditions under which the different introduced SORM methods perform well, but also impose recommendations on when to use FORM, when to use SORM methods, and when to use their own method (Zhao and Ono 1999c). However, Breitung (2002) has criticized some assumptions of the method by Zhao and Ono.

7.2.4 Hypersphere Division Method

Additional to the optimization procedure using Cartesian coordinates, the search for the minimum distance can also be carried out in spherical coordinates (see Fig. 7-7). After transformation into spherical coordinates, the angle and radius of a search vector are systematically adapted to find the minimum distance.

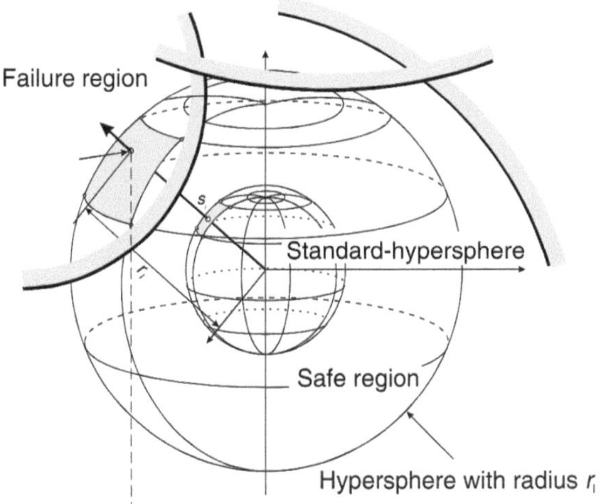

Fig. 7-7. Hypersphere division method

7.2.5 Response Surface Method

The probabilistic methods introduced so far required an analytically closed known limit state function $g(\mathbf{X})$. In other words, the limit state function should be one formula. However, in many cases, such as the computation of the ultimate load of an arch bridge, this requirement cannot be fulfilled. For example, consider a finite element program code. In those cases, the mathematical procedure has to be approximated by a surrogate, to reach to an acceptable computation when the aforementioned techniques like FORM and SORM are used. A procedure to develop such a surrogate is the response surface methodology. In this methodology, a simplified more dimensional function is computed based on some sample results of the extensive mathematical procedure originally established. The concept is easily understood when the complicated mathematical computations are substituted by some laboratory or field tests. There also, no function is known, but should be introduced. Based on the test results, a function will be introduced. Some general works about the concept can be found in Box and Draper (1987) and applications in structural engineering are mentioned in Bucher and Bourgund (1990) and Rajashekhar and Ellingwood (1993).

The concept can be described as follows. A certain function with the input variables $\mathbf{X}$ and some functional constants $\mathbf{K}$ is given with

$$g = f(\mathbf{X}, \mathbf{K}) \qquad (7\text{-}61)$$

but can only be pointwise solved. Therefore, an approximation function should be developed

$$\tilde{g} = f(\mathbf{X}, \mathbf{K}) . \qquad (7\text{-}62)$$

There are many different types of mathematical approximation functions. Probably the most applied methods are quadratic functions, either

$$\tilde{g} = a + \sum_{i=1}^{n} b_i \cdot x_i + \sum_{i=1}^{n} c_i \cdot x_i^2 \qquad (7\text{-}63)$$

or with mixed terms

$$\tilde{g} = A + \mathbf{X}^T \cdot \mathbf{B} + \mathbf{X}^T \cdot \mathbf{C} \cdot \mathbf{X} \qquad (7\text{-}64)$$

with A, $\mathbf{B}$ and $\mathbf{C}$ as constants.

$$\mathbf{B} = \begin{pmatrix} B_1 \\ B_2 \\ \vdots \\ B_n \end{pmatrix}, \quad \mathbf{C} = \begin{pmatrix} C_{11} & C_{12} & \cdots & C_{1n} \\ C_{21} & & & \\ \vdots & & & \\ C_{n1} & & & C_{nn} \end{pmatrix} . \qquad (7\text{-}65)$$

These constants can be computed based on the pointwise solutions. Depending on the number of pointwise solutions, there is an under determined number of solutions situation, an exact number, or an over determined number of solutions to compute $\mathbf{K}$. If an over determined number of solutions exist, then some minimum error methods should be applied. For the exact number of the solutions, the degrees of freedom of functions depending on the type of the function are shown in Table 7-2. Shapes of solutions points are shown in Fig. 7-8.

Table 7-2. Degrees of freedom for certain response surface functions (Weiland 2003)

Approximation function $\tilde{g}(x)$		Degrees of freedom
Linear regression	$\tilde{g}(x) = a + \sum_{i=1}^{n} b_i \cdot x_i$	$n+1$
Quadratic function including mixed terms	$\tilde{g}(x) = a + \sum_{i=1}^{n} b_i \cdot x_i + \sum_{i=1}^{n} \sum_{j=1}^{n} c_{ij} \cdot x_i \cdot x_j$	$\tfrac{1}{2} \cdot (n+1) \cdot (n+2)$
Quadratic function without mixed terms	$g(x) = a + \sum_{i=1}^{n} b_i \cdot x_i + \sum_{i=1}^{n} x_i^2$	$2n+1$
Polynomial third order with mixed terms	...	$\tfrac{1}{2} \cdot (2 + 3n + 3n^2)$
Polynomial third order without mixed terms	...	$3n+1$
Polynomial fourth order with mixed terms	...	$\tfrac{1}{2} \cdot (2 + 3n + 5n^2)$
Polynomial fourth order without mixed terms	...	$5n+1$

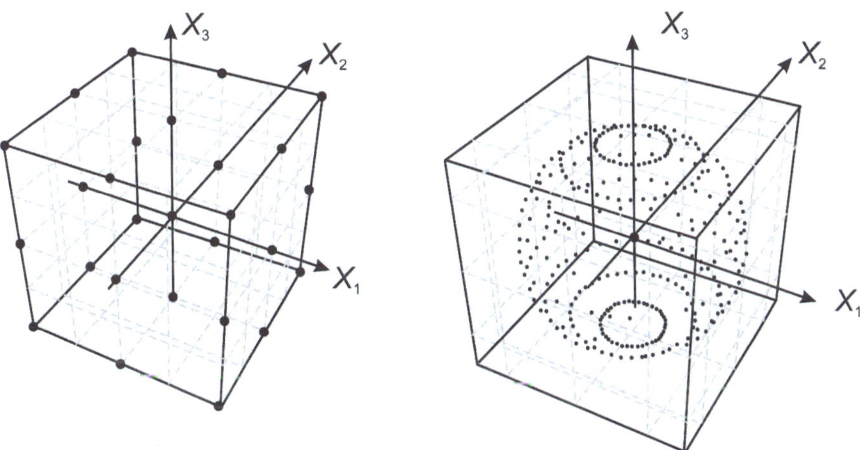

Fig. 7-8. Two examples of chosen solution points for three variables (Weiland 2003)

The approximated response surface can then be updated after the next probabilistic computation. This means that the response surface mainly acts as a local approximation and does not give a good approximation over the entire range of the original function. However, that is not required for FORM/SORM computations:

$$x_{D=m}^{(k+1)} = x_m^{(k)} + (x_D^k - x_m^{(k)}) \frac{g(x_m^{(k)})}{(g(x_m^{(k)}) - g(x_D^k))} \tag{7-66}$$

with x_m as centre point, x_D as design point based on a FORM computation using the response surface, and k as iteration counter. The iteration scheme is shown in Fig. 7-9.

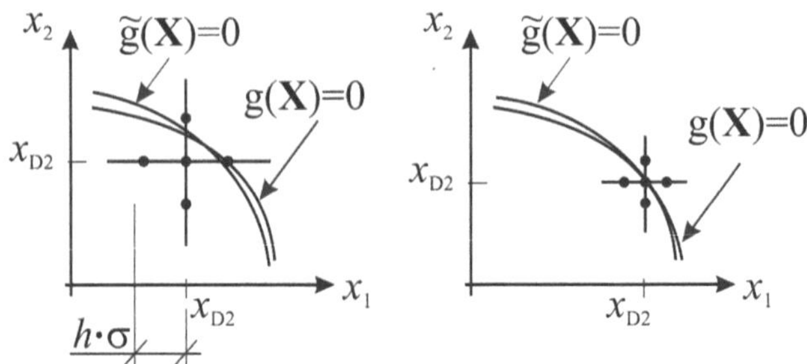

Fig. 7-9. Iterative improvement of the response surface (Klingmüller and Bourgund 1992)

The major advantage of this approximation is a simple application. The schema can easily be extended to existing finite element programs or other numerical tools. The computation is easily understandable and the number of computations is low.

The major disadvantage is a limited capability to find the extreme value of complicated functions. The number of iterations also depends on the number of random variables. Therefore, in high-dimensional cases, the response surface method may perhaps cause heavy computations. Furthermore, the approximation method only uses data points from one iteration cycle. However, it may perhaps be useful to keep the data for further investigations. There exist external programs that can carry out response surface computations afterwards by using all available data.

Since the limitations of the response surface method are known, in the last few years many new methods have been developed (Roos and Bayer 2008, Ross and Bucher 2003). An adaptive response surface method has been introduced by Most (2008).

7.2.6 Monte Carlo Simulation

7.2.6.1 Crude Monte Carlo Simulation

In contrast to the FORM and SORM methods, the Monte Carlo Simulation is an integration procedure, not an extreme value computation. Therefore, some assumptions required for the FORM/SORM are not relevant for the Monte Carlo Simulation. Also, Monte Carlo Simulation is extremely easy to program and to apply. However, if the probabilities of failure are extremely low and as required by codes, then Monte Carlo Simulation will require extensive computation power.

The general idea of Monte Carlo Simulation is, as the name already indicates, the application of pure random numbers into a computation flow. In its simplest description, Monte Carlo Simulation is an extensive version of trial and error. The only assumption for this technique is some quality requirements for random numbers. Since computers cannot provide real random numbers, they produce pseudo-random numbers based on purely deterministic causal computations; the period of the numbers should be big enough so that random numbers are not repeated in the Monte Carlo Simulation. There are many programs available to provide high-quality random numbers (NR 1992). The Monte Carlo Simulation is then

$$\int f \, dV \approx V \langle f \rangle \pm V \sqrt{\frac{\langle f^2 \rangle - \langle f \rangle^2}{N}} , \qquad (7\text{-}68)$$

where V represents the volume, $V \langle f \rangle$ stands for the mean value of the function f over the sample size N, and the $\pm$-term gives more or less an one standard deviation error estimator. The further functions are

$$\langle f \rangle \equiv \frac{1}{N} \sum_{i=1}^{N} f(x_i) \qquad \langle f^2 \rangle \equiv \frac{1}{N} \sum_{i=1}^{N} f^2(x_i) . \qquad (7\text{-}69)$$

The major advantage of the Monte Carlo Simulation concerning the dimensions is the fact that the statistical error is independent from the number of dimensions. For Monte Carlo Simulation, it does not matter if there is 1 or 50 random input variables, the statistical error will remain the same. This is completely different for Simpson's rule applied for integration, where the required computation grows exponentially with the number of dimensions.

However, the required sample size has a major influence on the error size in Monte Carlo Simulation. The next equation is an example to evaluate the required sample size n_{req} (Flederer 2001)

$$n_{req} = \frac{1}{1-P_\varepsilon} \cdot \frac{(1-P_f)}{P_f \cdot \varepsilon^2}$$

(7-70)

with P_f as probability of failure, P_ε as level of significance, and ε as statistical error. From the equation, it becomes understandable that with low probabilities of failure and low statistical error, high sample sizes are required. Macke (2000) has given a good example, where the probability of failure was 10^{-6} and the required statistical error was less than 50%. The required sample size was 4×10^6. If the statistical error should be less than 10%, then more than 10^8 samples are required. Based on these properties, Monte Carlo Simulation is a good method for high probabilities of failure and for high dimensions. If lower probabilities were to be investigated with Monte Carlo Simulation, the so-called variance reduction techniques should be applied. This is mainly done after a preliminary FORM/SORM computation.

7.2.6.2 Variance-reduced Monte Carlo Simulation

Since a great variety of variance reduction techniques exist, here only a few will be mentioned. Probably the most applied technique is importance sampling. Here, the original probability integral

$$P_f = \int \cdots \int_{g<0} f_x(\mathbf{x})d\mathbf{x}$$

(7-71)

will be transferred using an indicator or a weighting function to

$$P_f = \int \cdots \int_{all} I(x) f_x(\mathbf{x})d\mathbf{x} .$$

(7-72)

The weighting function is defined as

$$I(\mathbf{x}) = \begin{cases} 1 & g(\mathbf{x}) < 0 \\ 0 & g(\mathbf{x}) \geq 0 \end{cases}$$

(7-73)

This permits a transformation of the random distributions using a chosen distribution function $h_v(\mathbf{v})$:

$$P_f = \int \cdots \int_{all} I(\mathbf{v}) \frac{f_x(\mathbf{v})}{h_v(\mathbf{v})} h_v(\mathbf{v})d\mathbf{v} .$$

(7-74)

This function can be unbiasedly estimated with

$$\hat{P}_f = \frac{1}{m_c} \sum_{n=1}^{m_c} I(\mathbf{v}_n) \frac{f_x(\mathbf{v}_n)}{h_v(\mathbf{v}_n)} \,. \tag{7-75}$$

Furthermore, the variance can be computed as

$$\mathrm{Var}[\hat{P}_f] = \frac{1}{m_c - 1} \left[\frac{1}{m_c} \sum_{n=1}^{m_c} I(\mathbf{v}_n) \left(\frac{f_x(\mathbf{v}_n)}{h_v(\mathbf{v}_n)} \right)^2 - \hat{P}_f^2 \right]. \tag{7-76}$$

The major task in applying importance sampling is the search for a proper distribution function $h_v(\mathbf{v})$. If some prior information is available, for example by FORM/SORM computation, then the distribution function $h_v(\mathbf{v})$ can be selected very efficiently, yielding to an impressive drop of computational effort in the Monte Carlo Simulation. In general, importance sampling can be understood as a transformation of the random points toward interesting regions in the sampling space (Maes et al. 1993, Song 1997, and Ibrahim 1991).

The concept can be even further extended by updating the distribution function $h_v(\mathbf{v})$ after every sample step. This technique is called adaptive sampling (Bucher 1988 and Mori and Ellingwood 1993).

7.2.6.3 Quasi-random numbers

Another interesting technique is the application of quasi-random numbers instead of pseudo-random numbers. Quasi-random numbers are not random at all. They are constructed to fill a multidimensional space in a most efficient way. Also, they can be understood as a technique standing between the classical Simpson's rule for integration and the crude Monte Carlo Simulation. Since the numbers are deterministic, the computation error becomes related to the dimensions.

However, the major advantage of the application of quasi-random numbers with Monte Carlo Simulation is the fact that they can be applied to many finished programs. So, if Monte Carlo Simulation was chosen for a certain project and it turns out after some sample computations that the computation time is unacceptable, but the programming cannot be changed anymore, then perhaps quasi-random numbers can be produced externally and can then given to the program. This technique was applied in Flederer (2001). More details about possible application can be found in Curbach et al. (2002).

7.2.7 Combination of Safety Indexes

Whereas for Monte Carlo Simulation the number of limit state functions is irrelevant for the FORM and SORM methods, perhaps different limit state functions were considered separately and have to be merged into one single probability of failure or safety index. This can be seen at an arch bridge, where several different point of failure can be identified. The question that follows is whether these points are correlated or not.

If the different limit state functions are uncorrelated, then system probability of failure can be computed as

$$P_f = 1 - \prod_{j=1}^{n}(1 - P_{fj}).$$

(7-77)

If the single probabilities of failure are rather small, the values can be added instead of multiplying

$$P_f \approx \sum_{j=1}^{n} P_{fj}.$$

(7-78)

The error term is then not higher than

$$E(P_f) \leq \frac{1}{2}\left(\sum_{j=1}^{n} P_{fj}\right)^2.$$

(7-79)

If the probabilities of failure are expressed as safety indexes, the formulas become

$$\beta_{sys} = -\Phi^{-1}\left(1 - \prod_{j=1}^{n}\Phi(\beta_j)\right) \approx -\Phi^{-1}\left(\sum_{j=1}^{n}\Phi(-\beta_j)\right).$$

(7-80)

If the noncorrelation of the limit states does not hold true, then the correlations have to be considered and a multidimensional normal distribution can be applied. The correlations are expressed as

$$\rho_{jk} = \alpha_j^T \alpha_k = \alpha_{j1}\alpha_{k1} + \alpha_{j2}\alpha_{k2} + ... + \alpha_{jm}\alpha_{km}$$

(7-81)

and

$$\alpha_j^T = (\alpha_{j1} + \alpha_{j2} + ... + \alpha_{jm}).$$

(7-82)

With α_j^T are the weighting factors of the m random variables from the jth limit state function giving the safety index β_j. For the correlation matrix of the different limit state functions, one gets

$$R = \begin{pmatrix} 1 & \rho_{12} & \cdots & \rho_{1n} \\ \rho_{21} & 1 & & \vdots \\ \vdots & & \ddots & \vdots \\ \rho_{n1} & \rho_{n2} & \cdots & 1 \end{pmatrix} \tag{7-83}$$

and a vector for the safety indexes

$$\beta = \begin{pmatrix} \beta_1 \\ \beta_2 \\ \vdots \\ \beta_n \end{pmatrix}. \tag{7-84}$$

The system probability of failure is obtained by transformation in the standard normal space and linearization of the limit state functions:

$$P_f = 1 - P_s \tag{7-85}$$

$$= 1 - P\left(\bigcap_{j=1}^{n} (g_j(X) \geq 0) \right) \tag{7-86}$$

$$= 1 - P\left(\bigcap_{j=1}^{n} (h_j(Y) \geq 0) \right) \tag{7-87}$$

$$\approx 1 - P\left(\bigcap_{j=1}^{n} (l_j(Y) \geq 0) \right) \tag{7-88}$$

$$= 1 - P\left(\bigcap_{j=1}^{n} (-Z_j^* \leq \beta_j) \right) \tag{7-89}$$

$$= 1 - \Phi_n(\beta, \mathbf{R}), \tag{7-90}$$

with $\Phi_n(\beta, \mathbf{R})$ as standardized n-dimensional normal distribution.

$$\Phi_n(\beta, \mathbf{R}) = \frac{1}{(2\pi)^{n/2} |R|^{1/2}} \int_{-\infty}^{\beta_n} \cdots \int_{-\infty}^{\beta_1} \exp\left(-\frac{1}{2} \psi^T \mathbf{R}^{-1} \mathbf{y} \right) dy_1, dy_2, \cdots dy_n.$$

$$\tag{7-91}$$

Several different programs can be found for the evaluation of this function (Schervish 1984, Genz 1992, Drezner 1992, and Yuan and Pandey 2006).

To simplify the computation, it is often assumed that the correlations are equal between different limit state functions. The correlation matrix becomes

$$R = \begin{pmatrix} 1 & \rho & \cdots & \rho \\ \rho & 1 & & \vdots \\ \vdots & & \ddots & \vdots \\ \rho & \rho & \cdots & 1 \end{pmatrix} \tag{7-92}$$

and the multidimensional standard normal distribution changes to

$$\Phi_n(\beta, \mathbf{R}) = \int_{-\infty}^{+\infty} \varphi(x) \prod_{i=1}^{n} \Phi\left(\frac{\beta_i + \sqrt{\rho} \cdot x}{\sqrt{1-\rho}} \right) dx. \tag{7-93}$$

Based on this idea of a mean correlation matrix, the coefficients of the correlation matrix can be understood as products of single elements, like

$$R = \begin{pmatrix} 1 & \lambda_1 \cdot \lambda_2 & \cdots & \lambda_1 \cdot \lambda_n \\ \lambda_2 \cdot \lambda_1 & 1 & & \vdots \\ \vdots & & \ddots & \vdots \\ \lambda_n \cdot \lambda_1 & \lambda_n \cdot \lambda_2 & \cdots & 1 \end{pmatrix}, \ |\lambda_i| < 1,\ |\lambda_i| < 1,\ i,j = 1,\, 2,\, \dots,\, n \tag{7-94}$$

and the multidimensional standard normal distribution again changes

$$\Phi_n(\beta, \mathbf{R}) = \int_{-\infty}^{+\infty} \varphi(x) \prod_{i=1}^{n} \Phi\left(\frac{\beta_i + \lambda_i \cdot x}{\sqrt{1-\lambda_i^2}} \right) dx. \tag{7-95}$$

Suggestions of lower bounds for the correlations values are given as

$$\lambda_j \cdot \lambda_k \le \rho_{jk} \quad j \ne k \tag{7-96}$$

and for the upper bound as

$$\lambda_j \cdot \lambda_k \ge \rho_{jk} \quad j \ne k \tag{7-97}$$

and

$$-1 \le \lambda_j, \lambda_k \le 1 \tag{7-98}$$

should remain valid. An optimal solution would be achieved when

$$\lambda_j \cdot \lambda_k = \rho_{jk} \, . \tag{7-99}$$

That will only be possible in some rare cases. However, if the ρ_{jk} has approximately the same size and is positive, then the following recommendation for lower bounds

$$\lambda_j = \sqrt{\max_j \{\rho_{jk}\}} \quad j \neq k \tag{7-100}$$

and upper bounds

$$\lambda_j = \sqrt{\min_j \{\rho_{jk}\}} \quad j \neq k \, . \tag{7-101}$$

can be given.

If the contribution of the limit state functions to the system probability of failure can be roughly estimated, then the three limit state functions with the highest probability of failure or the lowest safety index can be chosen, and the following computation can be carried out:

$$\lambda_1 \cdot \lambda_2 = \rho_{12}, \quad \lambda_2 \cdot \lambda_3 = \rho_{23}, \quad \lambda_1 \cdot \lambda_3 = \rho_{13} \tag{7-102}$$

$$\lambda_1 = \sqrt{\frac{\rho_{12} \cdot \rho_{13}}{\rho_{23}}}, \quad \lambda_2 = \sqrt{\frac{\rho_{21} \cdot \rho_{23}}{\rho_{13}}}, \quad \lambda_3 = \sqrt{\frac{\rho_{31} \cdot \rho_{32}}{\rho_{12}}} \, . \tag{7-103}$$

For the remaining values, an upper bound

$$\lambda_j = \min_{k \leq j+1} \left\{ \frac{\rho_{jk}}{\lambda_k} \right\} \quad j = 4, 5... \tag{7-104}$$

and a lower bound can be estimated

$$\lambda_j = \max_{k \leq j+1} \left\{ \frac{\rho_{jk}}{\lambda_k} \right\} \quad j = 4, 5... \, . \tag{7-105}$$

Unfortunately, the requirement

$$-1 \leq \lambda_j, \, \lambda_k \leq 1 \tag{7-106}$$

can sometimes cause numerical problems.

For series system, further bounds can be given. In serious systems, the probability of failure increases by decreasing the correlation between the different elements or limit state functions. This can be understood as additionally random effects. The boundaries for a series system can be written as

$$\max_j P_{fj} \le P_f \le 1 - \prod_{j=1}^{n}(1 - P_{fj}) < \sum_{j=1}^{n} P_{fj} \tag{7-107}$$

or, in terms of the safety index

$$\min_j \beta_j \ge \beta_{sys} \ge -\Phi^{-1}\left(1 - \prod_{j=1}^{n}\Phi(\beta_j)\right) > -\Phi^{-1}\left(\sum_{j=1}^{n}\Phi(-\beta_j)\right). \tag{7-108}$$

Based on general additive theorems for probabilities, more precise bounds can be given:

$$P_f \le \min\left\{\begin{array}{l} 1 \\ \displaystyle\sum_{j=1}^{n} P(F_j) - \sum_{j=2}^{n}\max_{k<j} P(F_j \cap F_k) \end{array}\right. \tag{7-109}$$

$$P_f \ge P(F_1) + \sum_{j=2}^{n}\max\left\{\begin{array}{l} 0 \\ \displaystyle P(F_j) - \sum_{k=1}^{j-1} P(F_j \cap F_k). \end{array}\right. \tag{7-110}$$

The average volume of two probabilities of failure can be computed as

$$P(F_j \cap F_k) \approx \Phi_2(-\beta_j, -\beta_k; \rho_{jk}). \tag{7-111}$$

Using this for the above given boundaries, one achieves

$$P_f \le \min\left\{\begin{array}{l} 1 \\ \displaystyle\sum_{j=1}^{n}\Phi(-\beta_j) - \sum_{j=2}^{n}\max_{k<j}\Phi_2(-\beta_j, -\beta_k; \rho_{jk}) \end{array}\right. \tag{7-112}$$

$$P_f \ge \Phi(-\beta_1) + \sum_{j=2}^{n}\max\left\{\begin{array}{l} 0 \\ \displaystyle\Phi(-\beta_j) - \sum_{k=1}^{j-1}\Phi_2(-\beta_j, -\beta_k; \rho_{jk}). \end{array}\right. \tag{7-113}$$

The two-dimensional standard normal distribution can be approximated with

$$\Phi_2(x_2;x_2;\rho) = \int_{-\infty}^{+\infty} \Phi\left(\frac{x_1+\lambda_1\cdot w}{\sqrt{1-\lambda_1^2}}\right)\cdot\Phi\left(\frac{x_1+\lambda_1\cdot w}{\sqrt{1-\lambda_1^2}}\right)\cdot\varphi(w)\ dw \tag{7-114}$$

using

$$\lambda_1 = \lambda_2 = \sqrt{\rho},\ \rho>0 \tag{7-115}$$
$$\lambda_1 = \sqrt{-\rho},\ \lambda_2 = -\sqrt{-\rho},\ \rho<0.$$

The introduced methods have been programmed into FORTRAN 77 routines and are available free of charge to the reader of this book. Please simply contact the authors.

Arch bridges are usually considered to be a serious system: if one part fails, the entire bridge will collapse. However, some parts of bridge behave like parallel systems: one part will collapse and other parts will take more loads. Further works about the estimation of probabilities of failure for such types of systems can be found in Rackwitz and Hohenbichler (1981) or Gollwitzer and Rackwitz (1990), as seen in Fig. 7-10.

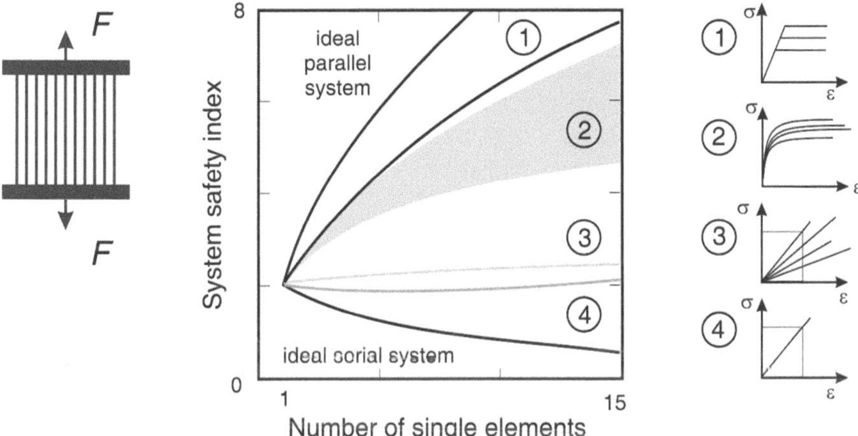

Fig. 7-10. Relationship between system safety index and material properties shown for a Daniel system

7.2.8 Limitation of the Presented Methods

The presented methods so far have described uncertainties of materials and loads by random distribution functions. They have not considered any correlations between the random variables that increase the numerical work. For the interested reader, the Rosenblatt transformation or the NATAF transformations are dealt with in Melchers (1999) or Liu and Der Kiureghian (1986). Copulas are an additional technique to transform the correlated random variables into noncorrelated variables.

Furthermore, these random distribution functions are usually kept constant over different distances or volumes. In contrast, it is well-known that, for example, material properties might not have a full correlation over a certain distance or volume. This correlation change over distance may be described by random fields (Vanmarcke 1983). In past years, random fields are increasingly applied in structural safety investigations to establish stochastic finite elements (Der Kiureghian and Ke 1988, Ghanem and Spanos 1991, and Pukl et al. 2006). The latest advances were shown by Bayer and Ross (2008).

Furthermore, not all presented techniques perform well under all conditions. Figures 7-11 and 7-12 give a good overview about the application conditions of the probabilistic techniques.

It is not intended by the authors to give here a full summary about current state of knowledge in the field of structural safety. A state-of-the-art report for computational stochastic mechanics was given by Berman et al. (1997), however, the latest developments should be considered. In contrast, the introduced methods can be easily programmed and applied to arch bridge problems by the reader. For more advanced studies, some commercial programs or programs from research institutes can be used.

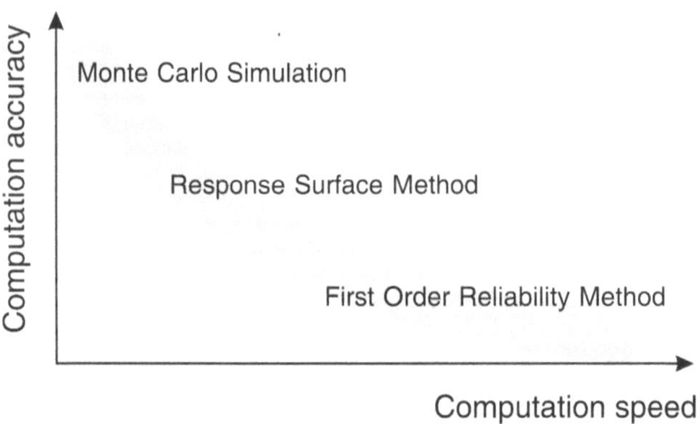

Fig. 7-11. Performance of methods for stochastic structural analysis (Bucher et al. 2000)

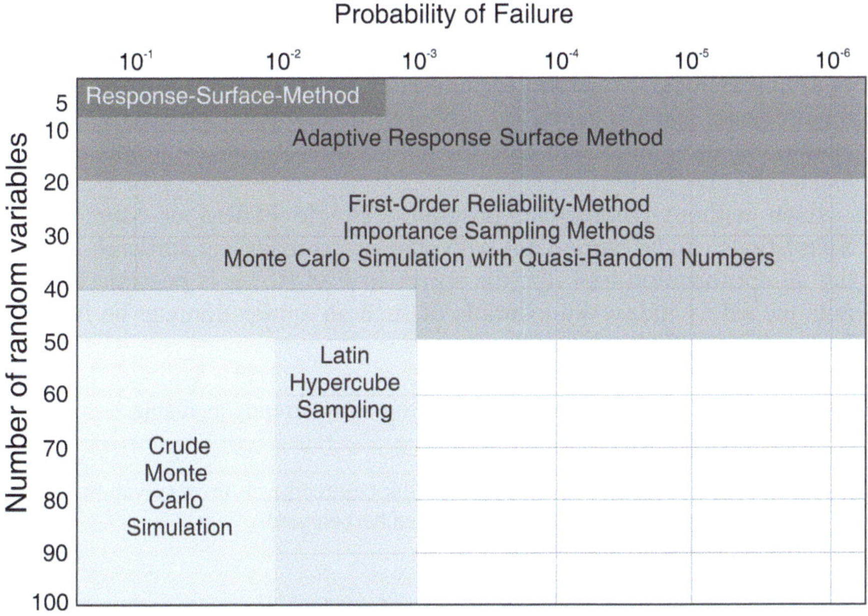

Fig. 7-12. Applicability of certain stochastic techniques subject to the number of random variables and the estimated probability of failure (Bayer 2008)

7.2.9 Commercial Programs

Although many universities have developed programs for the computation of probabilities of failure, in many cases these programs lack a sufficient manual, a graphical user interface, or simply an easy handling. Therefore, in many cases, commercial probabilistic programs were developed.

Currently, the following programs are known to the authors, however this list is subject to change: UNIPASS (Lin and Khalessi 2006), ProFES (Wu et al. 2006), Proban (Tvedt 2006), PHIMECA (Lemaire and Pendola 2006), PERMAS-RA/STRUREL (Gollwitzer et al. 2006), NESSUS (Thacker et al. 2006), COSSAN (Schueller and Pradlwarter 2006), CalRel/ FERUM/ OpenSees (Der Kiureghian et al. 2006), ANSYS PDS und DesignXplorer (Reh et al. 2006), ATENA/SARA/FREET (Pukl et al. 2006), VaP (Petschacher 1994), OptiSlang (Schlegel and Will 2007), RELSYS (Estes and Frangopol 1998), and the probabilistic toolbox ProBox (Schweckendiek and Courage 2006). Many of these programs can be downloaded free of charge for test runs (Table 7-3).

As mentioned above, parallel to the listed commercial programs, many further stand alone programs were or are under development in companies or at universities. Therefore, Epstein et al. (2008) suggested some general requirements and structures for probabilistic computer programs. The adaptation of these rules will ease the application of such programs and will extend the user group.

Even without commercial programs, simple FORM or Monte Carlo Simulations can be carried out with standard spreadsheet software. Including an optimization tool like the solver in EXCEL, it is possible to compute the safety index. An example of such an application can be found in Low and Teh (2000).

Table 7-3. List of different probabilistic programs currently available

Program	University	Homepage
VAP	ETH Zurich	http://www.ibk.baum.ethz.ch/proserv/vap.html
CALREL	University of California, Berkeley	http://www.ce.berkeley.edu
FERUM	University of California, Berkeley	http://www.ce.berkeley.edu/~haukaas/FERUM/ferum.html
NESSUS	Southwest Research Institute, San Antonia	http://www.nessus.swri.org/
PERMAS	INTES GmbH, Stuttgart	http://www.intes.de
SLANG	Bauhaus University, Weimar	http://www.uni-weimar.de/Bauing/ism/Slang
ISPUD	University Innsbruck	http://www.uibk.ac.at/c/c8/c810
COSSAN	University Innsbruck	http://www.uibk.ac.at/c/c8/c810
PROBAN	Det Norske Veritas Software	http://www.dnv.com
STRUREL	RCP GmbH	http://www.strurel.de
ANSYS	ANSYS Inc.	http://www.ansys.com
SARA	University Bruno,	
FREET	Cervenka Consulting	http://www.cervenka.cz
RACKV	University of Natural Resources and Applied Life Sciences, Vienna	

7.2.10 Goal Values of Safety Indexes

If a safety index or a probability of failure is used as a measure for safety of the structure, they have to be compared to a certain proof or goal value. Many references have published such goal probabilities of failure and examples are shown in Tables 7-4 to 7-10. However, most of the publications show nearly the same values; for new structures, maximum probability of failure is in the range of 10^{-6} per year or a minimum safety index of 3.8 per year. For existing structures, usually less stringent requirements are common. For example, a decrease of the safety index by 0.5 (Diamantidis et al. 2007). Furthermore, the probability of failure or the safety index can be adapted to more special conditions.

Table 7-4. Goal probability of failure per year according to the CEB (1976)

Average number of people endangered	Economical consequences		
	low	average	high
Low (< 0.1)	10^{-3}	10^{-4}	10^{-5}
Average	10^{-4}	10^{-5}	10^{-6}
High (> 10)	10^{-5}	10^{-6}	10^{-7}

Table 7-5. Goal probability of failures in some Scandinavian countries (Spaethe 1992)

Safety class	Failure consequences	Probability of failure for the limit state of ultimate load per year
Low	Low personal injuries Insignificant economical consequences	$1.0 \times \cdot 10^{-4}$
Normal	Some personal injuries Considerable economical consequences	1.0×10^{-5}
High	Considerable personal injuries Very high economical consequences	1.0×10^{-6}

Table 7-6. Goal probability of failures in the former East Germany (Franz et al. 1991)

Reliability class	Consequences	Probability of failure
I	Very high danger to the public Very high economical consequences Disaster	1.0×10^{-7}
II	High danger to the public High economical consequences High cultural losses	1.0×10^{-6}
III	Danger to some persons Economical consequences	1.0×10^{-5}

Reliability class	Consequences	Probability of failure
IV	Low danger to persons Low economical consequences	1.0×10^{-4}
V	Very low danger to persons Very low economical consequences	7.0×10^{-4}

Table 7-7. Goal probability of failures according to the GruSiBau (1981)

Safety class	Possible consequences of failure		Type of limit state	
	Limit state of ultimate load bearing	Limit state of serviceability	Ultimate load	Serviceability
1	No danger to humans and no economical consequences	Low economical consequences and low usage limitation	1.34×10^{-5}	6.21×10^{-3}
2	Some danger to humans and considerable economical consequences	Considerable economical consequences and strong limitation of further usage	1.30×10^{-6}	1.35×10^{-3}
3	High importance of the structure to the public	High economical consequences and high restriction to future usage	1.00×10^{-7}	2.33×10^{-4}

Table 7-8. Goal probability of failure according to the DIN 1055-100 (1999) and the Eurocode 1 (1994)

Limit state	Probability of failure	
	Lifetime	Per year
Ultimate load	7.24×10^{-5}	1.30×10^{-6}
Serviceability	6.68×10^{-2}	1.35×10^{-5}

Table 7-9. Goal safety indices according to ISO/CD 13822 (1999)

Limit state	Safety index
Serviceability	
Reversible	0.0
Irreversible	1.5
Fatigue	
Testable	2.3
Not testable	3.1
Ultimate load	
Very low consequences	2.3
Low consequences	3.1
Common consequences	3.8
High consequences	4.3

Table 7-10. Goal safety indices according to the JCSS Modelcode (2004)

Costs for safety measures	Low failure consequences	Average failure consequences	High failure consequences
Low	3.1	3.3	3.7
Average	3.7	4.2	4.4
High	4.2	4.4	4.7

The Eurocode also permits an adaptation of the safety index to some consequence classes in terms of failure consequence classes (CC) as shown in Table 7-11. Such consequence classes can then be related to some reliability classes (RC) listed in Tables 7-12 and 7-13.

Table 7-11. Graduation of failure CCs according to the Eurocode 1 (1994)

Failure CCs	Consequences	Examples
CC 3	High consequences to humans, the economy, social systems and the environment	Stands, public buildings, for example concert halls
CC 2	Average consequences to humans, the economy, social systems and the environment	Dwelling and office buildings, public buildings such as offices
CC 1	Low consequences to humans, the economy, social systems, and the environment	Agricultural structures or structures without regular persons' residence, for example barns, conservatories

Table 7-12. Graduation of RCs according to the Eurocode 1 (1994)

RC	Safety index per year	Safety index for 50 years
RC 3	5.2	4.3
RC 2	4.7	3.8
RC 1	4.2	3.3

Table 7-13. Adaptation factor for the partial safety index subject to the RC (Eurocode 1 1994)

Adaptation for the partial safety factors	RC		
	RC1	RC2	RC3
K_{FI}	0.9	1.0	1.1

The Eurocode furthermore can consider different types of production control of the building material in terms of changes of partial safety factors of the material. Still, this is for new structures only. Therefore, some other recommendations focus on existing structures. For example, in Tables 7-14 and 7-15, some adaptation factors are given. Furthermore, Strauss and Bergmeister (2005) have also introduced some factors.

Table 7-14. Adaptation of the safety index according to the CAN/CSA-S6-88 Canadian Limit States Design Standard (taken from Casas et al. 2001 and COST 345 2004)

$\beta = 3.5 - (\Delta_E + \Delta_S + \Delta_I + \Delta_{PC}) \geq 2.0$	value
Correction factor for element failure	Δ_E
Abrupt failure without warning	0.0
Abrupt loss of bearing capacity without warning with remaining capacity	0.25
Grateful failure with warning	0.50
Correction factor for system failure	Δ_S
Failure of one single element causes system failure	0.00
Failure of one single element does not cause system failure	0.25
Failure of one single element causes local failure only	0.50
Correction factor for monitoring	Δ_I
Element is not controllable	−0.25
Element is controlled regularly	0.00
Critical elements are controlled more frequently	0.25
Correction factor for live load	Δ_{PC}
All types of traffic without special permission	0.00
All types of traffic with special permission	0.60

Table 7-15. Adaptation of the safety index according to Schueremans and Van Gemert (2001)

$\beta = \beta_T - (\Delta_S + \Delta_R + \Delta_P + \Delta_I) \geq 2.0$	value
Adjustment for system behaviour	Δ_S
Failure leads to collapse, likely to impact occupants	0.00
Failure is unlikely to lead to collapse, or unlikely to impact occupants	0.25
Failure is local only, very unlikely to impact on occupants	0.50
Adjustment for risk category	Δ_R
High number of occupants (n) exposed to failure ($n = 100–1,000$)	0.00

$\beta = \beta_T - (\Delta_S + \Delta_R + \Delta_P + \Delta_I) \geq 2.0$	value
Normal occupancy exposed to failure ($n = 10\text{--}99$)	0.25
Low occupancy exposed to failure ($n = 0\text{--}9$)	0.50
Adjustment for past performance:	Δ_P
No record of satisfactory past performance	0.00
Satisfactory past performance or dead load measured	0.25
Adjustment for inspection:	Δ_I
Component not inspect able	−0.25
Component regularly inspected	0.00
Critical component inspected by expert	0.25

This adapted safety index can then be used to provide alternative safety measures in the semiprobabilistic safety concept.

Further background information about the development of goal probabilities of failure or safety indexes can be found in Proske (2008b) discussing different risk parameters and the current developments.

7.3 Semiprobabilistic Safety Concept

7.3.1 Introduction

The probabilistic safety concept for structures has introduced the measure of probability of failure as a measure of reliability and safety. Nevertheless, under everyday conditions, this measure is not practical and instead, simpler types of proof of safety have to be used for structures. Therefore, the probabilistic safety concept has to be transformed into a semiprobabilistic safety concept, which means nothing else than developing substitutes for the probability of failure proof, which are easier to handle. Such substitutes are the safety factors, characteristic values, and the design values. They are developed on some basic simplification.

One major advantage for these elements is the long tradition of the application of safety factors. It has been estimated that the first application of a global safety factor goes back up to 300 B.C. by Philo from Byzantium (Shigley and Mischke 2001). He introduced the global safety factor in terms of

$$\gamma = \frac{\text{resistance}}{\text{load}}. \tag{1-116}$$

Empirical geometrical rules remained valid over nearly the next two millenniums. Only in the last few centuries have the applications of safety factors become widespread. Over time, several different values were developed for different materials. In most cases, the values dropped significantly during the last century. As an example, in 1880 for brick masonry the safety factor of 10 was required, whereas only 10 years later the factor was chosen between 7 and 8. In the 20th century, the values ranged from factor 5 and 4, and now for the recalculation of historical structures with that material, factor 3 is chosen (Busch and Zumpe 1995, Schleicher 1949, Wenzel 1997, Mann 1987, and Tonon and Tonon 2006). This decline of safety factors could also be observed for other materials such as steel. The development of new materials especially led to more concerns about the safe application of those materials. The different developments for safety factors for materials led to the first efforts in the beginning of the 20th century to develop material-independent factors, as shown in Tables 7-16 and 7-17.

Table 7-16. Global safety factor according to Visodic (1948)

Safety factor	Knowledge of load	Knowledge of material	Knowledge of environment
1.2–1.5	Excellent	Excellent	Controlled
1.5–2.0	Good	Good	Constant
2.0–2.5	Good	Good	Normal
2.5–3.0	Average	Average	Normal
3.0–4.0	Average	Average	Normal
3.0–4.0	Low	Low	Unknown

Table 7-17. Global safety factor according to Norton (1996)

Safety factor	Knowledge of load	Knowledge of material	Knowledge of environment
1.3	Extremely well-known	Extremely well-known	Likewise tests
2	Good approximation	Good approximation	Controllable environment
3	Normal approximation	Normal approximation	Moderate
5	Guessing	Guessing	Extreme

As the tables show, a further decline of global safety factors seems to be limited; otherwise, the major requirement safety of structures might not be fulfilled anymore. Therefore, more advanced changes might be considered to meet the demanding requirements of economic and safe structures. Such a development would be special safety factors for the different columns in the table 7-16; for example, a safety factor for the load and a safety factor for the material. This is indeed the idea of the partial safety factor concept. It does not necessarily yield to lower safety factors, but it yields to a more homogenous level of safety.

The development of partial safety factors is strongly related to the development of the probabilistic safety concept. However, the practical application of partial safety factors took additional decades. First applications can be found in steel design, whereas a first example in the field of structural concrete could be ETV concrete—a concrete code in East Germany (ETV stands for Unified technical codes). ETV concrete was developed during the seventies of the 20th century and introduced in the beginning of the eighties. A comparison between ETV and the up-to-date German code DIN 1045-1 can be found in Wiese et al. (2005).

7.3.2 Partial Safety Factors

7.3.2.1 Introduction

The change from the global safety factor concept to the partial safety factor concept can be easily seen in the following equations. In general, the comparison of the resistance of a structure and the load or event remains

$$E_d \leq R_d. \tag{7-117}$$

But in contrast to the global safety factor format, where the safety factor can be separated like

$$E_d \leq R_d / \gamma_{Global}, \tag{7-118}$$

the safety factors accompany the parameters required for design. Then, the design load E_d is evaluated according to

$$E_d = \sum_{j \geq 1} \gamma_{G,j} \cdot G_{k,j} + \gamma_{Q,1} \cdot Q_{k,1} + \sum_{i>1} \gamma_{Q,i} \cdot \psi_{0,i} \cdot Q_{k,i} \tag{7-119}$$

and the design resistance R_d is based on

$$R_d = R\left(\alpha \cdot \frac{f_{ck}}{\gamma_c}; \frac{f_{yk}}{\gamma_s}; \frac{f_{tk,cal}}{\gamma_s}; \frac{f_{p0,1k}}{\gamma_s}; \frac{f_{pk}}{\gamma_s} \right). \tag{7-120}$$

A list of material partial safety factors is given in Table 7-18.

Table 7-18. Material partial safety factors

Material	Code or reference	Limit state of ultimate load	Accidental load conditions
Concrete (up to C 50/60)	DIN 1045-1	1.50	1.30
Concrete higher than C 50/60	DIN 1045-1	1.5/ $(1.1\text{-}f_{ck}/500)$	1.3/ $(1.1\text{-}f_{ck}/500)$
Non-reinforced concrete	DIN 1045-1	1.80	1.55
Non-reinforced concrete	DAfSt (1996)	1.25	
Pre-cast concrete	DIN 1045-1	1.35	
Collateral evasion	DIN 1045-1	2.00	–
Reinforcement steel	DIN 1045-1	1.15	1.00
Pre-stressing steel	DIN 1045-1	1.15	1.00
Steel yield strength	EN 4	1.10	1.00
Steel tensile strength	EN 4	1.25	1.00
Steel tensile strength	EN 4	1.00	1.00
Wood	EN 5	1.30	1.00
Masonry (Category A)	EN 6	1.7 (I)/2.0 (II)	1.20
Masonry (Category B)	EN 6	2.2 (I)/2.5 (II)	1.50
Masonry (Category C)	EN 6	2.7 (I)/3.0 (II)	1.80
Masonry – steel	EN 6	1.50/2.20	
Masonry	DIN 1053-100	1.50–1.875	1.30–1.625
Wall anchorage (C. A-C)	Mann (1999)	2.50	1.20
Floatglas/Gussglas	BÜV (2001)	1.80	1.40
ESV-Glas	BÜV (2001)	1.50	1.30
Siliconglas	BÜV (2001)	5.00	2.50
Carbon fibre	Onken et al.	1.20	
Carbon fibre	(2002) and	1.30[1]	
Carbon fibre cable	Bergmeister	1.20[1]	
Carbon fibre glue	(2003)	1.50[1]	
Cladding	DIN 18 516	2.00	
Aluminum yield strength	EN 9	1.10	
Aluminum tensile strength	EN 9	1.25	
Bamboo as building material	Bamboo (2005)	1.50	
Textile reinforced concrete	Own works	1.80[2]	
Concrete – multiaxial loading	Own works	1.35–1.70	
Granodiorit tensile strength	Own works	1.50–1.70	
Historical masonry arch bridges	UIC-Codex	2.00[3]	
Historical non-reinforced arch bridges	Bothe et al. (2004)	1.80[3]	

[1] Consider construction conditions (Bergmeister 2003).
[2] Recent researches indicate lower values.
[3] A partial safety factor for a system.

Unfortunately, the definition of the safety format for the resistance depends on the way of computing the forces in the structure. Eibl (1992) and Eibl and Schmidt-Hurtienne (1995) have pointed out the limitation of the partial safety factor concept in nonlinear force calculations. If the forces are computed in a nonlinear procedure, then an alternative definition has to be chosen, such as

$$R_d = \frac{1}{\gamma_R} R(f_{cR}; f_{yR}; f_{tR}; f_{p0,1R}; f_{pR}) \,. \tag{7-121}$$

Here, current work is being carried out. The reader should consult recent publications, such as Cervenka (2007), Allaix et al. (2007), Holicky (2007), and especially Pfeiffer and Quast (2003).

7.3.2.2 Design of partial safety factors

Several methods exist to develop partial safety factors for a certain material or a certain type of structure. But in general, all procedures rely on a statistical description of the material inherent uncertainty. Additionally, historical partial safety factors remain valid if they have proven to provide safe structures.

First, a historical procedure is introduced that permits the development of partial safety factors for concrete only on the coefficient of variation, an assumption about the type of probability distribution function, and the class of the structure (Murzewski 1974) (Tables 7-19 and 7-20). Usually the coefficient of variations for concrete depending on the production conditions lies around 10% (Spaethe 1992, Östlund 1991).

Table 7-19. Partial safety factor for a lognormal distributed strength

Class of structure	Coefficient of variation			
Class of structure	0.05	0.10	0.15	0.20
Dam	1.26	1.58	1.95	2.43
Bridge, theatre, cultural buildings	1.23	1.49	1.82	2.16
Residential buildings, office buildings	1.20	1.40	1.65	1.91
Lager, bunker, frames	1.15	1.31	1.46	1.62
Secondary buildings	1.10	1.17	1.22	1.25

Table 7-20. Partial safety factor for a normal distributed strength

Class of structure	Coefficient of variation			
	0.05	0.10	0.15	0.20
Dam	1.24–1.30	1.46–1.85		
Bridge, theatre, cultural buildings	1.21–1.26	1.41–1.67	1.60–2.47	
Residential or office buildings	1.18–1.22	1.35–1.53	1.50–2.00	1.65–2.86

	Coefficient of variation			
Lager, bunker, frames	1.15–1.16	1.27–1.37	1.38–1.61	1.49–1.92
Secondary buildings	1.10–1.11	1.17–1.19	1.21–1.25	1.23–1.28

The Eurocode (EN) is heavily based on works and suggestions by the Joint Committee of Structural Safety (JCSS). In the background document of the Eurocode provided by the JCSS (2004), an example for the development of a partial safety factor is presented. The general description for such a factor is

$$R = \Theta \cdot a \cdot x \tag{7-122}$$

with Θ as uncertainty factor for the calculation model, a as geometrical factor, and x as material strength. The resistance design value is then

$$R_d = \frac{\Theta_k \cdot a_k \cdot x_k}{\gamma_M}. \tag{7-123}$$

The three parameters might then be considered as independent random variables with a lognormal distribution. Please note that this is not true for many materials. The material partial safety factor can then be evaluated according to

$$\gamma_M = \exp(\alpha_R \cdot \beta \cdot (V_x + 0.4 \cdot V_a + 0,4 \cdot V_\Theta) - 1.64 \cdot V_x). \tag{7-124}$$

Additionally, simplified rules exist for the development of design values, including material partial safety factors based on testing of materials. Such procedures have been intensively discussed by Reid (1999) and have been applied for masonry (Curbach and Proske 2004). The procedures include some main assumptions; for example, the probability distribution function of the load and the resistance. Examples are the Australian Standard Procedure for Statistical Proof Loading, Australian Standard Procedure for Probabilistic Load Testing, or the Standard for Probabilistic Load tests. Most of the procedures consider the variance of the material strength, the variance of the load, some condition factors, the number of tests or a correction factor, and the required safety index.

Val and Stewart (2002) describe a procedure for the evaluation of partial safety factors for mainly the resistance of existing structures. The safety factor is split into two parts, the partial safety factor for the material strength and a factor for the consideration of additional uncertainties, $f_{k,est}$ is the estimated characteristic material strength. The design value for the strength is therefore

$$f_d = \frac{f_{k,est}}{\gamma_m \cdot \gamma_\eta} \cdot \qquad (7\text{-}125)$$

Tables 7-21 and 7-22 show some examples taken from the publication by Val and Stewart (2002). In Table 7-21, no *a priori* information was available, whereas in Table 7-22, *a priori* information could be used. It becomes clear that additional information is for existing structures of major importance for keeping safe and efficient structures. As already seen in the JCSS (2004) example, the consideration of modelling uncertainty, in which ever way, is a major part of many expansions of the traditional ways for the developing partial safety factors. Here it becomes visible that the clear rule for the application of statistics might sometimes be a drawback for a realistic estimation of the safety of a structure.

Table 7-21. Example of material partial safety factors without *a priori* and with six tested samples (Val and Stewart 2002)

Coefficient of variation of the samples	γ_m	Coefficient of variation for the calibration factor		
		0.05	0.10	0.15
0.04	1.1	1.07	1.19	1.27
0.08	1.4	1.05	1.17	1.24
0.12	1.6	1.04	1.15	1.21
0.16	1.9	1.03	1.10	1.19
0.20	2.2	1.02	1.07	1.15

Table 7-22. Example of material partial safety factors with *a priori* and with six tested samples (Val and Stewart 2002)

A priori coefficient of variation	γ_m
0.05	1.36
0.10	1.25
0.15	1.21

Melchers and Faber (2001) introduce a damage function, which is included in the following material safety factor:

$$\gamma_M = \gamma_{M0} \cdot \varphi_P \cdot \varphi_D \cdot \varphi_M \cdot \qquad (7\text{-}126)$$

The factors consider deterioration processes, protection against deterioration, and intensity of inspection.

Schneider (1999) has also investigated the development of partial safety factors. Partial safety factors for existing reinforced concrete structures have recently been discussed by Fischer and Schnell (2008) and Braml and Keuser (2008).

Besides the simplified techniques, the partial safety factor can also be computed using the results from full probabilistic computations (FORM). Then, the so-called weighting factors α_R of the random variables are used. The partial safety factor for material strength γ_m is computed as

$$\gamma_m = \frac{R_c}{R_d}.$$

(7-127)

The safety factor can be computed when design resistance R_d and characteristic resistance R_c are known. Since

$$R_d = \mu_E - \alpha_R \cdot \beta \cdot \sigma_R$$

(7-128)

and for a normal distribution with known mean value and standard deviation, the characteristic value is given as

$$R_c = \mu_E - 1.645 \cdot \sigma_R ,$$

(7-129)

hence the partial safety factor for the material strength can be computed. As an example, Fig. 7-13 shows the distribution of the weighting factor quadrates for historical arch bridges, either for road or for railway traffic based on FORM computations. This is a common type of diagram since the quadrates of the weighting factors have to sum up to one. The figure clearly shows why in railway codes the partial safety factor for the traffic load is often lower; for example, 1.3 compared to road traffic bridges with 1.5. Here it can be seen that the weighting factor is significantly lower for railway traffic than for road traffic.

Table 7-23 shows the computation of partial safety factors for live and dead load subjected to an adaptable goal safety index for arch bridges. Table 7-24 lists system partial safety factors for stone arch bridges subject to different masonry types. The investigation is based on own probabilistic computations.

In general, the issue of partial safety factors for historical stone arch bridges is still under discussion since they show a highly nonlinear behaviour and such structures can only be understood as a system and not on the cross-section layer.

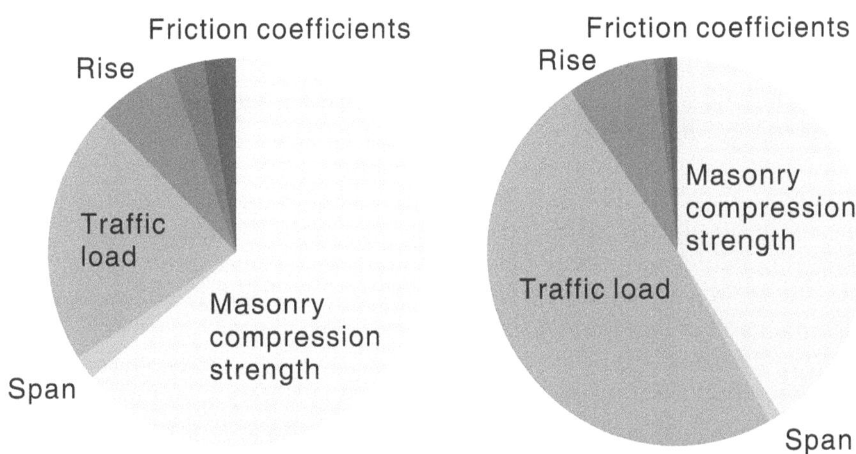

Fig. 7-13. Distribution of the square of the weighting factors for an arch bridge using a linear-elastic model for railway traffic (*left*) and road traffic (*right*)

Table 7-23. Partial safety factor for dead and live load considering the adaptation of the safety index according to Schueremans and Van Gemert (2001)

$\Delta_S + \Delta_R + \Delta_P + \Delta_I \leq 1.5$	Partial safety factor for		Factor of combination
	Dead load	Live load	
−0.25	1.42	1.56	0.69
0.00	1.35	1.50	0.70
0.25	1.28	1.45	0.71
0.50	1.22	1.39	0.72
0.75	1.16	1.34	0.73
1.00	1.10	1.29	0.74
1.25	1.05	1.25	0.75
1.50	1.00	1.20	0.76

Table 7-24. Own suggestions for partial safety factors for historical masonry arch bridges subject to different masonry types in the arch

Masonry types

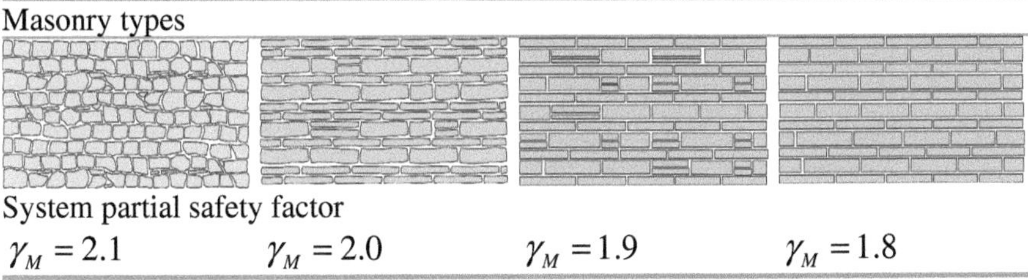

System partial safety factor

$\gamma_M = 2.1$ $\gamma_M = 2.0$ $\gamma_M = 1.9$ $\gamma_M = 1.8$

7.3.3 Characteristic Values

Characteristic values are probability fixed defined values of a probability distribution. They relate the probability distribution property—for example, strength—to a certain probability. A characteristic strength value f_k of structural materials of 5% of the overall population is very common. That means that only 5% of the overall population will experience a lesser strength, and 95% of the population will show a higher strength than this 5% fractile material strength. The terms percentile and quantile can often also be found in literature instead of fractile. The chosen value of 5% is arbitrary, but the general idea about characteristic values are the proofs in the state of serviceability carried out without partial safety factor (one). Since characteristic values and partial safety factors interact, the characteristic value simply has to be chosen to provide the partial safety factor of one for the limit state of serviceability. Therefore, the assumption of the 5% fractile value can be found as a general requirement, for example, in the Eurocode 1 (1994) or in the German DIN 1055-100, Sect. 6.4 (1999). As an example, some material-related codes are listed here to show the wide application of the 5% fractile value assumption:

- Reinforcement steel according to DIN 488 (90% confidence interval)
- Concrete compression strength according to DIN 1045, DIN 1048
- Masonry after testing according to DIN 1053 (75% confidence interval) (Schubert 1995)
- Outer wall panelling according to DIN 18516 (75% confidence interval)
- Masonry according to DIN 18152 (1987) with 90% confidence interval
- Reinforcement steel according to ENV 100080 (90% confidence interval)
- Wooden structures according to DIN V ENV 1995 (84,1%-confidence interval, coefficient of variation greater or equal 0.1, more than 30 samples)
- Masonry-aerated concrete (95%-confidence interval)
- Artificial brick stones (Schubert 1995)
- Natural stones (90% confidence interval) (Schubert 1995)

To compute a 5% fractile value of a certain property, statistical information is required. Using such statistical information, a probability distribution function can be estimated. There exists a wide variety of such probability functions as shown in Table 7-25. Many distribution functions can be related to each other (Fig. 7-14). Other distribution functions describe certain distribution families or certain conditions; for example, Piersons differential equation and the Fleishmann system. An overview of distribution families can be found in Plate (1993), Bobee and Ashkar (1991), and Fischer (1999). Besides that for extreme value distributions, the work by Castillo et al. (2008) is mentioned. For practical reasons, however, only a limited number of distributions is considered for construction material: the normal distribution, the lognormal distribution, and the Weibull distribution (Fischer 1999, Eurocode 1 1994, and GruSiBau 1981). The normal distribution has found wide application and can be easily explained by the central limit state theorem. This theorem states that a sum of certain random variables will yield to a normal distributed random variable if certain conditions are fulfilled (Van der Werden 1957). Such conditions are, for example, that no single random variable dominates the results. And indeed, many material properties can be seen as the sum of certain other properties—for example, the strength of a natural stone can be seen as sum of the strength of the single elements of the stone. Therefore, many publications assume a normal distribution for the concrete compression strength (Rüsch et al. 1969). A disadvantage of the normal distribution is possible negative values. Therefore, instead of the normal distribution, the lognormal distribution is often used if the average value of a material property is low and experiences a high standard deviation, such as the tensile strength of concrete or masonry. The lognormal distribution does not feature negative values and therefore no negative tensile strengths are then possible. The lognormal distribution can also be related to the central limit state theorem, if the data are logarithmic. However, this implies the multiplication of the single input random variables—in other terms, the logarithmic distribution fulfils the central limit state theorem for the case of multiplication. A further often-used distribution for construction material properties is the Weibull distribution (Weibull 1951). This distribution belongs to the group of extreme value distributions and describes a chain interaction of single elements. If the weakest part of the chain fails, then the entire chain fails, which indeed fulfil's the requirements of an extreme value distribution. The properties of brittle materials, such as glass, can be described with this distribution (Button et al. 1993, Güsgen et al. 1998). Of course, many materials do not comply with the chain rule for a serial system but show rather a mixed parallel-serial system. Further considerations are then needed. Some theoretical works dealing with this issue have been taken out by Rackwitz and Hohenbichler (1981), Gollwitzer and Rackwitz (1990), and Kadarpa et al. (1996) for brittle materials; Chudoba et al. (2006) and Chudoba and Vorechovsky (2006) for glass yarns.

Table 7-25. Certain probability functions

	Name of distribution		Name of distribution
1	χ-Distribution	29	Laplace distribution
2	General Pareto distribution	30	Logarithmic Pearson typ-3 distribution
3	Arcsin distribution	31	Logarithmic–logistic distribution
4	Beta distribution	32	Logistic distribution
5	Binomial distribution	33	Lognormal distribution
6	Birnbaum–Saunders distribution	34	Lorenz distribution
7	Breit–Wigner distribution	35	Maxwell distribution
8	Cauchy distribution	36	Neville distribution
9	Erlang distribution	37	Pareto distribution
10	Exponential distribution	38	Pearson, typ-3, gamma distribution
11	Extreme value distribution typ I max	39	Pearson, typ-3 distribution
12	Extreme value distribution typ I min	40	Poisson distribution
13	Extreme value distribution typ II max	41	Polya distribution
14	Extreme value distribution typ II min	42	Potential distribution
15	Extreme value distribution typ III max	43	Power normal distribution
16	Extreme value distribution typ III min	44	Rayleigh distribution
17	Fisher distribution	45	Uniform distribution
18	Fréchet distribution	46	Reverse Weibull distribution
19	F distribution	47	Rossi distribution
20	Gamma distribution (G-distribution)	48	Simpson or triangle distribution
21	Gauss order normal distribution	49	Sinus distribution
22	General extreme value distribution	50	Snedecor distribution
23	General Pareto distribution	51	Student-t-distribution
24	Geometric distribution	52	Tukey Lambda distribution
25	Gumbel distribution	53	Wakeby distribution
26	Hypergeometric distribution	54	Weibull distribution
27	Krickij–Menkel distribution	55	Wishart distribution
28	Landau distribution	56	Z-Distribution

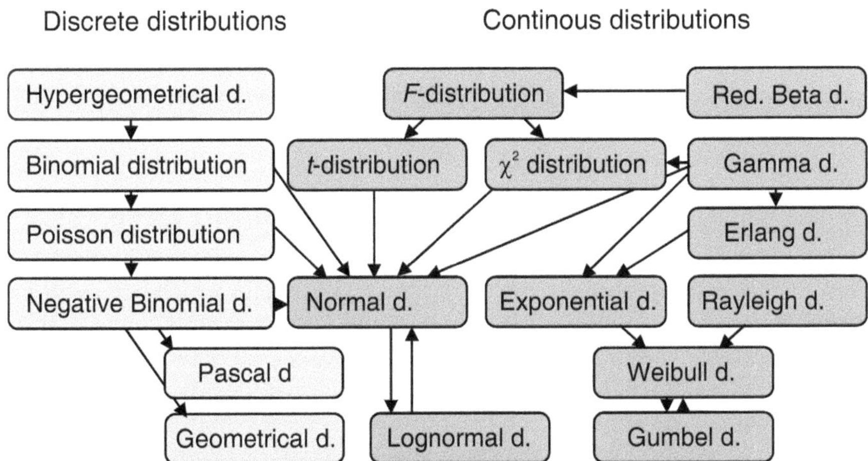

Fig. 7-14. Distribution families according to Fischer (1999)

However, theoretical considerations often do not prove the type of probability distribution. Therefore, usually the statistical data are investigated, according to DIN 53804 (1981). The most robust and fast converting parameters of the random data are general tendency estimators such as the arithmetical mean, harmonically mean, geometrical mean, generalized mean, quadratic mean and median (50% fractile), and mode (most probable value). However, more interesting from a statistical point of view, are deviation measures or uncertainty measures such as variance, standard deviation (same unit as mean value), coefficient of variation, span, modified interquartile range, and mean deviation. Besides such measures, the skewness and the kurtosis are also of interest (Fig. 7-15).

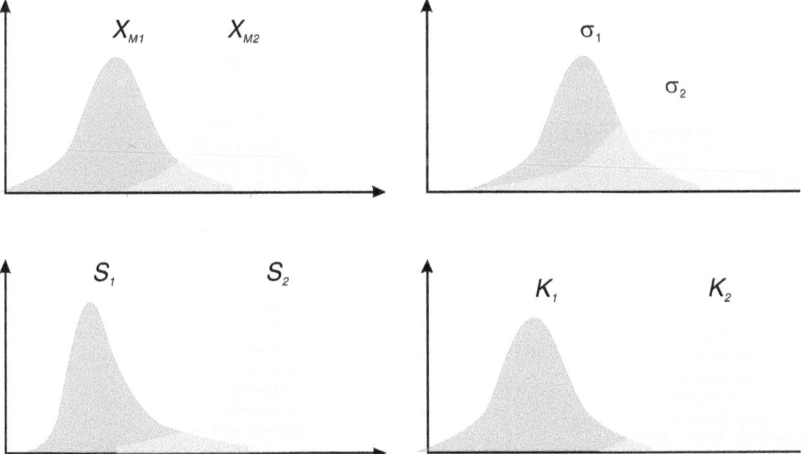

Fig. 7-15. Meaning of the statistical parameter arithmetic mean (X_m), standard deviation (σ), skewness (S), and kurtosis (K)

For historical masonry, one should keep in mind that such statistical parameters may be insufficient due to corrupted data (Fig. 7-16). Statistical data from historical masonry with natural stone may feature outliers, censored data, and multimodal data. Outliers are samples that do not belong to the population. However, since we do not know the population, it is difficult to identify outliers. Here, Dixons test, Barnett–Lewis test, Chauvenets criteria, Grupps test, the David-Hartley-Pearson test and other criteria may be used for outlier identification (McBean and Rovers 1998, Fischer 1999, and Bartsch 1991). Censored data describe a condition in which access to the original population data is filtered by a process. For example, investigating the strength of historical mortar may yield to censored data since the drilling process using high-pressure cooling water may destroy mortar inside the masonry. Furthermore, sawing the test specimen may further destroy material. Therefore, the compression test results of the mortar may indicate high average compression strength, however the tests do not consider the failure of all weak material in the preparation process. Therefore, the data are censored and have to be corrected. Then, for example, Cohens and Aitchison's method can be applied for data correction (McBean and Rovers 1998). Finally, it may be the case that the test results do not come from one population. For example, different stone types may be used for masonry or the natural stone material may come from different stone quarries. Then the compression strength can show multimodal behaviour. Using optimization methods, it is possible to disintegrate different distributions from such multimodal data (Proske 2003).

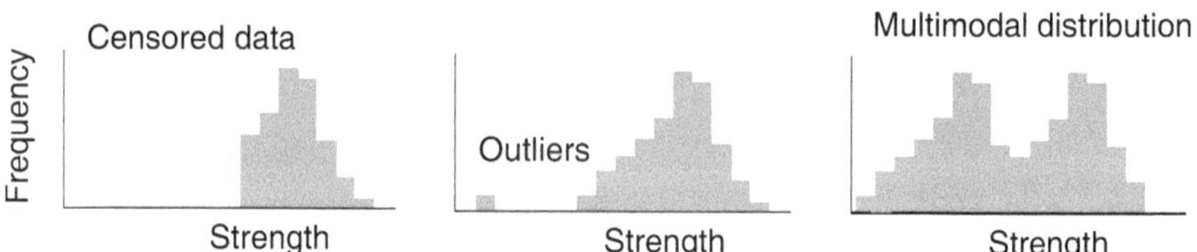

Fig. 7-16. Examples of corrupted data

After the evaluation of the single parameters, the probability distribution function usually has to be chosen. Several statistical techniques exist to investigate statistical data and recommend a distribution type:

- Relating the coefficient of variance and the type of distribution
- Relating skewness and kurtosis and the type of distribution (Fig. 7-17)
- The minimum sum square error based on histograms
- The χ^2 test and $n\omega^2$ test
- The Kolmogoroff-Smirnoff test
- The Shapiro-Wilk or Shapiro-Francia test

- Probability plots
- Quantile-correlation values

Such statistical tests are often called goodness-of-fit tests. However, as described, statistical testing is only one part of the investigation. The understanding of the material extraction, the testing techniques, and the material itself is compelling to interpret the results of material testing and statistical investigations.

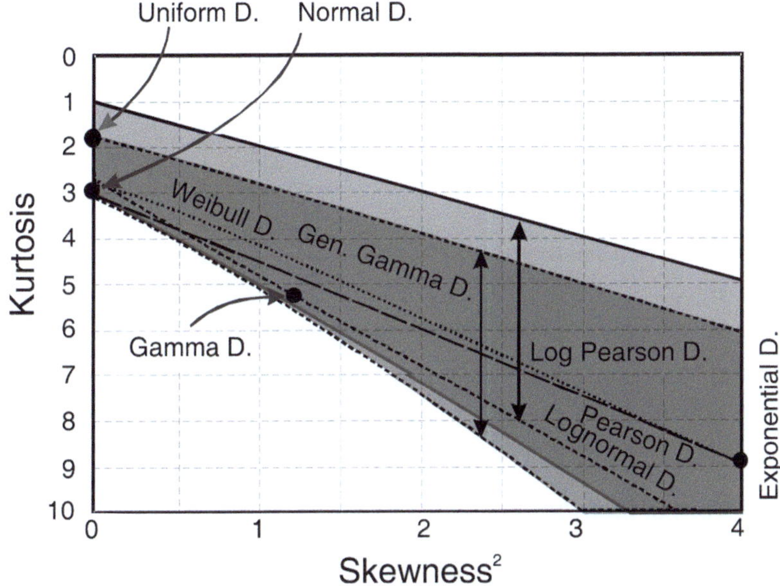

Fig. 7-17. Relation between skewness and kurtosis and the type of distribution (Plate 1993) D. = distribution

Besides the behaviour of single random variables, correlation between the random variables is often of great interest in identifying deterministic formulas. However, the identification of different correlation coefficients is rather laborious due to the required high sample sizes. Figure 7-18 shows that measuring a correlation coefficient of 0.5 for ten samples shows a value for the population between –0.5 and 0.8. Furthermore, Pearson's coefficient of correlation may not be adequate, other correlation coefficients such as from Spearman, Kendall, and Hoeffding may be helpful. Additionally, nonlinear regression using the Levenberg-Marquardt method can be considered.

Besides such limitations, the next sections will discuss the computation of characteristic 5% fractile values for certain probability distribution functions.

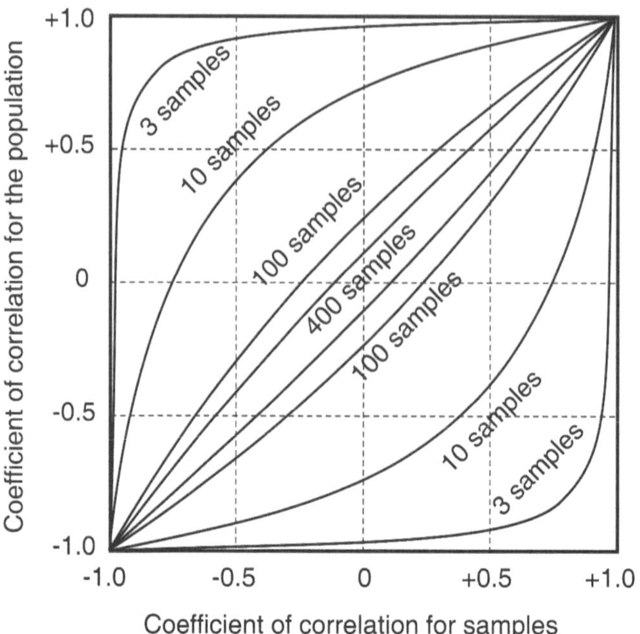

Fig. 7-18. Ninty-five percent confidence interval for coefficient of correlations (Steel and Torrie 1991)

7.3.3.1 Normal distribution

The 5% fractile value f_k of a normal distributed material strength can be computed as (Storm 1988)

$$f_k = f_m - k \cdot \sigma , \tag{7-130}$$

where f_m is the mean value and σ is the standard deviation. If the mean value and the standard deviation are known, then the k-factor corresponding to the 5% fractile value amounts to 1.645. However, usually only some suggestions for the mean value and the standard deviation are known due to a limited number of samples. Although the mean value converges very fast with a low number of samples, the problem remains for the evaluation of the standard deviation. The uncertainty in the empirical statistical parameters is usually considered in the choice of the k-factor.

Assuming, for example, a normal distributed property with 15 samples and a confidence interval of 95%, the k-factor becomes 1.76 based on a student-t distribution (Table 7-26). For very high sample numbers, the k-value converges to 1.645 again.

Table 7-26. k-factors based on a student's t-distribution (degrees of freedom = sample number – 1)

Degree of freedom	Fractile value		Degrees of freedom	Fractile value	
	5%	2.5%		5%	2.5%
1	6.314	12.706	20	1.725	2.086
2	2.920	4.303	21	1.721	2.080
3	2.353	3.182	22	1.717	2.074
4	2.132	2.776	23	1.714	2.069
5	2.015	2.571	24	1.711	2.064
6	1.943	2.447	25	1.708	2.060
7	1.895	2.365	26	1.706	2.056
8	1.860	2.306	27	1.703	2.052
9	1.833	2.262	28	1.701	2.048
10	1.812	2.228	29	1.699	2.045
11	1.796	2.201	30	1.697	2.042
12	1.782	2.179	40	1.684	2.021
13	1.771	2.160	60	1.671	2.000
14	1.761	2.145	80	1.664	1.990
15	1.753	2.131	100	1.660	1.984
16	1.746	2.120	200	1.653	1.972
17	1.740	2.110	500	1.648	1.965
18	1.734	2.101	1000	1.646	1.962

The Eurocode 1 suggests slightly different k-factors. For example, according to the Eurocode 1, 15 samples would yield a k-factor of 1.84.

A simple example should illustrate the application of a normal distribution. About 500 compression test results of Posta sandstone are available for an investigation. The mean value of the compression strength is 58.05 MPa and the standard deviation is given with 10.32 MPa. Assuming a normal distribution for 500 samples, the k-value becomes 1.648. The characteristic compression strength then is computed as

$$f_{st,k} = 58.05 \text{ MPa} - 1.648 \cdot 10.32 \text{ MPa} = 41.04 \text{ MPa} . \tag{7-131}$$

This value is compared with other references discussing the compression strength of Posta sandstone. Since the sample size is unknown, the k-value is kept constant with 1.645.

Grunert (1982) $f_{st,k} = f_m - 1.645 \cdot \sigma = 45.6 - 1.645 \cdot 11.6 = 26.52 \text{ MPa}$

Grunert et al. (1998) $f_{st,k} = f_m - 1.645 \cdot \sigma = 31.6 - 1.645 \cdot 6.8 = 20.41 \text{ MPa}$

Peschel (1984) $f_{st,k} = f_m - 1.645 \cdot \sigma = 41.6 - 1.645 \cdot 11.6 = 22.52 \text{ MPa}$

7.3.3.2 Lognormal distribution

As mentioned above, the lognormal distribution can be reasoned based on a product formulation of the central limit theorem. The computation of the 5% fractile value is analogous to the normal distribution, except that the data have to be transformed by the logarithm:

$$L' = \overline{y} - k \cdot s^* = \ln(L) \tag{7-132}$$

$$\overline{y} = \sum_{i=1}^{n} \frac{y_i}{n} \text{ with } y_i = \ln(x_i) \tag{7-133}$$

$$s^* = \sqrt{\frac{\sum_{i=1}^{n}(y_i - \overline{y})^2}{n-1}} \, . \tag{7-134}$$

The following example should show the application. Four compression strength tests from natural stone masonry have been carried out (Table 7-27).

Table 7-27. Test data

Sample	Compression strength in MPa	Logarithm
1	6.70	1.90
2	6.20	1.82
3	5.70	1.74
4	6.60	1.89
Mean value	6.30	1.84
Standard deviation	0.39	0.06

The characteristic value assuming a lognormal distribution can be computed as follows (for 2.353 see Table 7-26):

$$f_{mw,k} = \exp(1.84 - 2.353 \cdot 0.06) = 5.47 \text{ MPa} \, . \tag{7-135}$$

If a normal distribution is assumed, then the characteristic value becomes

$$f_{mw,k} = 6.30 \text{ MPa} - 2.353 \cdot 0.39 \text{ MPa} = 5.37 \text{ MPa} \, . \tag{7-136}$$

7.3.3.3 Weibull distribution

For the Weibull distribution, the 5% fractile value can be estimated with

$$f_k = \left(-\frac{1}{\lambda} \ln(1-q) \right)^{\frac{1}{k}} . \tag{7-137}$$

The value q describes the probability that is chosen in our case as 0.05 for the 5% fractile value. The factors λ and k are parameters of the Weibull distribution, which can be computed based on statistical data. The mean value and the standard deviation can be computed as

$$f_m = f_0 + \lambda^{-1/k} \cdot \Gamma\left(1+\frac{1}{k}\right), \tag{7-138}$$

$$\sigma = \lambda^{-1/k} \sqrt{\Gamma\left(1+\frac{2}{k}\right) - \Gamma\left(1+\frac{1}{k}\right)^2} . \tag{7-139}$$

Since the mean value and the standard deviation of the test data can usually be easily computed, these formulas are useful in computing the factors λ and k.

Again, an example shows the application. Consider the test results of the compression strength of natural stone masonry shown in Table 7-27. Using the mean value of 6.30 MPa and the standard deviation of 0.39 MPa, the k value can be computed with 20.01 and the λ becomes 5.84×10^{-17}. The 5% fractile value is then given as

$$f_k = \left(-\frac{1}{5.84 \cdot 10^{-17}} \ln(0.95) \right)^{\frac{1}{20.01}} = 5.58 \text{ MPa} . \tag{7-140}$$

7.3.3.4 Leicester method

The Leicester method estimates the 5% fractile value without the selection of a probability function (Hunt and Bryant 1996). The 5% fractile value is computed as

$$f_k = A \cdot \left(1 - \frac{2.7 \cdot v}{n} \right) \tag{7-141}$$

with
n as number of samples preferred higher than 30
v as coefficient of variation preferred smaller than 0.5
A as empirical 5% fractile of the data. This value is often computed by linear interpolation.

The simple application is shown in the following example. Using 500 stone compression strength tests, the empirical 5% fractile value of the sample data is estimated with 42.3 MPa. Such estimations can, for example, be done with many spreadsheet programs such as Excel. In Excel, the "rang and quantile" function can be used. The coefficient of variation is 0.177. Then, the corrected 5% fractile value is given by

$$f_k = 42.3 \text{ MPa} \cdot \left(1 - \frac{2.7 \cdot 0.177}{500}\right) = 42.26 \text{ MPa} . \tag{7-142}$$

7.3.3.5 Öferbeck method

A further technique is the Öfverbeck Power Limit (Hunt and Bryant 1996). Using the constant ε, subject to the number of samples and the number of used samples q, the 5% fractile value is given as

$$f_k = x_q^{1-\varepsilon} \prod_{i=1}^{q-1} x_i^{\varepsilon/(q-1)} . \tag{7-143}$$

The relevant constants are given in Table 7-28.

Table 7-28. Relevant constants q and ε subject to the overall sample size

Sample size n	Number of used samples q	Öfverbeck constant ε
5	2	5.93
6	2	5.35
7	2	4.85
8	2	4.42
9	2	4.03
10	3	3.31
11	3	3.12
12	3	2.96
13	3	2.80
14	3	2.66
15	3	2.53
20	4	2.22
30	5	1.80
40	6	1.58
50	7	1.44

Again, the data of masonry compression strength are used as illustration. Only four samples are available. Using Table 7-28, q is 2 and 6.00 is chosen for ε by extrapolation. The samples are given by 5.7, 6.2, 6.6, and 6.7 MPa. The 5% fractile value is then

$$f_k = x_q^{1-\varepsilon} \prod_{i=1}^{q-1} x_i^{\varepsilon/(q-1)} = x_2^{1-6.00} \prod_{i=1}^{2-1} x_i \tag{7-144}$$

$$= 6.2^{1-6} \cdot 5.7^{6/(2-1)} = \frac{5.7^6}{6.2^5} = 3.74 \text{ MPa} .$$

7.3.3.6 Jaeger and Bakht method

Jaeger and Bakht (1990) present a technique to estimate the 5% fractile based on a combination of different probability functions. They call this artificial function a log arc sinh normal polynomial distribution. Assuming a quadratic polynomial in the distribution, and the consideration of only the lowest, average and highest test value, the fractile value can be computed with the following steps:

1. Sort the data.
2. Compute the mean value. If the sample number is uneven, then use the median. If the sample number ($n=2 \times k$) is even, use the following formula:

$$f_m = \frac{(f_k + f_{k+1})}{2} . \tag{7-145}$$

3. Transform the data according to

$$y_i = \frac{f_i^2 - f_m^2}{2 \cdot f_i \cdot f_m} . \tag{7-146}$$

4. Chose the characteristic value—in our case, the 5% fractile value. Then chose the representing k-value of a normal distribution, here $k = z^* = -1{,}645$.
5. Chose a z_1 according to Table 7-29.

Table 7-29. Representative z_1 value subject to the sample number

Sample number n	z_1	
10	1.34	
11	1.38	As approximation, the following equation
12	1.43	can be recommended:
13	1.47	$z_1 = -0.8004 - 0.0649 \cdot n + 0.0011 \cdot n^2$
14	1.50	
15	1.53	
16	1.56	

Sample number n	z_1
17	1.59
18	1.62
19	1.65
20	1.67

6. Compute the 5% fractile value for the transformed data:

$$y^* = \left(\frac{y_1 - y_n}{2}\right)\left(\frac{z^*}{z_1}\right) + \left(\frac{y_1 + y_n}{2}\right)\left(\frac{z^*}{z_1}\right)^2 . \tag{7-147}$$

7. Compute the 5% fractile value for the original data:

$$f_k = f_m(y^* + \sqrt{1 + (y^*)^2} . \tag{7-148}$$

The introduced list is only one possible technique mentioned by Jaeger and Bakht (1990); however, they consider this one as the numerically most robust one. The other techniques consider other data points, such as the smallest, the second smallest, and the mean value, or other orders of the polynomial, and yield slightly different 5% fractile values.

Again, an example illustrates the approach. Fifteen bending test results of granite stones are used. The test results are then related to one specimen height and listed in Table 7-30.

Table 7-30. Test results and transformed data

No.	Bending tensile strength in MPa	Percent	y_i
1	14.710	100.00	0.35106
2	12.802	92.80	0.20674
3	12.700	85.70	0.19858
4	12.506	78.50	0.18291
5	11.150	71.40	0.06719
6	10.829	64.20	0.03793
7	10.822	57.10	0.03729
8	10.426	50.00	0.00000
9	10.208	42.80	–0.02113
10	10.002	35.70	–0.04153
11	9.919	28.50	–0.04987
12	9.664	21.40	–0.07597
13	9.412	14.20	–0.10250
14	6.281	7.10	–0.52875
15	4.233	0.00	–1.02851

According to Step 2, the mean value is given as f_m=10.426 MPa. The transformed data (Step 3) are already included in Table 7-30. Furthermore, the 5% fractile value should be computed, therefore $z^* = -1.645$ and $z_1 =$ 1.53. Then the transformed 5% fractile value is given as

$$y^* = \left(\frac{-1.029-0.351}{2}\right)\left(\frac{-1.645}{-1.53}\right) + \left(\frac{-1.029+0.351}{2}\right)\left(\frac{-1.645}{-1.53}\right)^2 \qquad (7\text{-}149)$$

$$y^* = -1.133$$

and for the original data

$$f_k = 10.426 \text{ MPa} \cdot \left(-1.133 + \sqrt{1+(-1.133)^2}\right) = 3.94 \text{ MPa}. \qquad (7\text{-}150)$$

7.3.3.7 Binomial distribution

Mehdianpour (2006) has used diagrams of binomal distribution to estimate characteristic compression strength values. Details can be found in the original reference.

All mentioned techniques consider the uncertainty of material and loading data in a stochastic way. However, such a consideration does not necessarily end in the preparation of fractile values and partial safety factors. Increasingly, full probabilistic computations of structures and also of historical stone arch bridges can be found in the literature. Therefore, Chapter 8 discusses the results of those numerical investigations.

References

Allaix DL, Carbone VI & Mancini G (2007) Global Safety Factors for Reinforced Concrete Beams. Proceedings of the 5th International Probabilistic Workshop, L Taerwe & D Proske (Eds), Ghent, Belgium

Bamboo (2005) The Hong Kong Polytechnic University and International Network for Bamboo and Rattan: Bamboo Scaffolds in Building Construction – Design Guide. www.inbar.int/publication/txt/INBAR_Technical_Report_No23.doc

Bartsch HJ (1991) Mathematische Formeln. 23. Auflage, Fachbuchverlag Leipzig

Bayer (2008) Efficient Modelling and Simulation of Random Fields. Presentation at the 6th International Probabilistic Workshop, Darmstadt

Bayer V & Ross D (2008) Efficient Modelling and Simulation of Random Fields. Proceedings of the 6th International Probabilistic Workshop, CA Graubner, H Schmidt & D Proske (Eds), Darmstadt, pp 3363–380

Bergman LA, Shinozuka M, Bucher CG, Sobczyk K, Dasgupta G, Spanos PD, Deodatis G, Spencer Jr. BF, Ghanem RG, Sutoh A, Grigoriu M, Takada T, Hoshiya M, Wedig WV, Johnson EA, Wojtkiewicz SF, Naess A, Yoshida I,

Pradlwarter HJ, Zeldin BA, Schuëller GI & Zhang R (1997) A state-of-the-art report on computational stochastic mechanics. Probabilistic Engineering Mechanics, 12 (4), pp 197-321

Bergmeister K (2003) Kohlenstofffasern im Konstruktiven Ingenieurbau. Verlag Ernst und Sohn, Berlin

Bobee B & Ashkar F (1991) The Gamma Family and Derived Distributions Applied in Hydrology. Water resources publications, Littleton

Bothe E, Henning J, Curbach M, Bösche T & Proske D (2004) Nichtlineare Berechnung alter Bogenbrücken auf der Grundlage neuer Vorschriften. Beton- und Stahlbetonbau 99, Heft 4, pp 289–294

Box GEP & Draper NR (1987) Empirical Model-Building and Response Surfaces. John Wiley & Sohn

Braml T & Keuser M (2008) Structural Reliability Assessment for Existing Bridges with In Situ Gained Structural Data Only – Proceeding and Assessment of Damaged Bridges. Proceedings of the 6th International Probabilistic Workshop, CA Graubner, H Schmidt & D Proske (Eds), Darmstadt, pp 149–165

Breitung K (1984) Asymptotic approximations for multinormal integrals. Journal of the Engineering Mechanics Division, 110 (3), pp 357–366

Breitung K (2002) SORM: Some common misunderstandings and errors. In: Application of Statistics and Probability in Civil Engineering, A Der Kiureghian, S Madanat & JM Pestana (Eds), Millpress, Rotterdam 2003, pp 35–40

Bucher C & Bourgund U (1990) A fast and efficient response surface approach for structural reliability problems. Structural Safety, 7, pp 57–66

Bucher C (1988) Adaptive sampling – An iterative fast Monte Carlo procedure. Structural Safety, 5, pp 119–126

Bucher C, Hintze D & Roos D (2000) Advanced analysis of structural reliability using commercial FE-codes. European Congress on Computational Methods in Applied Sciences and Engineering (ECCOMAS), CIMNE, Barcelona, 11–14 September 2000

Busch P & Zumpe G (1995) Tragfähigkeit, Tragsicherheit und Tragreserven von Bogenbrücken. 5. Dresdner Brückenbausymposium, 16.3.1995, Fakultät Bauingenieurwesen, Technische Universität Dresden, pp 153–169

Button D et al. (1993) Glass in Building – A Guide to Modern Architectural Glass Performance. Pilkington Glass Ltd. Flachglas AG, Pilkington

BÜV (2001) Empfehlung für die Bemessung und Konstruktion von Glas im Bauwesen. veröffentlicht in: Der Prüfingenieur 18 April 2001, pp 55–69

Cai GQ & Elishakoff I (1994) Refined second-order reliability analysis. Structural Safety, 14, pp 267–276

Casas JR, Prato CA, Huerta S, Soaje PJ & Gerbaudo CF (2001) Probabilistic assessment of roadway and railway viaducts. Safety, Risk, Reliability – Trends in Engineering, Malta 2001, pp 1001–1008

Castillo E, Castillo C & Mínguez R (2008) Use of extreme value theory in engineering design. Safety, Reliability and Risk Analysis: Theory, Methods and Applications, S Martorell et al. (Eds), Taylor & Francis Group, London, pp 2473–2488

CEB (1976) Comité Euro-International du beton: International system of unified standard – codes of practice for structures. Volume I: Common unified rules for different types of construction and material (3rd Draft, Master Copy), Bulletin d'information 116 E, Paris, November 1976

Cervenka V (2007) Global Safety Format for Nonlinear Calculation of Reinforced Concrete. Proceedings of the 5th International Probabilistic Workshop, L Taerwe & D Proske (Eds), Ghent, Belgium

Chudoba R & Vořechovský M (2006) Stochastic modeling of multi-filament yarns: II. Random properties over the length and size effect, International Journal of Solids and Structures, 43 (3-4), pp 435–458.

Chudoba R, Vořechovský M & Konrada M (2006) Stochastic modeling of multi-filament yarns. I. Random properties within the cross-section and size effect, International Journal of Solids and Structures, 43 (3-4), pp 413–434

COST-345 (2004) European Commission Directorate General Transport and Energy: COST 345 – Procedures Required for the Assessment of Highway Structures: Numerical Techniques for Safety and Serviceability Assessment – Report of Working Groups 4 and 5

Curbach M & Proske D (2004) Zur Ermittlung von Teilsicherheitsfaktoren für Natursteinmaterial. 12. November 2004, 2. Dresdner Probabilistik Symposium. Technische Universität Dresden. pp 99–128

Curbach M, Michler H & Proske D (2002) Application of Quasi-Random Numbers in Monte Carlo Simulation. 1st International ASRANet Colloguium. 8–10th July 2002, Glasgow, auf CD

DAfSt – Deutscher Ausschuß für Stahlbeton (1996) Richtlinie für Betonbau beim Umgang mit wassergefährdenden Stoffen, Beuth-Verlag: Berlin

Der Kiureghian A & Ke J-B (1988) The stochastic finite element method in structural reliability. Probabilistic Engineering Mechanics, 3 (2), pp 105–112

Der Kiureghian A, Haukaas T & Fujimura K (2006) Structural reliability software at the University of California, Berkeley. Structural Safety 28, pp 44–67

Diamantidis D, Holicky M & Jung K (2007) Assessment of existing structures – On the applicability of the JCSS recommendations. In: Aspects of Structural Reliability – In Honor of R. Rackwitz, MH Faber, T Vrouwenvelder, K Zilch (Eds), Herbert Utz Ver-lag, München

DIN 1045-1 (2001) Tragwerke aus Beton, Stahlbeton und Spannbeton, Teil 1: Bemessung und Konstruktion, Juli 2001

DIN 1053-100 (2004) Mauerwerk – Berechnung auf der Grundlage des semiprobabilistischen Sicherheitskonzeptes. NABau im DIN e.V.: Berlin

DIN 1055-100 (1999): Einwirkungen auf Tragwerke, Teil 100: Grundlagen der Tragwerksplanung, Sicherheitskonzept und Bemessungsregeln, Juli 1999

DIN 18 516 (1999) Außenwandbekleidungen

DIN 53 804 (1981) Teil 1: Statistische Auswertungen – Meßbare (kontinuierliche) Merkmale. September 1981

Drezner Z (1992) Computation of the multivariate normal integral. ACM Transaction on Mathematical Software, 18 (4), pp 470–480

Eibl J & Schmidt-Hurtienne B (1995) Grundlagen für ein neues Sicherheitskonzept, Bautechnik, 72, Nr. 8, pp 501–506

Eibl J (1992) Nichtlineare Traglastermittlung/Bemessung. Beton- und Stahlbetonbau 87, Heft 6, pp 137–139

EN 1994 (2006) Eurocode 4: Design of composite steel and concrete structures

EN 1995 (2008) Eurocode 5: Design of timber structures

EN 1996 (2006) Eurocode 6: Design of masonry structures

EN 1999 (2007) Eurocode 9: Design of aluminium structures

Epstein S, Rauzyb A & Reinhart F M (2008) The Open PSA Initiative for Next Generation Probabilistic Safety Assessment. Proceedings of the 6th International Probabilistic Workshop, CA Graubner, H Schmidt & D Proske (Eds), Darmstadt

Estes AC & Frangopol DM (1998) RELSYS: A computer program for structural system reliability. Structural Engineering and Mechanics, 6 (8), pp 901–919

Eurocode 1 (1994) (ENV 1991 –1): Basis of Design and Action on Structures, Part 1: Basis of Design. CEN/CS, August 1994

Fießler B, Hawranek R & Rackwitz R (1976) Numerische Methoden für probabilistische Bemessungsverfahren und Sicherheitsnachweise. Technische Universität München, Berichte zur Zuverlässigkeitstheorie, Heft 14

Fischer A & Schnell J (2008) Determination of partial safety factors for existing structures. Proceedings of the 6th International Probabilistic Workshop, CA Graubner, H Schmidt & D Proske (Eds), Darmstadt, pp 133–147

Fischer L (1999) Sicherheitskonzept für neue Normen – ENV und DIN-neu, Grundlagen und Hintergrundinformationen. Teil 3: Statistische Auswertung von Stichproben im eindimensionalen Fall. Bautechnik 76 (1999), Heft 2, pp 167–179, Heft 3, pp 236–251, Heft 4, pp 328–338

Flederer H (2001) Beitrag zur Berechnung durchlaufender Stahlverbundträger im Gebrauchszustand unter Berücksichtigung streuender Eingangsgrößen. Dissertation, Technische Universität Dresden, Lehrstuhl für Stahlbau

Franz G, Hampe E & Schäfer K (1991) Konstruktionslehre des Stahlbetons. Band II: Tragwerke, Zweite Auflage, Springer Verlag, Berlin

Freudenthal AM (1947) Safety of Structures, Transactions ASCE, V. 112, 1947, pp 127–180

Genz A (1992) Numerical computation of multivariate normal probabilities, Journal of Computational and Graphical Statistics, 1, pp 141–149

Ghanem G & Spanos PD (1991) Stochastic Finite Element: A Spectral Approach, publisher, Springer, New York

Gollwitzer S & Rackwitz R (1990) On the reliability of Daniels systems. Structural Safety, 7, pp 229–243

Gollwitzer S, Kirchgäßner B, Fischer R & Rackwitz R (2006) PERMAS-RA/STRUEL system of programs for probabilistic reliability analysis. Structural Safety 28, pp 108–129

Grunert B, Grunert S & Grieger C (1998) Die historischen Baustoffe der Marienbrücke zu Dresden. Wissenschaftliche Zeitschrift der Technischen Universität Dresden, 47, Heft 5/6, pp 11–20

Grunert S (1982) Der Sandstein der Sächsischen Schweiz als Naturressource, seine Eigenschaften, seine Gewinnung und Verwendung in Vergangenheit und Gegenwart. Dissertation, Technische Universität Dresden

GruSiBau (1981) Normenausschuß Bauwesen im DIN: Grundlagen zur Festlegung von Sicherheitsanforderungen für bauliche Anlagen. Beuth Verlag

Güsgen J, Sedlacek G & Blank K (1998) Mechanische Grundlagen der Bemessung tragender Bauteile aus Glas, Stahlbau 67, pp 281–292

Hasofer AM & Lind MC (1974) Exact and invariant second moment code format, Journal of Engineering Mechanics Division, 100, pp 111–121

Holicky M (2007) Probabilistic Approach to the Global Resistance Factor for Reinforced Concrete Members. Proceedings of the 5th International Probabilistic Workshop, L Taerwe & D Proske (Eds), Ghent, Belgium

Hunt RD & Bryant AH (1996) Statistical Implications of Methods of Finding Characteristic Strengths. Journal of Structural Engineering, 122 (2), pp. 202–209

Ibrahim Y (1991) Observations on applications of importance sampling in structural reliability analysis, Structural Safety, 9, pp 269–281

ISO/CD 13822 (1999) Bases for design of structures – Assessment of existing structures. International Organisation for Standardization, Geneva

Jaeger LG & Bakht B (1990) Lower Fractiles of Material Strength from limited test data. Structural Safety, 7, pp 67–75

Jäger W (2005) Zur Einführung der DIN 1053-100, Mauerwerk – Berechnung auf der Grundlage des semiprobabilistischen Sicherheitskonzepts. Mauerwerk 9, 1, pp 21–23

JCSS (2004) – Joint Committee of Structural Safety. Probabilistic Modelcode. www.jcss.ethz.ch

Kandarpa S, Kirkner DJ & Spencer BF (1996) Stochastic damage model for Brittle materials subjected to monotonic loading. Journal of Engineering Mechanics, 122 (8), Aug. 1996, pp 788–795

Klingmüller O & Bourgund U (1992) Sicherheit und Risiko im Konstruktiven Ingenieurbau. Braunschweig/Wiesbaden: Vieweg

Köylüoglu HU & Nielsen SRK (1994) New approximations for SORM integrals. Structural Safety, 13, pp 237–246

Lemaire M & Pendola M (2006) PHIMECA-Soft. Structural Safety, 28, pp 130–149

Lin H-Z & Khalessi MR (2006) General outlook of UNIPASS V5.0: A general purpose probabilistic software system. Structural Safety, 28, pp 196–216

Liu P & Der Kiureghian A (1986) Multivariate distribution models with prescribed marginals and covariances. Probabilistic Engineering Mechanics, 1, pp 105–112

Low BK & Teh CI (2000) Probabilistic Analysis of Pile Deflection under Lateral Loads. In: Application of Statistics and Probability ICASP 8, RE Melchers & MG Stewart (Eds), Volume 1, Balkema, Rotterdam, pp 407–414

Macke M (2000) Variance reduction in Monte Carlo Simulation of dynamic systems. Applications of Statistics and Probability (ICASP 8). RE Melchers & MG Stewart (Eds), Balkema, Rotterdam, Band II, pp 797–804

Maes MA, Breitung K & Dupuis DJ (1993) Asymptotic importance sampling. Structural Safety, 12, pp 167–186

Mann W (1987) Überlegungen zur Sicherheit im Mauerwerksbau. Mauerwerks-Kalender. Verlag Ernst & Sohn, Berlin, pp 1–5

Mann W (1999) Mauerwerk in Europa. Der Prüfingenieur. 14. April 1999, VPI, pp 53–59

Mayer M (1926) Die Sicherheit der Bauwerke und ihre Berechnung nach Grenzkräften anstatt nach zulässigen Spannungen. Verlag von Julius Springer, Berlin

McBean EA & Rovers FA (1998) Statistical procedures for analysis of environmental monitoring data & risk assessment. Prentice Hall PTR Environmental Management & Engineering Series, Volume 3, Prentice Hall, Inc., Upper Saddle River

Mehdianpour M (2006) Tragfähigkeitsbewertung aus Versuchen – Probenanzahl versus Aussagesicherheit. Proceedings of the 4th International Probabilistic Symposium, D Proske, M Mehdianpour & L Gucma (Eds), Berlin, 2006

Melchers RE & Faber MH (2001) Aspects of Safety in Design and Assessment of Deteriorating Structures, Proceedings to the International IABSE Conference on Safety, Risk and Reliability - Trends in Engineering, March 21-23, Malta 2001, pp. 161-166

Melchers RE (1999) Structural Reliability Analysis and Prediction. 2nd Edition, John Wiley & Sons

Mori Y & Ellingwood BR (1993) Time-dependent system reliability analysis by adaptive importance sampling. Structural Safety, 12, pp 59–73

Most T (2008) An adaptive response surface approach for reliability analyses of discontinuous limit state functions. Proceedings of the 6th International Probabilistic Workshop, CA Graubner, H Schmidt & D Proske (Eds), Darmstadt, pp 381–396

Murzewski J (1974) Sicherheit der Baukonstruktionen. VEB Verlag für Bauwesen, Berlin, DDR

Norton RL (1996) Machine Design – An Integrated Approach. Prentice-Hall, New York

NR (1992) Numerical Recipes in FORTRAN 77: The Art of Scientific Computing. Cambridge University Press

Onken P, vom Berg W & Neubauer U (2002) Verstärkung der West Gate Bridge, Melbourne, Beton- und Stahlbetonbau, Heft 2, 97 Jahrgang, pp 94–104

Östlund L (1991) An Estimation of gamma-Values. An application of a probabilistic method. In: Reliability of Concrete Structures. Final report of Permanent Commission I. CEB Bulletin d' Information. Heft 202, pp 37–51

Peschel A (1984) Natursteine der DDR – Kompendium petrographischer, petrochemischer, petrophysikalischer und gesteinstechnischer Eigenschaften. Beilage zur Dissertation (B) an der Bergakademie Freiberg

Petschacher M (1994) VaP a Tool for Practicing Engineers. Proceedings of ICOSSAR'93, GI Schueller et al. (Ed), 6th International Conference on Structural Safety and Reliability in Innsbruck, Balkema, pp 1817–1823

Pfeiffer U & Quast U (2003) Nichtlineares Berechnen stabförmiger Bauteile. Beton- und Stahlbetonbau 98, Volume 9, pp 529–538

Plate EJ (1993) Statistik und angewandte Wahrscheinlichkeitsrechnung für Bauingenieure, Ernst & Sohn Verlag, Berlin

Polidori DC, Beck JL & Papadimitriou C (1999) New approximations for reliability integrals. Journal of Engineering Mechanics, April 1999, pp 466–475

Proske D (2003) Ein Beitrag zur Risikobeurteilung von alten Brücken unter Schiffsanprall. Ph.D. Thesis, Technische Universität Dresden

Proske D (2008a) Definition of Safety and the Existence of "Optimal Safety". Proceedings of the ESREL 2008 conference, Valencia

Proske D (2008b) Catalogue of risks – Natural, technical, health and social risks. Springer, Berlin – Heidelberg

Proske D, Michler H & Curbach M (2002) Application of quasi-random numbers in Monte Carlo Simulation. ASRANET, Glasgow, on CD

Pukl R, Novak D, Vorechovsky M & Bergmeister K (2006) Uncertainties of Material Properties in Nonlinear Computer Simulation. Proceedings of the 4th International Probabilistic Symposium, D Proske, M Mehdianpour und L Gucma (Eds), Berlin, pp 127–138

Pukl R, Novák D, Vořechovský M & Bergmeister K (2006) Uncertainties of material properties in nonlinear computer simulation. D Proske, M Mehdianpour & L Gucma (Eds), Proceedings of the 4th International Probabilistic Symposium, 12th–13th October 2006, Berlin

Rachwitz R & Hohenbichler M (1981) An Order Statistics Approach to Parallel Structural Systems, Berichte zur Zuverlässigkeitstheorie der Bauwerke, SFB 96, Technische Universität München, Heft 58

Rackwitz R & Hohenbichler M (1981) An Order Statistics Approach to Parallel Structural Systems, Berichte zur Zuverlässigkeitstheorie der Bauwerke, SFB 96, Technische Universität München, Heft 58

Rajashekhar MR & Ellingwood BR (1993) A new look at the response surface approach for reliability analysis, Structural Safety, 12, pp 205–220

Reh S, Beley J-D, Mukherjee S & Khor EH (2006) Probabilistic finite element analysis using ANSYS. Structural Safety, 28, pp 17–43

Reid SG (1999) Load–testing for design of structures with Weibull–distributed strengths. Application of Statistics and Probability (ICASP 8), Sydney, 1999, Band 2, pp 705–712

Roos D & Bayer V (2008) An efficient robust design optimization approach using advance surrogate models. Proceedings of the 6th International Probabilistic Workshop, CA Graubner, H Schmidt & D Proske (Eds), Darmstadt

Ross D & Bucher C (2003) Adaptive response surfaces for structural reliability of nonlinear finite element structures. NAFEMS Seminar: Use of Stochastics in FEM Analyses. May 7–8 2003 Wiesbaden

Rüsch R, Sell R & Rackwitz R (1969) Statistische Analyse der Betonfestigkeit. Deutscher Ausschuß für Stahlbeton, Heft 206, Berlin

Schervish MJ (1984) Multivariate normal probabilities with error bound. Applied Statistics, Royal Statistical Society, 33 (1), pp 81–94

Schlegel R & Will J (2007) Sensitivitätsanalyse und Parameteridentifikation von bestehenden Mauerwerkstrukturen. Mauerwerk 11, Heft 6, pp 349–355

Schleicher F (Ed) (1949) Taschenbuch für Bauingenieure, Springer-Verlag

Schneider J (1999) Zur Dominanz der Lastannahmen im Sicherheitsnachweis. Festschrift zum 60. Geburtstag von Prof. Dr. Edoardo Anderheggen. Institut für Baustatik und Konstruktion, Eidgenössische Technische Hochschule Zürich, pp 31–36

Schubert P (1995) Beurteilung der Druckfestigkeit von ausgeführtem Mauerwerk aus künstlichen Steinen und Naturstein. Mauerwerkskalender 1995, pp 687–701

Schueller GI & Pradlwarter HJ (2006) Computational stochastic structural analysis (COSSAN) – a software tool. Structural Safety, 28, pp 68–82

Schueremans L & Van Gemert D (2001) Assessment of existing masonry structures using probabilistic methods - state of the art and new approaches. STRUMAS 2001, 18.–20. April 2001, Fifth International Symposium on Computer Methods in Structural Masonry

Schweckendiek T & Courage W (2006) Structural reliability of sheet pile walls using finite element analysis. Proceedings of the 4th International Probabilistic Symposium, D Proske, M Mehdianpour & L Gucma (Eds), Berlin, pp 97–111

Shigley JE & Mischke CR (2001) Mechanical Engineering Design. 6th Edition McGraw Hill. Inc., New York

Song BF (1997) A technique for computing failure probability of structure using importance sampling. Computers & Structures, 72, pp 659–665

Spaethe G (1992) Die Sicherheit tragender Baukonstruktionen, 2. Neubearbeitete Auflage, Wien, Springer Verlag

Spaethe G (1992) Die Sicherheit tragender Baukonstruktionen, 2. Neubearbeitete Auflage, Wien, Springer Verlag

Steel RD & Torrie JH (1991) Prinsip dan Prosedur Statistika, Edition 2, Gramedia Pustaka Utama, Jakarta, 1991

Storm R (1988) Wahrscheinlichkeitsrechnung, mathematische Statistik und statistische Qualitätskontrolle. VEB Fachbuchverlag Leipzig

Strauss A & Bergmeister K (2005) Safety concepts of new and existing structures. Vortrag zur fib Commission 2 Sitzung "Safety and performance concepts" am 23. Mai 2005 in Budapest

Thacker BH, Riha DS, Fitch SHK, Huyse LJ & Pleming JB (2006) Probabilistic engineering analysis using the NESSUS software. Structural Safety 28, pp 83–107

Tichý M (1976) Probleme der Zuverlässigkeit in der Theorie von Tragwerken. Vorträge zum Problemseminar: Zuverlässigkeit tragender Konstruktionen. Weiterbildungszentrum Festkörpermechanik, Konstruktion und rationeller Werkstoffeinsatz. Technische Universität Dresden – Sektion Grundlagen des Maschinenwesens. Heft 3/76

Tonon F & Tonon L (2006) Renovation of the 17th-Century Ponte Lungo bridge in Chioggia, Italy. Journal of Bridge Engineering, 11 (1), January-February 2006, pp 13–20

Tvedt L (1988) Second-order reliability by an exact integral. In: Proceedings of the 2nd IFIP Working Conference on Reliability and Optimization on structural Systems, P Thoft-Christensen (Ed), Springer, pp 377–84

Tvedt L (2006) Proban – probabilistic analysis. Structural Safety, 28, pp 150–163

Val DV & Stewart MG (2002) Safety factors for assessment of existing structures. Journal of Structural Engineering, 128 (2), pp 258–265

Van der Waerden BL (1957) Mathematische Statistik. Springer Verlag, Berlin

Vanmarcke E (1983) Random fields: Analysis and synthesis. MIT Press

Visodic JP (1948) Design Stress Factors. Vol. 55, May 1948, ASME International, New York

Weibull W (1951) A Statistical Distribution of wide Applicability, J. App. Mech., Vol. 18, 1951, pp 253

Weiland S (2003) Das Antwort-Flächen-Verfahren. 1st Dresdner Probabilistik-Symposium – Sicherheit und Risiko im Bauwesen. D Proske (Ed), TU Dresden, 2nd Edition 2006

Wenzel F – (Ed) (1997) Mauerwerk – Untersuchen und Instandsetzen durch Injizieren, Vernadeln und Vorspannen. Erhalten historisch bedeutsamer Bauwerke. Empfehlungen für die Praxis. Sonderforschungsbereich 315, Universität Karlsruhe

Wiese H, Curbach M, Al-Jamous A, Eckfeldt L & Proske D (2005) Vergleich des ETV Beton und DIN 1045-1. Beton- und Stahlbetonbau 100, Heft 9, pp 784–794

Wu Y-T, Shin Y, Sues RH & Cesare MA (2006) Probabilistic function evaluation system (ProFES) for reliability-based design. Structural Safety, 28, pp 164–195

Yuan X-X & Pandey MD (2006) Analysis of approximations for multinormal integration in system reliability computation. Structural Safety, 28 (4), pp 361–377

Zhao Y-G & Ono T (1999a) New approximations for SORM: Part 1. Journal of Engineering Mechanics, ASCE 125 1, pp 79–85

Zhao Y-G & Ono T (1999b) New approximations for SORM: Part 2. Journal of Engineering Mechanics, ASCE 125 1, pp 86–93

Zhao Y-G & Ono T (1999c) A general procedure for first/second-order reliability method (FORM/SORM). Structural Safety, 21 (2), pp 97–112

8 Examples

8.1 Introduction

Only a few examples can be found in literature dealing with the probabilistic safety assessment of historical arch bridges. Whereas in other technical fields such as car design, turbine design, or design of new structures, probabilistic techniques are now common, especially in fields with highly nonlinear material behaviour such as masonry, and difficult numerical expression of the overall load-bearing capacity such as found for arch bridges, the success of probabilistic methods was so far limited. However, the number of examples rises and a few example are mentioned here.

8.2 Examples in Literature

Modena and Sonda (1998) have investigated a reinforced concrete arch constructed in 1948 with a span of 60 m by probabilistic computation. First, they investigated the safety index using only existing data, and received 3.78 (7.93×10^{-5}) and 4.44 (4.41×10^{-6}). After extraction of updated material properties, the safety index for different cross sections of the arch increased to 5.80 (3.27×10^{-9}) and 6.68 (1.17×10^{-11}). Please note the strong increase of the safety index by only considering more realistic material parameters.

Casas (1999) has published probabilistic computations for several historical arch bridges. He has used two different types of models: a linear-elastic model that describes failure as the origin of a new hinge and a nonlinear model that considers failure either by overload of the compression strength of the masonry or by development of a four-hinge mechanism.

The most simple model uses the formulae

D. Proske, P. van Gelder, *Safety of Historical Stone Arch Bridges*, DOI 10.1007/978-3-540-77618-5_8,
© Springer-Verlag Berlin Heidelberg 2009

$$g = H - \frac{N}{f_c \cdot B} - 2 \cdot \frac{M}{N} \tag{8-1}$$

with

- H as arch thickness described by a Gauss distribution with a coefficient of variation of 10%
- B as arch width described by a Gauss distribution with a coefficient of variation of 5%
- f_c as compression strength of the masonry described by a Gauss distribution with a coefficient of variation of 20%
- M as a bending moment described by a Gumbel distribution with a coefficient of variation of 10% for the live load and a Gauss distribution with a coefficient of variation of 5% for the dead load
- N as normal force described by a Gumbel distribution with a coefficient of variation of 10% for the live load and a Gauss distribution with a coefficient of variation of 5% for the dead load

Based on these input data and using the linear-elastic model Casas (1999) computed for the Magarola Bridge the following safety indexes (Table 8-1).

Table 8-1. Safety indexes for the Magarola Bridge for the linear-elastic model (Casas 1999)

Cross section at	No correlation $r = 0$	Strong correlation $r = 0.9$
Springing	1.54	1.64
Quarter point	4.81	4.82
Crown	1.11	1.14

The results in Table 8-1 give rather low values and indicate an insufficient safety. The consideration of correlations decreases the uncertainty of the input data and therefore usually increases the safety index, but in this case not sufficiently. However, if the mechanical model is changed from type 1 to type 2 using nonlinear failure considerations, then the safety index improves strongly. Casas (1999) obtained safety indexes for the bridge under live load between 10.6 and 14.3 and without live load between 11.6 and 15.5.

For two further bridges, the Jerge Arch Bridge and the San Rafael bridge, the results are shown in Tables 8-2 and 8-3.

Table 8-2. Safety indexes for the Jerge arch bridge (Casas 1999)

Cross section at	No correlation	Average correlation	Strong correlation
Springing	3.64	3.71	4.26
Quarter point	3.60	3.61	4.32
Crown	3.85	3.84	4.75

Table 8-3. Safety indexes for the San Rafael arch bridge (Casas 1999)

Cross section at	No correlation
Springing	5.49
Quarter point	7.99
Crown	7.39

In Table 8-4, all results from Casas (1999) are summarized. In general, it can be seen that linear-elastic models often yield to unacceptable safety indexes. However, for the computation, the current traffic model by the Eurocode 1 was assumed and it is understandable that historical bridges were not designed for such loads. Therefore, cases where the bridges do not meet the safety requirements in terms of a safety index are understandable. However, nonlinear models yield to acceptable safety indexes in all cases. This confirms that linear-elastic models underestimate the ultimate load of arch bridges significantly. Obviously, the historical design of the bridges was based on linear-elastic models, and therefore the bridges were permanently over designed. However, using more advanced models can extend the usage of such bridges to the much higher load that we face today.

Table 8-4. Summary of the probabilistic results from Casas (1999) in terms of the safety index

Bridge	Material	Span in m	β linear-elastic	β nonlinear
Magarola	Brick	20.0	1.6	13.0
Jerte	Granite stone	22.5	3.5	6.0
San Rafael	Concrete	24.0	5.5	17.6
Duenas	Limestone	15.0	0.24	10.0
Quintana	Concrete	19.0	7.0	
Torquemada	Concrete	40.0	4.5	

Schueremans (2001), Schueremans, Smars and Van Gemert (2001), and Schueremans and Van Gemert (2001, 2004) have heavily investigated the safety of arches in terms of safety factors and probabilistic measures. In one example, they give the static safety factor of 2.39 and the geometrical safety factor of 1.23 when using average material parameters.

Quan (2004) presents a probabilistic computation of a reinforced concrete arch. He receives safety indexes between 3.53 and 4.31 and considers the values as too low. Quan (2004) has used an independently developed program.

Žak et al. (2003) introduce a probabilistic computation of a historical arch bridge. The mechanical model was done in the finite element program ANSYS, however no probabilistic results are included in the publication. Busch (1998) and Busch and Zumpe (1995) have carried out an intensive probabilistic investigation of the Marien Bridge in Dresden, Germany. Möller et al. (1998) and Möller et al. (2002) have also used probabilistic investigations of historical arch bridges and vaults. Ng and Fairfield (2002) have carried out a Monte Carlo Simulation to investigate the safety of an arch bridge in probabilistic terms. Recently, Brencich et al. (2007) carried out probabilistic computations. Tschötschel (1989) has carried out intensive probabilistic computations of masonry, however he has not dealt with arch bridges.

In the next section, probabilistic computations of arch bridges carried out by the authors will be discussed.

8.3 Own Examples

8.3.1 Bridge 1

Bridge 1 is situated near Würzburg in the south of Germany. The bridge is a six-arch bridge with a span of approximately 25 m. It was constructed between 1872 and 1875, and it is made of regular coursed ashlar stone work. The material is red main sandstone, a high-quality coloured sandstone with a uniaxial compression strength of up to 140 MPa. Toward the end of World War II, one pier was blasted. This pier was rebuilt in 1945 and 1946 using concrete. During the reconstruction phase, the so-called explosion chambers were built inside the pier.

Figure 8-1 shows a view of the site nowadays. A detailed description of the bridge can be found in Proske (2003). Attention has been drawn to this bridge because it was hit by ships in 1999.

The investigation for the bridge was split into three steps. First, the bridge regions most severely stressed during the impact were detected by simple numerical calculations and then selected for drilling. Since Bridge 1 consists of different materials due to the partial replacement of blasted piers at the end of World War II, the transfer of results from one structural element to the same element in another position was not possible. Moreover, Bridge 1 has a unique foundation. With due regard to these specific factors, 26 drillings with a length of up to 15 m were planned and carried out on this bridge. The drillings had an overall length of 150 m: 90 m in masonry and 60 m in concrete.

Fig. 8-1. Picture of Bridge 1

The drilling produced a comparatively large amount of bridge material that was used for material testing. More than 500 material tests were carried out, including compressive and tensile strength tests of the sandstone, mortar, and concrete; Young's modulus tests; height and width measurements of the sandstone; density measurements; and measurements of the shear strength of the masonry. With the data from the tests, it was possible to describe the material input parameters in terms of random distributions. The choice of the distribution type has been discussed intensively in Proske (2003). Several statistical techniques have been used to determine the type of statistical distribution for the investigated material properties. In addition, the strength of the concrete compressive strength samples was multimodal. Therefore, this distribution has been decomposed into original distributions. This statistical effect could also be identified visually on the testing specimen and historically with documents from the reconstruction after the war. The statistical properties of the input variables for the numerical model are given in Table 8-5.

Table 8-5. Statistical properties input variables for Bridge 1

Parameter	Distribution	x_m [a]	s [b]	Unit
Sandstone compressive strength	Lognormal	75.40	21.30	MPa
Concrete compressive strength	Lognormal	47.90	22.28	MPa
Sandstone splitting strength	Lognormal	4.72	1.30	MPa
Concrete tensile stress	Lognormal	1.15	0.69	MPa
Young's modulus sandstone	Lognormal	28,534	7,079.6	MPa
Young's modulus concrete	Lognormal	22,552	8,682.1	MPa
Density sandstone	Normal	2.27	0.15	kg/dm^3

Parameter	Distribution	x_m[a]	s[b]	Unit
Density concrete	Normal	2.26	0.10	kg/dm^3
Mortar compressive strength	Lognormal	11.00	7.25	MPa
Ship impact force (frontal)	Lognormal	2.04	1.5	MN
Ship impact force (lateral)	Lognormal	0.61	0.385	MN
Sandstone height	Normal	0.7	0.13	m
Sandstone width	Lognormal	0.8	0.08	m
Mortar joint height	Lognormal	0.037	0.048	m
Impact height	Normal	3.0	0.5	m

[a] x_m – empirical mean, [b] s – empirical standard deviation

In the second step, the mechanical behaviour of the bridge under traffic loads and under ship impact has been modelled with a finite element program (ANSYS). The model of Bridge 1 was particularly complex due to the inhomogeneous structural system. Figure 8-2 permits a view inside the bridge. Figure 8-3 shows the finite element model using area and volumes (unmeshed), and Fig. 8-4 shows the principal compression stress inside a hit pier. The typical element size was about 0.5 m, but smaller elements were used in some regions. Cracks observed on the bridge, and caused by dead and live loads, could be approved with the used models.

To achieve such results, a realistic numerical model for the description of the load-bearing behaviour of natural stone masonry had to be incorporated into the finite element program. There exist several different techniques to describe the load-bearing behaviour of natural stone structures under normal and shear forces. The models from Mann, Hilsdorf, Berndt and Sabha for one layer walls, and the models from Warnecke and Egermann for several layer walls (Warnecke et al. 1995) have been investigated (Proske 2003). The von Berndt (1996) model was chosen after intensive numerical investigations to describe the load-bearing behaviour of the natural stone piers of the bridges. This model is valid for normal forces and shear forces and therefore able to describe the load conditions under impact. Also, this model has proven to reach acceptable results in the comparison of the load-bearing behaviour with a wide range of experimental data. In addition, the implementation into the finite element program was convenient.

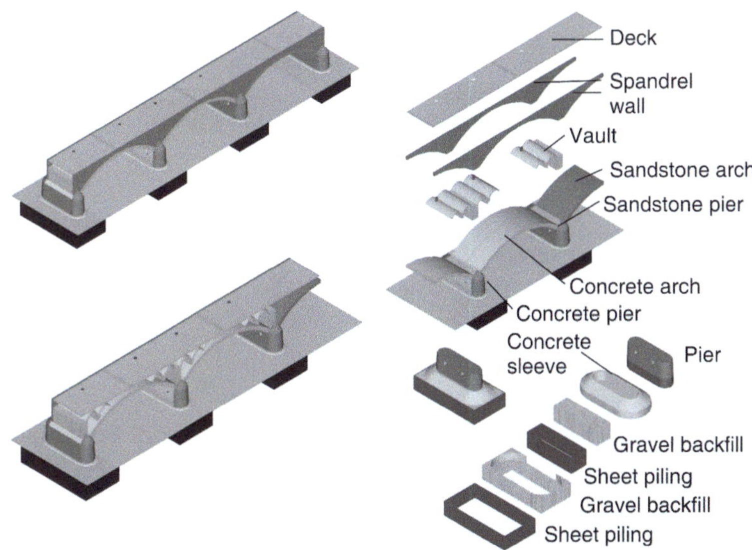

Fig. 8-2. Assembly of Bridge 1

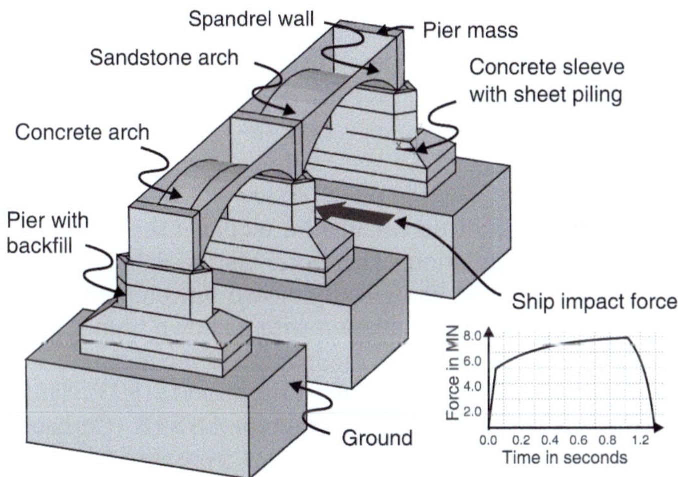

Fig. 8-3. Example of a finite element model of Bridge 1 (not meshed)

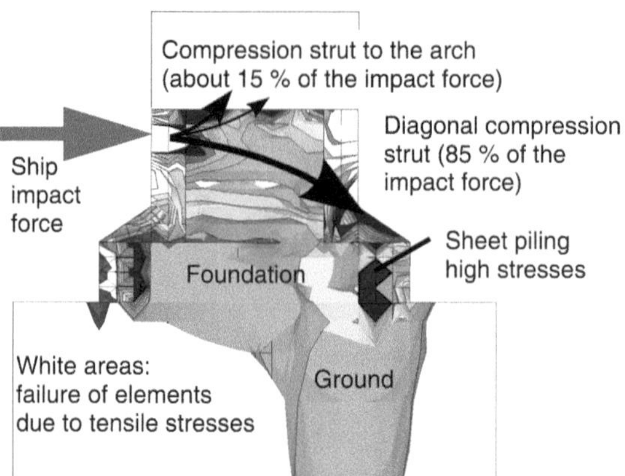

Fig. 8-4. Principal compression stress in longitudinal section of frontal hit pier of Bridge 1

Using this model, the calculation of one deterministic dynamic impact against Bridge 1 on an IBM workstation with a power II processor took about one hour. Of course, simple linear-elastic static and dynamic models for the bridge have also been used to check the results of the sophisticated models.

To incorporate the random variables in the third step, a probabilistic calculation was done using the FORM and SORM methods. After the FORM (Spaethe 1992) calculation, different SORM methods were applied to improve the quality of the first. Importance sampling has also been used to back the results (Spaethe 1992). All the methods used gave results that were comparable from an engineering point of view.

The criteria for the probabilistic calculation included the results of the dynamic finite element calculation. Therefore, the so-called limit state function was not available in an analytically closed form. One way to obtain results with the probabilistic calculation with a known limit state function that is not analytically closed is the application of the response surface methodology; for example, Rajashekhar and Ellingwood (1993). This procedure was included in the finite element program ANSYS (Curbach and Proske 1998). The FORM and SORM techniques were also incorporated into the finite element program ANSYS using the customized capabilities of the program. For that purpose, the techniques had to be provided as FORTRAN subroutines and could then be compiled and linked into the program.

8.3.2 Bridge 2

Bridge 2 has been chosen for comparison reasons with Bridge 1. Bridge 2 was built in 1893 and consists of a steel frame superstructure with natural stone piers. Parts of the bridge were destroyed during World War II. The static system of Bridge 2 is a four-field beam with a span of approximately 39 m. Due to the different static systems of Bridges 1 and 2, Bridge 2 will show a different behaviour under impact. In contrast to the excellent natural stone material of Bridge 1, the material in Bridge 2 has a lower strength (Table 8-6). Figure 8-5 shows a view of the site nowadays. A detailed description of the bridge can also be found in Proske (2003). This bridge was chosen to represent typical historical German bridges with steel superstructures and masonry piers over inland waterways.

Fig. 8-5. Picture of Bridge 2

Table 8-6. Statistical properties input variables for Bridge 2

Parameter	Distribution	x_m [a]	s [b]	Unit
Natural stone compressive strength	Normal	21.2	2.4	MPa
Natural stone splitting strength	Normal (lognormal)	0.38	0.094	MPa
Mortar compressive strength	Normal	15.5	3.58	MPa
Ship impact force (frontal)	Lognormal	2.04	1.5	MN
Ship impact force (protection)	Lognormal	0.046	0.8368	MN
Ship impact force (lateral)	Lognormal	0.61	0.385	MN
Impact height	Normal	3.0	0.5	m
Normal load	Normal	0.242	0.0242	MPa

[a] x_m – empirical mean, [b] s – empirical standard deviation

First, the maximum possible impact forces were investigated. In the next evaluation step, the probabilistic investigation was accomplished. The results of the probabilistic investigation of Bridges 1 and 2 are shown in Table 8-7. Several structural solutions to increase the load-bearing capacity of the bridges under ship impact were also investigated. They are visualized in Fig. 8-6. The results are shown as the probability of failure, either per impact $P(V|A)$ or per year $P(V \cap A)$. The value per year also includes the probability of the ship impact event. To show that the models of the bridges are comparable, the probability of failure under dead and live load conditions were also evaluated. Lines 6 and 14 from Table 8-7 show approximately the same value. Only these two lines refer to the failure of the piers under normal stress, all other lines refer to the shear failure of the piers or the arch. To allow a better comparison of both bridges, Table 8-8 summarizes the major properties of Bridges 1 and 2. The maximum permitted probability of failure per year is about 1.3×10^{-6} (E DIN 1055-100). Due to unsatisfactory results (see Table 8-7), the description of safety in terms of risk has been extended. However, the risk assessment is not discussed here.

Table 8-7. Probability of failure for different structural versions

I	II	III	IV	V	VI	VII	VIII	IX	X	XI
#	Bridge	Load	Element	Version	$P(V|A)\cdot10^{-6}$	$\beta(V|A)$	$P(V\cap A)\cdot10^{-6}$	$\beta(V\cap A)$	$P(V\cap A)\cdot10^{-6}$	$\beta(V\cap A)$
					per impact		per year		per year	
1	2		Pier 2	Damage (crack) [a]	313,667.7	0.4854	5,018.7	2.5745		
2		Frontal impact	Pier 2	No damage (no crack) [b]	154,256.0	1.0183	2,468.1	2.8111		
3			Pier 2	Impact fender system	1,540.5	2.9595	24.6	4.0605		
4			Pier 2	Pier increased × 2.3 [c]	11,843.4	2.2621	189.5	3.5541		
5			Pier 2	Tension strength × 2 [d]	43,179.2	1.7149	690.9	3.1985		
6		Dead and Live load	Normal stress [f]		240,0	3.4919	4.8	4.4259		
7		Lateral Pier 2		No damage [b]	328,986.4	0.4427	5,263.8	2.5580		
8		impact Pier 2		Impact fender system	84,539.3	1.3751	1,352.6	2.9994		
9	1		Pier II	Explosion chamber [e]	80,760.0	1.4002	1,292.2	3.0132	596.0	3.2410
10		Frontal impact	Pier II	No explosion chamber [f]	23,300.0	1.9904	372.8	3.3722	172.0	3.5799
11			Pier II	Pre-stressed [g]	340.0	3.3977	5.4	4.3958	2.5	4.5639
12			Pier II	Reinforced concrete [h]	32.0	3.9976	0.5	5.0370	0.2	5.0370
13			Pier II	GEWI-elements [i]	28.0	4.0290	0.4	5.0370	0.2	5.0370
14		Dead and Live load	Normal stress [f]		203.0	3.5363	4.1	4.4619	4.1	4.4619
15		Frontal Pier III		Explosion chamber [e]	35,930.0	1.8004	578.8	3.2511	265.2	3.4651
16		impact Pier III		No explosion chamber [f]	28,720.0	1.9004	459.5	3.3143	212.0	3.5249
17		Pier III		Pre-stressed [g]	30.0	4.0128	0.5	5.0493	0.2	5.0493
18		Arch			1,500.0	2.9681	24.0	4.0652	11.1	4.2421
19		Lateral Pier III		No explosion chambers [f]	25,670.0	1.9491	410.7	3.3454	54.9	3.8678
20		impact Pier III		No explosion chambers [f]	10,720.0	2.3006	171.5	3.5809	36.1	3.9688

Note: For both bridges, there exist different probabilities of impact – row VIII and IX for Bridge 2 and row X and XI for Bridge 1. To compare both bridges, Bridge 1 has in row VIII and IX the same probabilities of impact applied as for Bridge 2. Note: The probabilities are given as multiples of 10^{-6}

[a] The pier has been found with a 3 m crack
[b] Assumption of closing the crack
[c] Size of the pier increased by factor 2.3
[d] Hypothetical material with higher tensile strength
[e] Explosion chamber was found inside the piers
[f] Closing of the explosion chamber
[g] Pre-stressing of the pier with no-bond tendons (2 × 2 MN and 2 × 4 MN, respectively)
[h] Reinforced concrete replacement type piles (2 × 3 Ø 1,5 m) inside the pier and closing the explosion chamber
[i] Use of threaded rods (GEWI) inside the piers (2 × 4) and closing explosion chamber
[f] Not considering an impact

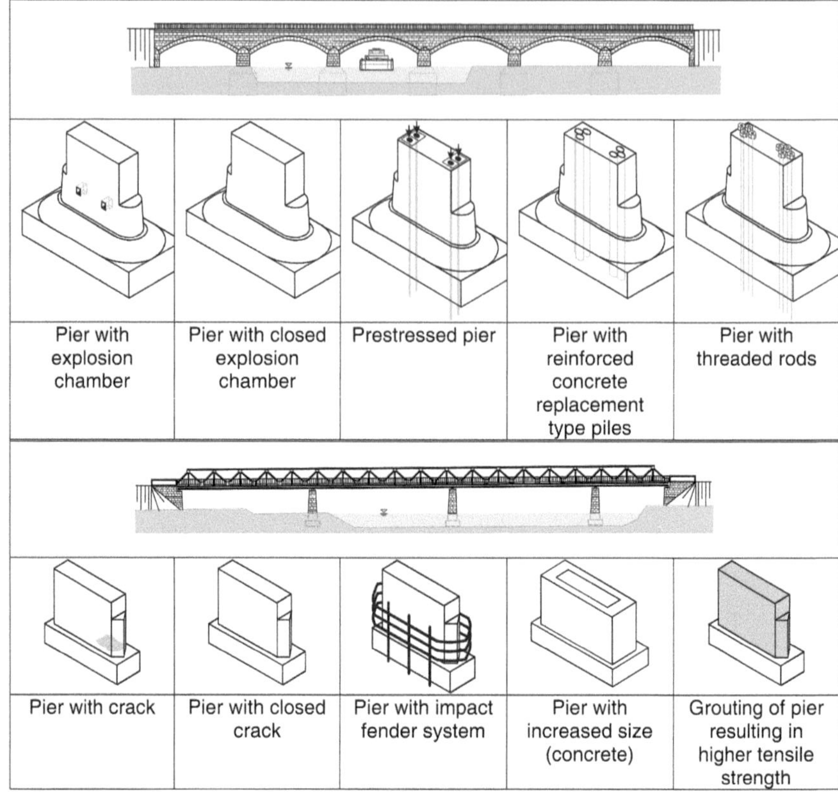

Fig. 8-6. Visualization of the investigated different strengthening technologies for Bridge 1 (*top*) and Bridge 2 (*bottom*)

Table 8-8. Comparison of the properties of Bridge 1 and Bridge 2

	Bridge 1		Bridge 2	
Probability of failure per impact	0.023	(1.0)[a]	0.15	(6.5)[a]
Probability of failure under dead and live loads	$2.030 \cdot 10^{-4}$	(1.0)[a]	$2.400 \cdot 10^{-4}$	(1.2)[a]
Area of the pier in m^2	48	(1.9)[a]	25	(1.0)[a]
Dead load in MN	37	(7.4)[a]	5	(1.0)[a]
Existing normal stress in MPa	0.84	(3.2)[a]	0.26	(1.0)[a]
Acceptable normal stress in MPa	25	(2.5)[a]	10	(1.0)[a]
Acceptable shear stress in MPa	0.8	(2.7)[a]	0.3	(1.0)[a]
Maximal dynamic impact force in MN	13.0	(2.9)[a]	4.5	(1.0)[a]
Quantile value of the impact force in %	99.99		97.00	

[a] Numbers in brackets give the ratio between the two bridges

8.3.3 Bridge 3

The third example deals with a railway bridge constructed between 1911 and 1913. The bridge consists of three arch spans built of brick masonry with concrete backfill. In the beginning, different statical models were developed, starting with simple beam models and increasing towards two- and three-dimensional finite element models with nonlinear material behaviour. Figure 8-7 shows a two-dimensional model using ATENA. Because the railroads are not symmetrically located at the bridge, a three-dimensional model was also created (Fig. 8-8). The modelling of the arch was done in shell elements and the shell elements were fixed to the piers. Input data were based on material tests. The material was provided by core drilling. A nonlinear computation on an IBM working station with 2 GB RAM required approximately one hour. Besides the deterministic computation, the probability of failure was also computed.

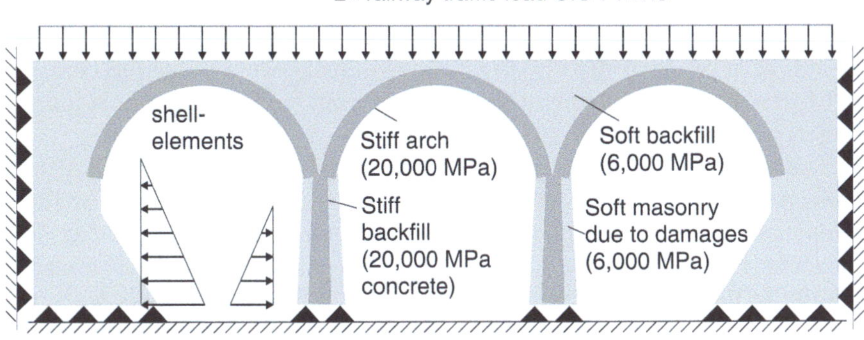

Fig. 8-7. Finite element model of Bridge 3 using ATENA

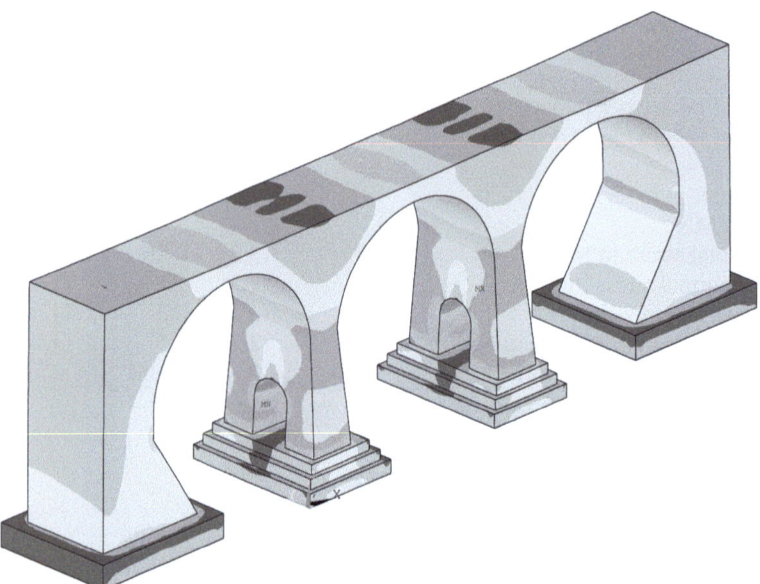

Fig. 8-8. Finite element model of Bridge 3 using the program ANSYS

8.3.4 Bridge 4

Bridge 4 is a railway masonry arch bridge made of natural stones (Fig. 8-9). The bridge has an overall length of 423 m and consists of 34 arches and 33 piers. The arches have a span of approximately 10 m. The piers are distinguished into regular piers and group piers. Group piers simply have a greater width (2.8 m) in comparison to normal piers (1.5 m).

The bridge was constructed in 1871 and was refurbished in 1929–1930. Originally the bridge was designed for two railway tracks, however only one track is in use. In general, the bridge seems to be in good condition, but some crack damage and deformations of the spandrel walls were found. As a result of the investigation, the cracks were related to centrifugal forces and unsymmetrical loads of the train since the bridge is located in a curvature.

The numerical load-bearing investigation started by constructing a fixed arch, a two-hinged arch, and a three-hinge arch beam model. For those simple models, the acceptable stress values were exceeded. Besides the first models, drillings were also carried out at the bridge. Based on the results in terms of material properties, the arch beam models were improved (Figs. 8-10 and 8-11) and nonlinear two-dimensional (Fig. 8-12) and three-dimensional models were created (Fig. 8-13). Considering the backfill and the spandrel walls, the load-bearing capacity could be temporarily proved. If one considers that the original suggestion was the destruction of the bridge and the construction of a new reinforced concrete bridge, the result can be seen as a success. It was recommended to improve the backfill and add a reinforced concrete slab, however the owner agreed only to the concrete slab. The probability of failure was also computed.

Fig. 8-9. Picture of Bridge 4

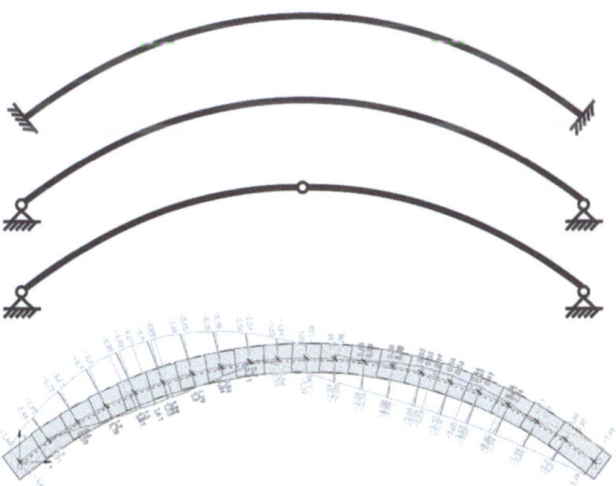

Fig. 8-10. Simple arch beam models without backfill

Fig. 8-11. Simple arch beam model considering the backfill

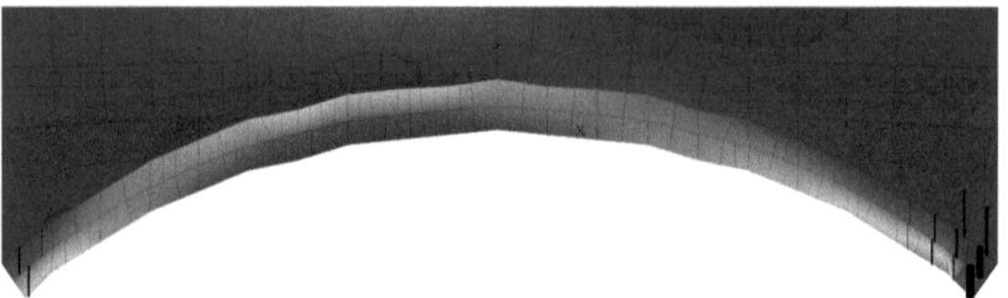

Fig. 8-12. Two-dimensional model using ATENA

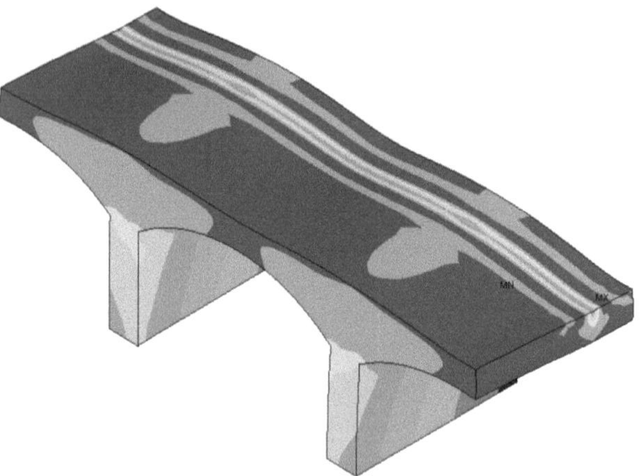

Fig. 8-13. Three-dimensional model using ANSYS

8.3.5 Bridge 5

The fifth bridge is a thought-up bridge using average geometrical and material parameters based on references. The bridge was then modelled using the finite element programs ATENA (Fig. 8-14) and ANSYS (Fig. 8-15). The probability of failure per year was then computed with 8.8×10^{-6}.

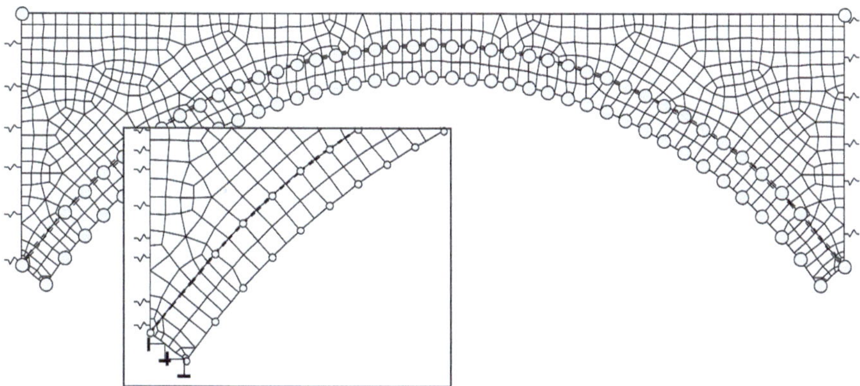

Fig. 8-14. Finite element model using ATENA

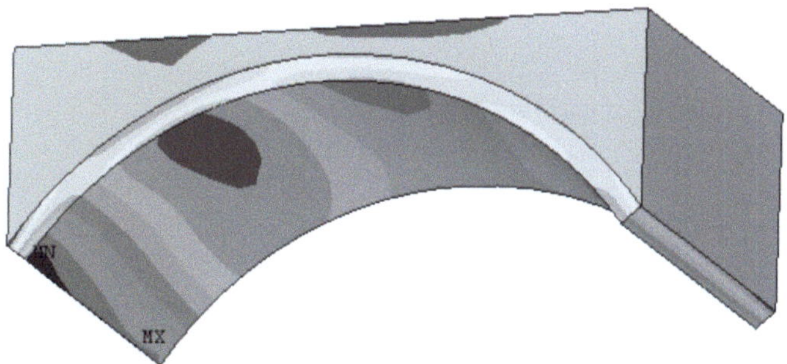

Fig. 8-15. Finite element model using ANSYS

8.3.6 Bridge 6

Bridge 6 is an arch bridge built of tamped concrete using granite stones in the arch. Drilling cores were taken from the structure, and a beam model was developed by Bothe et al. (2004) using the software package SOFISTIK (Fig. 8-16). A probabilistic computation was then carried out to define a partial safety factor of the arch system.

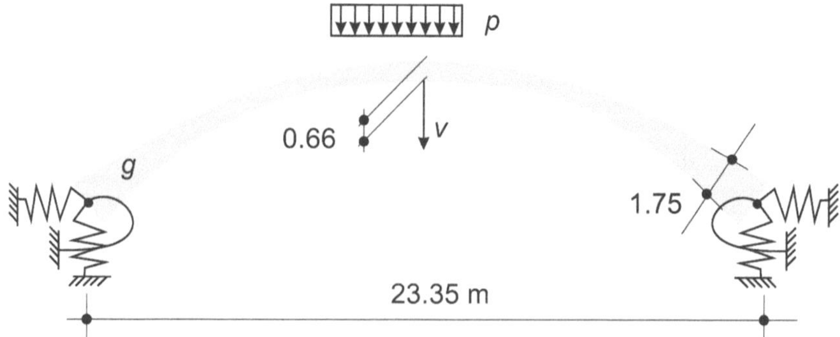

Fig. 8-16. Beam model for Bridge 6 using the software SOFISTIK (Bothe et al. 2004)

8.3.7 Summary

Several bridge structures investigated by deterministic and probabilistic methods have been introduced. Table 8-9 compares the results with other probabilistic calculations of historical bridges based on references or calculated by different authors. The authors are well aware of the influence of different mechanical models or different input data. However, the historical bridges have existed under real world load conditions for a long time and the results of a probabilistic calculation must reflect this fact in terms of comparability.

Table 8-9. Probabilities of failure for different historical arch bridges

Proof under dead and live loads	Chosen value	per	Reference
Mulden Bridge Podelwitz in 1888[a]	$591.50 \cdot 10^{-6}$	Year	Möller et al. (1998)
Flöha Bridge Olbernhau	$0.04 \cdot 10^{-6}$	Year	Möller et al. (2002)
Syraltal Bridge Plauen in 1905	$360.00 \cdot 10^{-6}$	Year	Möller et al. (2002)
Bridge 1 (Bernd model) in 1875	$4.10 \cdot 10^{-6}$	Year	Proske (2003)
Bridge 2 in 1893	$4.80 \cdot 10^{-6}$	Year	Proske (2003)
Bridge 5	$8.8 \cdot 10^{-6}$	Year	
Marien Bridge Dresden in 1846	$1{,}279.0 \cdot 10^{-6}$	Load	Busch (1998)
Bridge 1 (Mann model) in 1875	$33{,}430. \cdot 10^{-6}$	Load	Proske (2003)
Bridge 1 (Berndt model) in 1875	$248.0 \cdot 10^{-6}$	Load	Proske (2003)
Bridge 2 constructed in 1893	$203.0 \cdot 10^{-6}$	Load	Proske (2003)
Bridge 3	$343.0 \cdot 10^{-6}$	Load	
Bridge 6	$159.0 \cdot 10^{-6}$	Load	
Artificial Bridge	$130.0 \cdot 10^{-6}$	Year	Schueremans et al. 2001
Magarola Bridge	$5.48 \cdot 10^{-2} - 10^{-12}$	Year	Casas (1999)
Jerte Bridge	$2.33 \cdot 10^{-4} - 10^{-10}$	Year	Casas (1999)

Proof under dead and live loads	Chosen value	per	Reference
San Rafael Bridge	$1.90 \cdot 10^{-8} - 10^{-12}$	Year	Casas (1999)
Duenas Bridge	$4.05 \cdot 10^{-1} - 10^{-12}$	Year	Casas (1999)
Quintana Bridge	$1.28 \cdot 10^{-12}$	Year	Casas (1999)
Torquemada Bridge	$3.40 \cdot 10^{-6}$	Year	Casas (1999)

[a] Limit state of serviceability

8.4 Further Examples

The authors have not only investigated arch bridges using probabilistic models, but also many other historical and modern structures. To show the application to other historical structures, two further examples are illustrated.

8.4.1 Historical Stone Beam Bridges

In the Lausitz part of Saxony, Germany, many simple stone beam bridges built of Lausitz granite are found (Fig. 8-17). Such bridges often have an age of more than 150 years and therefore do not comply with current structural codes. To conserve these historical structures, a proof concept was developed that not only considers special properties of the material, but also complies with the new code generation using partial safety factors.

For the development, several tests on real-scale stone bridges were carried out. The bending tests not only included original stones but also strengthened stones. The mass of the stones reached up to 1 tonne. Besides tests on the entire stones, compression and splitting tensile tests were also carried out to predict the bending tensile strength indirectly. Using the test data, statistical data investigation was carried out and characteristic bending tensile strength values, as well as partial safety factors, were given (Curbach et al. 2004). The characteristic bending tensile strength can be evaluated as a function of the stone height. The concept also considers the nondestructive tests in the partial safety factor for the stone material. If ultrasonic or radar tests are carried out and confirm a damage-free stone material, then the partial safety factor can be decreased from 1.7 to 1.5.

Fig. 8-17. Stone bridges in Saxony, Germany

8.4.2 Anchoring Chamber of the Blue Wonder Bridge

The next example considers only a part of a bridge: the numerical investigation of the anchoring chamber of the Blue Wonder Bridge in Dresden (Fig. 8-18). The bridge experienced a major flood in 2002. Although the bridge looks like a steel framework bridge, it is actually a suspension bridge. The top chord of the framework functions like a suspension cable and the framework inclusive of the roadway hangs on this element. Even further, the frameworks are constructed like independent slabs. This was done to design a statical determined system, which was easier to calculate during the time of construction at the end of the 19th century. The different slabs are separated by hinges. Since a few of the hinges do not function very well anymore, different statical systems have to be considered nowadays for recomputation. Whereas the dead load is still applied to the statical determined system, traffic and temperature loads have to be applied to a statical indeterminate system. However, all loading cases function only if the tensile forces in the top chord can be transferred to the foundations. This is mainly done in the anchoring chamber where a counterweight is located. During the flood in 2002, the question come up as to which water level the anchoring chamber can bear the water pressure and when the anchoring chamber should at least partially be flooded to limit the maximum water pressure. If the anchoring chamber fails completely, then the counterweight goes nearly completely underwater and is put under buoyant force. However, the load-bearing capacity of the bridge then decreases.

Since the anchoring chamber is built of major stone elements with lean concrete and has a complicated three-dimensional body, a finite element model was used for the computation (Fig. 8-19). The nonlinear computation of a loading case ran up to one day on a PC. The input data for the computation were partially based on material tests using test specimen from core drillings.

As a deterministic result, the normal load-bearing capability of the bridge can be provided up to 7 m of Dresden water level. The computation confirmed a partial filling of the anchoring chamber during the 2002 flood with a Dresden water level of more than 9 m. In relation to the return period of such water levels, the probability of failure for the bridge under the flooding load and without traffic restriction during flooding has been estimated.

Fig. 8-18. Blue Wonder Bridge in Dresden, Germany

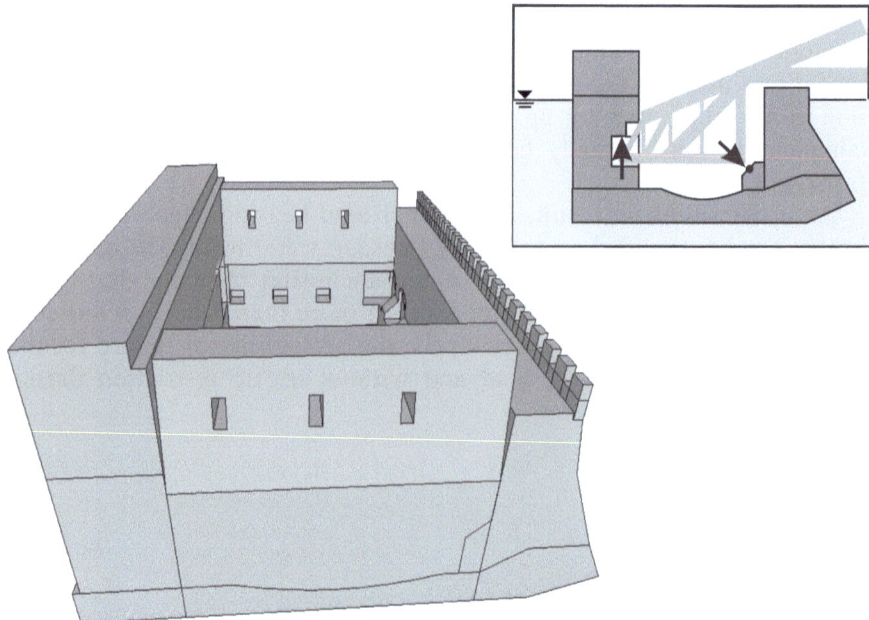

Fig. 8-19. Anchoring chamber finite element model and load during flooding

8.5 Conclusion

Humans can never describe the reality in one single model since all models simplify and abstract. This is valid for social, chemical, physical, or mathematical models. The limits of the models follow the general laws and limits in certain areas of science. However, in contrast to scientific thinking, humans behave differently under everyday life conditions. They permanently change the limits of model borders independent from any administrative scientific types or pedagogic rules. That means, when a model does not fit to describe real-world behaviour, the model will be changed independent from the amount of knowledge we have. For example, we do not have permanent scientific studies about the traffic density of roads on hand, but we have to decide which road we would take and we are able to do that. Models have to be developed, and have to be effective and applicable. Therefore, all the models introduced here do have their meaning. There may be cases where a linear-elastic model fits best to the conditions. In other cases, a full probabilistic computation may be required. Even further, in the case of historical arch bridges, we have had a development of models for probably more than 2,000 years and even now, sometimes the early simple rules may be of use. Such a development can be seen all over the engineering fields. For example, for steel bridges, often we apply advanced finite element models, however simple beam models are still used (Unterweger 2002).

However, all humans develop their models based on their gained information over a lifetime and their knowledge. That means that every human develops different models based on his genetic code, his own experience, and history. If humans reach the same results despite that diversity using their own models, we can assume that the models function well and are objective. The term "objectivity" can be differently defined. For example, objectivity is the independence of the goal of an investigation and the chosen model. In other words, objective decisions are conscious decisions. Another definition of objectivity is based on the length of causal relationships.

The model for the investigation of an arch bridge should also be objective. However, as stated above, the models are related to the individual. Engineers often have different opinions and discuss different parameters, different models, and different assumptions. Some readers may not agree with the presented models or statements, while others do. The final question is then are we able to describe the safety of arch structures objectively or will we have under all conditions some subjective parts, and if so, are we able to transfer the subjective parts into codes or rules?

We assume that research in the field of arch bridges will continue to experience progress by the objective description of the load-bearing capacity of arch bridges, such as that shown in Schubert (2003), Purtak et al. (2007), ICOMOS (2001), Ciria (2006), and Schiemann (2005).

However, this is only one part. We will underestimate the subjective part when we do not consider subjective elements of bridges. In many books, arch bridges are only dealt with in a subjective way when discussing the beauty of such bridges (Widmer 1996, Stiglat 1997, Sprotte 1977, Pearce and Jobson 2002, Mühlen 1969, Dietrich 1998, Cruppers 1965, and Bonatz and Leonhardt 1953). The objective modelling of complex systems is limited: complex systems under all conditions include properties that can only be assessed subjectively (Proske 2008).

Arch bridges are, for many people, not only a tool to cross a river or a valley; they are part of the human history, culture, and human effort. They make proud, they often fit perfectly into the landscape, and they are so long-lasting. They are probably, besides temples, the technical product with the longest usage time, sometimes reaching 2,000 years. We rather doubt that we can currently build bridges that can be used in the year 4,000 A.D.

In everything that man pushed by his vital instinct, builds and raises,
for me, nothing is more beautiful or more precious than bridges.
Bridges are more important than houses,
more sacred because they are more useful than temples.
They belong to everybody and they are the same for everybody,
always built in the right place
in which the major part of human necessity crosses,
more durable than all other constructions.

Ivo Andrić 1892–1975: "Na Drini ćuprija",
taken from Armaly, Blasi and Hannah (2004)

References

Armaly M, Blasi C & Hannah L (2004) Stari Most: rebuilding more than a historic bridge in Mostar. Museum International, Blackwell Publishing, No. 224, 56 (4), pp 6–17

Berndt E (1996) Zur Druck- und Schubfestigkeit von Mauerwerk – experimentell nachgewiesen an Strukturen aus Elbsandstein. Bautechnik 73 (1996), Heft 4, pp 222–234

Bonatz P & Leonhardt F (1953) Brücken. Verlag Karl Robert Lange Wiesche. Königstein im Taunus

Bothe E, Henning J, Curbach M, Bösche T & Proske D (2004) Nichtlineare Berechnung alter Bogenbrücken auf der Grundlage neuer Vorschriften. Beton- und Stahlbetonbau 99, Heft 4, pp 289–294

Brencich A, Bambarotta L & Sterpi E (2007) Stochastic distribution of compressive strength: effect on the load carrying capacity of masonry arches. Proceedings of the Arch 07 – 5th International Conference on Arch Bridges, PB Lourenço, DV Oliveira & A Portela (Eds), 12–14 September 2007, Madeira, pp 641–648

Busch P & Zumpe G (1995) Tragfähigkeit, Tragsicherheit und Tragreserven von Bogenbrücken. 5. Dresdner Brückenbausymposium, 16.3.1995, Fakultät Bauingenieurwesen, Technische Universität Dresden, pp 153–169

Busch P (1998) Probabilistische Analyse und Bewertung der Tragsicherheit historischer Steinbogenbrücken: Ein Beitrag zur Angewandten Zuverlässigkeitstheorie. Dissertation. Technische Universität Dresden

Casas JR (1999) Assessment of masonry arch bridges. Experiences from recent case studies. Current and future trends in bridge design, construction and maintenance in Bridge design construction and maintenance, PC Das, DM Frangopol & AS Nowak (Eds). Thomas Telford, London, pp 415–423

Ciria (2006) Masonry and brick arch bridges (RP692), www.ciria.org

Cruppers H (1965) Die Trierer Römerbrücken. Verlag Philipp von Zabern, Mainz, 1965

Curbach M & Proske D (1998) Module zur Ermittlung der Versagenswahrscheinlichkeit. 16. CAD-FEM Nutzermeeting. Grafing

Curbach M, Günther L & Proske D (2004) Sicherheitskonzept für Brücken aus Granitsteindeckern. Mauerwerksbau, Heft 4, pp 138–145

Dietrich RJ (1998) Faszination Brücken – Baukunst, Technik, Geschichte. Verlag Georg D. W. Callwey, München

ICOMOS (2001) Recommendations for the analysis, conservation and structural restoration of architectural heritage, International Scientific Committee for Analysis and Restoration of Structures of Architectural Heritage, Draft Version n°11, September 2001

Modena C & Sonda D (1998) Safety evaluation and retrofitting of an r.c bridge. Arch Bridges: History, Analysis, Assessment, Maintenance and Repair. Proceedings of the Second International Arch Bridge Conference, A Sinopoli (Ed), Venice 6.–9. October 1998, Balkema, Rotterdam, pp 223–229

Möller B, Beer M, Graf W, Schneider R & Stransky W (1998) Zur Beurteilung der Sicherheitsaussage stochastischer Methoden. Sicherheitsrisiken in der

Tragwerksmodellierung, 2. Dresdner Baustatik-Seminar. 9. Oktober 1998, Technische Universität Dresden, pp 19–41

Möller B, Graf W, Beer M & Bartzsch M (2002) Sicherheitsbeurteilung von Naturstein-Bogenbrücken. 6. Dresdner Baustatik-Seminar. Rekonstruktion und Revitalisierung aus statisch-konstruktiver Sicht. 18. Oktober 2002. Lehrstuhl für Statik + Landesvereinigung der Prüfingenieure für Bautechnik Sachsen + Ingenieurkammer Sachsen. Technische Universität Dresden, pp 69–91

Mühlen F (1969) Technische Kulturdenkmale des Wasser-, Straßen- und Brückenbaus in Westfalen. Deutsche Kunst und Denkmalpflege Jahrgang 1969, Sonderdruck

Ng K-H & Fairfield CA (2002) Monte Carlo Simulation for Arch Bridge Assessment. Construction and Building Materials 16, pp 271–280

Pearce M & Jobson R (2002) Bridge Builders. John Wiley & Sohns, Ltd. London

Proske D (2003) Ein Beitrag zur Risikobeurteilung von alten Brücken unter Schiffsanprall. Ph.D. Thesis, Technische Universität Dresden

Proske D (2008) Catalogue of Risks – Natural, technical, health and social risks. Springer, Berlin – Heidelberg

Purtak F, Geißler K & Lieberwirth P (2007) Bewertung bestehender Natursteinbogenbrücken. Bautechnik 8, Heft 8, pp 525–543

Quan Q (2004) Modeling current traffic load, safety evaluation and SFE analysis of four bridges in China. Risk Analysis IV, CA Brebbia (Ed). WIT-Press, Southampton 2004

Rajashekhar MR & Ellingwood BR (1993) A new look at the response surface approach for reliability analysis. Structural Safety, 12, pp 205–220

Schiemann L (2005) Tragverhalten historischer gemauerter Bogenbrücken (Dissertation in Bearbeitung), http://www.lt.arch.tu-muenchen.de/

Schubert P (2003) Übersicht über abgeschlossene und laufende Forschungsvorhaben im Mauerwerksbau: Realistische Tragfähigkeitseinstufung gemauerter Bogenbrücken mit Hilfe numerischer Berechnungen. Mauerwerk Kalender, Jahrgang 2004, p 646

Schueremans L & Van Gemert D (2001) Assessment of existing masonry structures using probabilistic methods – state of the art and new approaches. STRUMAS 2001 18.-20. April 2001, Fifth International Symposium on Computer Methods in Structural Masonry

Schueremans L & Van Gemert D (2004) Assessing the Safety of Existing Structures: Reliability Based Assessment Framework and Applications. WTA Conference Prague November 23.–24., 2004

Schueremans L (2001) Probabilistic evaluation of structural unreinforced masonry, December 2001, Dissertation, Katholieke Universiteit Leuven, Faculteit Toegepaste Wetenschappen, Departement Burgerlijke Bouwkunde, Heverlee, Belgien,

Schueremans L, Smars P & Van Gemert D (2001) Safety of arches - a probabilistic approach. 9th Canadian masonry symposium, Frederiction: University of New Brunswick, Canada, 12 pp

Spaethe G (1992) Die Sicherheit tragender Baukonstruktionen, 2. Neubearbeitete Auflage, Wien, Springer Verlag

Sprotte B (1977) Aus der Geschichte der Tauberbrücken. Eigenverlag, 6893 Kreuzwetheim, Brückenstraße 30

Stiglat K (1997) Brücken am Weg. Frühe Brücken aus Eisen und Beton in Deutschland und Frankreich. Ernst und Sohn, Berlin

Tschötschel M (1989) Zuverlässigkeitstheoretisches Konzept zur Bemessung von Mauerwerkskonstruktionen. Dissertation, Technische Hochschule Leipzig

Unterweger H (2002) Globale Systemberechnung von Stahl- und Verbundbrücken – Leistungsfähigkeit einfacher Stabmodelle, Habilitation, TU Graz

Warnecke P, Rostasy FS & Budelmann H (1995) Tragverhalten und Konsolidierung von Wänden und Stützen aus historischem Natursteinmauerwerk. Mauerwerkskalender 1995, Ernst & Sohn, Berlin, pp 623–685

Widmer M (1996) Eisenbahnbrücken. Schweiz, Deutschland, Österreich. Transpress Verlagsgesellschaft, Stuttgart

Žak J, Florian A & Hradil P (2003) A contribution to the analysis of historical structures using LHS method. Structural Studies, Repairs and Maintenance of Heritage Architecture VIII, CA Brebbia (Ed), WIT-Press 2003, Southampton, pp 252–256

Index

D. Proske, P. van Gelder, *Safety of Historical Stone Arch Bridges*, DOI 10.1007/978-3-540-77618-5_BM2,
© Springer-Verlag Berlin Heidelberg 2009